Climate Crisis Unmasked

Unraveling The Web Of Betrayal And Greed

Jude S. Ngu' Ewodo

Rwenzori Society

ISBN: 978-1-83556-095-2 (PAPERBACK)

ISBN: 978-1-83556-096-9 (EBOOK)

Book Design by HMDPUBLISHING

Climate Crisis Unmasked

Unraveling The Web Of Betrayal And Greed

Jude S. Ngu' Ewodo

Rwenzori Society

EXPRESSION OF GRATITUDE

Within the pages of this tome, I humbly present a sincere tribute – a testament – to the memory of my beloved father, Mengue Nkoa Clement Georges. He traversed this earthly realm before me, imprinting indelible marks upon the tapestry of time. His legacy entwines with the myriad lives that have shaped our existence, a rich mosaic that lays the cornerstone upon which we now stand.

With warmth as abundant as the sun's embrace, I extend this dedication to my precious daughter, Shana Charlayne, and my cherished son, Menelik Chaka. In them, I discern the promise of tomorrow. Through their eyes, I witness the unfolding chapters of a narrative yet unwritten – a narrative where aspirations and dreams meld into a portrait of limitless potential. In their hearts, the seeds of tomorrow find haven, nurtured by the cumulative ambitions of generations yet to be.

Moreover, within the bounds of this literary pursuit, I am compelled to express deep-seated gratitude to my enduring life companion, mother to our children, and steadfast friend, Becky. Her steadfast love and unwavering support have been my lodestar – a source of fortitude. Through her, I glimpse the countless souls who diligently labor to weave the intricate tapestry of a better world. Their resilience and compassion illuminate the path we tread, reminding us that our unity fuels the power to sculpt a destiny of brilliance.

The vision of adorning each chapter with artwork inspired by a beloved friend has arisen as a radiant guiding star, illuminating the path for both you, esteemed readers, and me, through the intricate labyrinth of the climate change narrative. This friend's wisdom and insight illuminated the enchanting world of art during our numerous conversations and visits to exhibitions. With heartfelt appreciation, I express my gratitude to this friend for enriching my life with this newfound appreciation.

The concept of beginning each chapter with a philosophical, spiritual, or educational thought, acting as the lodestar that guided me through that chapter's journey, also belongs to the essence of this book. May you find both of these concepts inspirational as you explore these pages.

To all these souls – those who have etched their mark upon history, those who stand alongside us today, and those who await the threshold of tomorrow – I extend a resounding chorus of gratitude and devotion. May this modest offering encapsulate their essence and reverberate as a harmonious chord within humanity's shared odyssey.

In remembrance, in hope, and in profound gratitude, let these words cascade through the annals of time – an enduring testimony to the profound influence of those who have graced our lives and to the resounding impact of those who have yet to join our journey.

With heartfelt appreciation.

Jude S. Ngu' Ewodo

Contents

CHAPTER

One

A Homecoming To Remember: Tears On The Infested Shores

"The one who plants trees, knowing that he will never sit in their shade, has at least started to understand the meaning of life."
– Malawian proverb

I. Vow Renewed: A Father's Second Demise

In August 2018, an extraordinary journey unfolded for me and my beloved daughter – a voyage that would forever intertwine the threads of our lives with the fate of a planet in peril. Our destination was the beautiful and flamboyant City of Douala in Cameroon, my birthplace, where my ailing mother awaited our embrace. A sacred pilgrimage, we knew this trip might be the last chance to bask in the simple pleasures of family, laughter, and ancestral blessings. As an African native, this homecoming held profound significance.

Surrounded by the warmth of familial reunions, rekindling old friendships, and navigating the intricacies of business, I entrusted my sisters with the care of my 18-year-old daughter. A chance for her to walk in my footsteps, I arranged quality time with me, hoping to pass on the essence of my roots. As we meandered through the paths of my childhood, each step an homage to memories long cherished, pride swelled within me.

One destination, however, held unparalleled significance – the beach of my youth. Beyond sun, sand, and waves, it was a living testament to my late father. I recalled him carrying me on his shoulders during warm evenings, the distant flickering lights of ships tantalizing us with tales of faraway lands and endless adventure. The beach was inseparable from memories of my father, etched deep in my soul.

Yet, the moment our feet touched that sacred shoreline, the idyllic imagery I clung to shattered. The once majestic beach, a pristine vessel of cherished memories, had succumbed to an ugly plague of plastic. Sand and waves entwined with the remnants of our disposable culture – the legacy of thoughtlessness. My heart sank, not just for the ravaged beach but for the profound loss of my father. This time, not by the hand of cancer, but by the relentless onslaught of plastics.

In that heart-wrenching moment, my mind became a portal to the past – the beach of my youth, a tapestry of cherished memories woven together. I remembered the family picnics the aroma of grilled corn filling the air, and the tantalizing spices of braised fish that permeated the atmosphere. It was a sensory symphony, a culinary dance that brought together generations.

Oh, how the young and old lovers reveled in each other's company, just like in the song of Georges Brassens, "les amoureux sur les bancs publics", blissfully oblivious to the passing of time. The beach was a sanctuary of love, where romance unfurled its tender wings, and passion bloomed like wildflowers in the spring.

The sunsets, oh the sunsets – they were a spectacle of unparalleled beauty. People stood in awe, gazing at the skies for hours as if they were worshipping pagan gods. The beach was alive with laughter and joy, with volleyballs flying and other beach sports, where beauty came in all shapes and sizes, and judgment had no place. It was a place where the soul was nourished, where the heart found solace, and where the spirit soared.

Yes, we were not affluent, but riches lay in the simple joys of life on that beautiful beach. You couldn't buy the happiness that permeated the air, the warmth of community, and the connectedness with nature. In that humble simplicity, we found abundance – the richness of love, laughter, and shared experiences.

But now, as I cast my eyes upon that once-cherished shore, I saw a different landscape – an eerie transformation of vibrant colors into a monochrome of grey dirt. The remnants of faraway lands invaded our sacred space, as bottles and cans from Western brands such as Nestlé, Coca-Cola, Fanta, and Pepsi littered the once-pristine sands. How far had they traveled on their journey before landing here, I wondered?

The invading plastic containers, a symbol of our throwaway culture, had taken residence on the beach of my memories, displacing the echoes of laughter and the whispers of love that once graced the shores. A heart-wrenching irony – it was a landscape now inhabited by the ghosts of consumerism, polluting not only the environment but also the essence of our shared experiences.

As I stood there, the salty tears mingled with the crashing waves, my heart heavy with grief. The beach of my youth, once a sanctuary of harmony and joy, had become a battlefield for the insidious invasion of plastics. And with it, my father's memory, my ancestors' wisdom, and the essence of our collective heritage were fading into the horizon.

Standing beside me was my daughter – a true embodiment of her German upbringing, molded to revere the values of "Sauberkeit und Ordnung," cleanliness and order, as they call it in Germany, her homeland. In her eyes, I saw not just disappointment, but a storm of emotions – anger, indignation, and a profound sense of betrayal, or "Betroffenheit," as they say in German.

As we faced the desecration of these shores, I felt an overwhelming sense of disgrace – a gnawing pain that echoed the sentiments of our collective negligence. This was the land of my ancestors, a place once teeming with natural abundance and sacred memories. Yet, here it stood, tainted by the relentless tide of pollution, a testament to the disregard that had befallen our once-pristine environment.

In that disheartening moment, it felt as though I lost my father all over again. Not to the passage of time, but to the slow destruction of the very landscape that shaped my fondest memories. The beach, once a beacon of joy and connection, had become a haunting reminder of our planet's vulnerability. The relentless tide of pollution had consumed not only the environment but the legacy of my father.

As tears welled in my eyes, a profound resolve stirred within me. The journey to write a book on climate change took on an intensely personal dimension. It wasn't just a reaction to a global crisis; it was an intimate mission. Armed with determination, I knew I had to raise my voice, to tell the stories of a changing world. Through the pages of my book, I would depict the consequences of our actions – the plastic-strewn shores, the vanishing ice caps, the imperiled wildlife, and the communities teetering on the brink of survival. It became a testament to the beauty of our planet and an urgent call for collective responsibility.

In every word I penned, in every story I shared, I would strive to re-kindle the spirit of that beach – the love, the laughter, and the shared moments of bliss. I would bear witness to the desecration of our sacred spaces, urging others to awaken to the consequences of our actions.

For the beauty of that beach was not lost in the sands of time – it lived on in the memories etched in the hearts of those who once walked its shores. And as I gazed upon the relentless tide of plastic pollution, I pledged to honor the legacy of my father, to safeguard the precious memories of the past, and to reclaim figuratively the purity of that beach – for my daughter, my son, my family, my ancestors, myself, future generations, and for the soul of humanity.

II. Artistic Echoes Of Climate Change: The Epoch Of The Solemn Pledge

CHAPTER

Two

Untangling Illusions: Taming The Wild Beast Of Madam Climate Hoax

"He who does not clean his environment will be consumed by it."
– Igbo proverb (Nigeria)

I. Shadows of Veridium: A Fictional Saga of Climate Doubt and Awakening

Once upon a time, in a fictional metropolis known as Veridium City, a group of imagined characters called "The Consortium" had a vested interest in maintaining their control over the city's vast fossil fuel industry. For years, they had amassed unimaginable wealth and influence by exploiting the city's natural resources, heedless of the environmental consequences.

As scientific studies began to reveal the alarming impact of human activities on the climate, the Consortium grew uneasy. They feared that the mounting evidence of climate change would lead to widespread demands for stricter regulations and a shift towards renewable energy sources, endangering their profits and status.

The leader of The Consortium, the enigmatic billionaire named Damien Blackthorn, devised a cunning plan. He enlisted the help of a shrewd public relations expert, Maxwell Sterling, who had a knack for manipulating public opinion. Their goal was to sow doubt and confusion about climate change, framing it as a contentious topic among experts rather than a scientifically established fact. Yes, they intended to gaslight the public. Maxwell and his team meticulously crafted a disinformation campaign. They cherry-picked scientific studies that seemed to contradict the consensus on climate change, funding a few fringe scientists to produce reports that downplayed the significance of human influence on global warming.

At the same time, they used their immense wealth to influence media outlets, ensuring that climate change skeptics received prominent coverage and platformed them as "experts" on the topic. Their well-coordinated messaging aimed to create the illusion of a balanced debate, even though the overwhelming majority of climate scientists agreed on the reality of climate change.

To further cement their narrative, they backed political candidates sympathetic to their cause, pouring money into their campaigns and lobbying efforts. These candidates, indebted to The Consortium, would parrot the climate change denial talking points, giving them a veneer of legitimacy in the eyes of the public.

As the years passed, The Consortium's disinformation campaign began to take root. Doubt was sown, and the general public became divided on the issue of climate change. The media landscape was flooded with contradictory messages, leaving people unsure of what to believe.

Despite the mounting evidence from extreme weather events, melting ice caps, and rising sea levels, a sizable portion of the population remained skeptical about climate change. The Consortium's strategy had succeeded in creating an atmosphere of confusion and inaction, effectively delaying any meaningful efforts to address the urgent climate crisis.

But unbeknownst to The Consortium, there were those who saw through their deception. A small group of intrepid investigative journalists, environmental activists, and concerned citizens began to dig deeper, uncovering the truth behind the disinformation campaign. Their tireless efforts to expose The Consortium's plot became a beacon of hope in the darkness of deception.

As the public became increasingly aware of the truth, protests erupted in Veridium City and beyond. People demanded accountability from the powerful elite who had willfully endangered the planet for their own gain. The Consortium's façade of power and influence started to crumble under the weight of public scrutiny.

In a stunning turn of events, the tides shifted. Political leaders once swayed by The Consortium's wealth and influence now faced the wrath of an informed electorate. Calls for action on climate change grew louder and more urgent.

Finally, the city of Veridium woke up to the reality of climate change. Renewable energy initiatives flourished, and the once mighty fossil fuel industry began to crumble. The Consortium's power waned as the world embraced sustainable practices, leaving them isolated in their denial.

In the end, the story of The Consortium served as a cautionary tale about the dangers of prioritizing short-term profits over the well-being of the planet. It highlighted the importance of critical thinking, scientific consensus, and the need for transparency in an era when misinformation could sway the fate of the entire world. And so, the powerful people's plot to push the idea of climate change as a hoax was ultimately overcome by the collective will of the people to face the truth and take action.

Now, as we turn the page, let us gracefully depart from the realms of fiction and entertainment, and venture forth into the enlightening domain of knowledge, science, and facts. Within this chapter, we shall earnestly delve into the profound matters surrounding the science of climate change, exploring the very essence of its existence and implications. With minds open and hearts receptive, we shall navigate through the depths of understanding, seeking to confront the genuine challenges that climate change poses to our world and all its inhabitants.

II. Climate Action: A Dire Necessity To Combat The Killing Deception

In the realm of our planet's uncertain future, the urgency of the climate crisis reverberates like a relentless siren. Its impacts, both present and impending, gnaw at the very fabric of eco-systems, economies, and human societies. The dire consequences that await us, should we choose to turn a blind eye, are nothing short of cataclysmic.

Delving into the depths of the multifaceted phenomenon of Climate Change, the Intergovernmental Panel on Climate Change (IPCC) emerges as the torchbearer of knowledge, a collective of eminent climate scientists from across the globe. Their meticulous research has birthed a series of reports that encapsulate the essence of our understanding of climate change.

Subsequently, according to the authoritative voice of the Intergovernmental Panel on Climate Change (IPCC) in their gripping report of 2021, global temperatures have already soared by a startling 1.1°C above pre-industrial levels (IPCC, 2021). As we hurtle forward on this ominous trajectory, the critical threshold of 1.5°C looms on the horizon, threatening to ensnare us by around 2040. The specter of such a reality conjures images of unrelenting chaos: scorching heatwaves, parched lands gripped by drought, unforgiving floods, tempestuous storms, and rising tides that swallow our shores. The consequences ripple across our delicate tapestry of existence, from the scarcity of sustenance and water to the upheaval of mass migration and the eruption of conflict.

In a spellbinding study published in the esteemed journal Environmental Research Letters (Screen et al, 2018), the chilling effects of global warming were laid bare, focusing on the, teetering precipice that German cities now find themselves upon. A vivid tapestry woven by climate models unveiled a disquieting reality: if we continue down the treacherous path of business-as-usual, where greenhouse gas emissions go unchecked, Berlin's annual heatwave days may skyrocket from a mere three to a mind-numbing thirty by the end of this very century. As the mercury soars, lives teeter on the brink, with heat-related deaths surging from 67 to over 1,000 per year. The same study foretold a future where extreme precipitation events, those malevolent downpours, could gain an appalling 40% in frequency and 60% in intensity, transforming Berlin into a watery labyrinth of floods and destruction.

Alas, these haunting predictions do not confine themselves to the lands of the Old World. Across African nations, a study etched in the annals of Climatic Change (Corner, 2014) forewarned of the perils that climate change holds for agriculture in the region. With meticulous precision, climate models unveiled an unsettling reality: by the middle of this century, up to half of the fertile croplands in certain African countries may wilt away, rendered inhospitable for the very crops that sustain millions. A grim tapestry of food shortages and famine may become an indelible mark upon the continent's legacy. Echoing this dire chorus, a prophetic study chronicled in the journal Nature Climate Change (2018) stitched together the threads of climate models and economic analysis. It projected a harrowing decline in agricultural productivity across vast swaths of the globe, stretching from Europe's verdant fields to North America's fertile plains and Australia's sunburnt earth. The repercussions reverberate far beyond mere sustenance, igniting economic losses and social upheaval on an unprecedented scale.

Drawing from the IPCC's Fifth Assessment Report (AR5), a resounding truth reverberates: human activities, with their insatiable thirst for fossil fuels, have emerged as the primary architects of the climate change unraveling before our eyes. This conclusion, backed by an overwhelming probability of 95–100%, is not hinged on mere conjecture but is fortified by a wealth of evidence. As our planet warms, the undeniable rise in global temperatures stands as a testament, supported by an array of temperature records spanning over a century, alongside the stark retreat of glaciers and the vanishing Artic Sea ice (IPCC, 2013). The symphony of change, orchestrated by climate models, dances in sync with reality, amplifying our understanding as greenhouse gas concentrations surge.

Yet, the evidence is not confined to temperature alone; it permeates every facet of our climate system. The pages of the IPCC reports bear witness to the shifting patterns of precipitation, the fierce onslaught of extreme weather events, all in harmonious synchrony with a planet brimming with heat (2014). It is a tale of cause and effect, a tapestry where humanity's insatiable appetite for fossil fuels intersects with the complex workings of our climate, setting the stage for transformative shifts.

A wealth of evidence solidifies the narrative, painting a vivid portrait of human influence on our climate's destiny. Atmospheric greenhouse gas concentrations surge relentlessly, their fingerprints visible in the isotopic signature of carbon hovering in the air we breathe . The very timing of our planet's warming mirrors the rise of these greenhouse gases (GHGs), a dance choreographed with scientific precision that sets human activities squarely in the spotlight.

The repercussions of this unfolding drama are not confined to the realms of nature alone. They reverberate through the fabric of our societies, threatening to reshape the landscapes we call home. The IPCC reports unfurl a tapestry of change, where heatwaves grow bolder, droughts deepen their grasp, floods unleash their fury, and storms roar with newfound vigor. Rising sea levels encroach upon our shores, while ocean acidification erodes the delicate balance of underwater eco-systems. Biodiversity, that intricate web of life, unravels under the weight of humanity's influence.

The scientific chorus that resounds in unison does not waver within the hallowed halls of the IPCC alone. Institutions like the National Aeronautics and Space Administration (NASA, 2022) and the National Oceanic and Atmospheric Administration (NOAA, 2022) add their resounding voices to the symphony, endorsing the irrefutable reality of climate change and its human origins.

Thus, we find ourselves at the crossroads of knowledge, confronted by a resounding consensus. The clarion call of science reverberates, weaving a tale that transcends borders and disciplines. It is unequivocally clear: human activities, driven by a relentless thirst for fossil fuels, have propelled us into an era where the impacts of climate change shall be far-reaching, shaping the very foundations of our natural world and the societies we inhabit.

Step into the world of temperature fluctuations, where the rise in global temperatures emerges as a haunting truth, etched into the annals of history. The diligent work of James Hansen and his cohorts in 2010 offers a glimpse into the past, a portal through which we can witness the evolution of our planet's thermostat over the last 130 years. Their scientific symphony of historical temperature records and computer modeling reveals a chilling reality: our world has warmed by approximately 0.8°C (1.4°F) since the late 1800s, with the lion's share of this warming unfolding in recent decades (Hansen et al., 2010).

Who bears the brunt of responsibility for this climatic transformation? The answer lies in the mirror, reflecting the choices of humanity. Human activities, like the relentless burning of fossil fuels, the rapacious deforestation, and the relentless reshaping of land, have taken center stage as the primary culprits behind this warming spectacle (Hansen et al., 2010; Hansen et al., 1988).

This discovery, while breathtaking, is not an isolated revelation. A chorus of scientific studies and esteemed organizations, including the revered IPCC, has lent their resounding voices to this resolute truth. The IPCC's Fifth Assessment Report (AR5) leaves no room for doubt, asserting that human activities have assumed the role of the "extremely likely" drivers of the warming witnessed since the mid-20th century (IPCC, 2014).

The implications of this temperature ascent are profound, painting a canvas of ecological upheaval and societal tremors. Behold the metamorphosis of our world: melting glaciers and ice sheets, rising sea levels, and the oppressive embrace of heatwaves growing in frequency and ferocity. Rainfall dances to a different tune, as precipitation patterns waltz into unfamiliar rhythms. Yet, let us not be deceived by the modest 0.8°C increase in global temperatures. It is merely an average, concealing the disproportionate heat that ravages certain corners of our planet. The Arctic, ever the harbinger of change, has witnessed temperatures soaring at a pace twice that of the global average, a chilling omen of the turmoil that lies in store.

Alas, the stage is set for a climatic performance of unparalleled consequence. If we fail to arrest the rising temperatures, the repercussions will reverberate across the fabric of our existence. Brace yourself for a world besieged by even fiercer extreme weather events, where food and water supplies falter under the relentless assault of nature's fury. The intricate tapestry of life, our precious biodiversity, hangs in the balance, teetering on the precipice of irreparable loss. The IPCC's projections cast a sobering shadow: a temperature increase of 1.5°C above pre-industrial levels bears the weight of profound impacts, while a leap beyond the 2°C threshold plunges us into a realm of cataclysmic proportions.

Remember, as we grapple with the implications of this warming trend, that the 0.8°C rise in global temperatures discussed here is but a relative measure. We have already surpassed the 1°C milestone, the world now enveloped in a 1.1°C warmer embrace than the pre-industrial era. With each passing day, our planet marches further down this perilous path, beckoning us to act with urgency, to heed the clarion call of responsibility. The stakes could not be higher, for it is the very essence of our world that hangs in the balance (Chapman, 2019; Reiner, 2023).

Step now into the realm where the ceaseless tides clash against the very fabric of our coastal existence. Sea level rise emerges as a harbinger of profound consequences, threatening the delicate balance between humanity and the relentless embrace of the ocean. As we confront this pressing challenge, it becomes crucial to fathom the potential ramifications on specific cities, peering through the lens of urgency to comprehend the magnitude of the issue that looms before us.

To gauge the impact of sea level rise on coastal cities, we turn to the artistry of simulation modeling. Through these digital realms, studies have sought to predict the fate of our coastal brethren in the face of rising tides. The mesmerizing work of Kulp and Strauss in 2019 stands as a testament to this endeavor, as they deployed a high-resolution digital elevation model to simulate the potential consequences of sea level rise on cities spanning the globe. Their symphony of data conjured a stark reality: if we persist with emissions at current rates, up to 190 million souls could find themselves ensnared by the clutches of sea level rise by 2100, with over 1,200 cities teetering on the precipice of partial inundation (Kulp et al., 2019).

Cast your gaze upon the pantheon of vulnerable cities, perched precariously upon low-lying coastal realms. Paris, London, Amsterdam, Hamburg, New York, Miami, Bangkok, Shanghai – the list unfurls like an atlas of peril. They bear witness to the encroaching specter of rising seas, where the watery boundary breaches with greater frequency, leaving a trail of flooding and coastal erosion in its wake.

New York City, a bustling metropolis ensconced by the Atlantic's embrace, knows all too well the perils of this climatic dance. Rising sea levels have orchestrated a symphony of coastal flooding, leaving a trail of destruction in its wake – infrastructure battered, properties desecrated. Meanwhile, Miami, another coastal gem in the United States, finds itself trapped in a watery labyrinth, besieged by the relentless encroachment of sea level rise, an unwelcome guest intruding upon the livelihoods of local businesses and homeowners.

Travel afar, to the vibrant heart of Lagos, Nigeria. Here, the sea level rise amplifies the ferocity of storm surges, lashing out at low-lying areas, exacerbating the risk of flooding and plunging vulnerable communities into the depths of uncertainty. Across the seas, in Douala, Cameroon, the insidious erosion of coastal realms unfolds as sea level rise takes its toll on vital infrastructure, threatening the very foundations upon which communities rely.

Peer into the looking glass of our future, and one cannot help but shudder at the impending cataclysm. Sea level rise, a menacing specter on the horizon, carries the potential to unleash a tempest of economic and societal consequences. The echoes of Union of Concerned Scientists (UCS) reverberate through the halls of certainty, warning that by 2045, approximately 300,000 coastal homes in the United States alone face the specter of chronic flooding, with a staggering $136 billion loss in property value hanging ominously in the balance (2018). The tremors of this economic upheaval will reverberate far and wide, with the insurance industry left grappling with skyrocketing premiums, threatening to cast homeowners adrift in a tempest of uncertainty.

But the tempest's toll extends far beyond economic realms, for the rhythm of daily life is destined to be disrupted. In the coastal regions, where saltwater intrusion knows no bounds, the delicate balance of freshwater resources teeters on the precipice. Crops wither, livestock languish, and food security unravels, leaving vulnerable populations.

Let us now delve into the intricate web of climate change's impact on human systems, as we unravel the threads that connect our health and economic well-being in this ever-changing world.

First, let us cast our gaze upon the realm of health impacts, where the consequences of climate change reverberate with a force that cannot be ignored. Within this realm, both direct and indirect impacts intertwine to shape the landscape of human well-being. Direct impacts manifest in the form of extreme weather events, such as scorching heatwaves and devastating hurricanes, which leave a trail of injury and death in their wake. But it is the indirect impacts that often lurk in the shadows, silently weaving their intricate tapestry of health woes.

Consider the toll exacted by soaring temperatures. In the year 2019 alone, France mourned over 14,000 lives lost to the unforgiving grip of heat-related fatalities, while the United States grappled with over 1,400 souls claimed by the relentless heat (Watts et al., 2019). And let us not forget the intricate dance between climate change and vector-borne diseases, as the warming climate alters the landscapes where disease-carrying creatures thrive. Malaria and dengue fever, once confined to specific regions, now spread their wings of transmission, projected to infiltrate new territories with fervor (IPCC, 2014).

But amidst this mosaic of health impacts, we must not overlook the most vulnerable among us – our seniors and residents of elderly care facilities. Age has a way of rendering our defenses frail, leaving these individuals particularly susceptible to the ravages of extreme weather events and the diseases they beckon. Heat-related illnesses cast a menacing shadow, as seniors and care facility residents bear the burden of physiological changes and underlying health conditions that amplify their vulnerability. And as the climate shifts, so too do the patterns of infectious diseases, with the likes of West Nile virus and Lyme disease altering their course, threatening to intrude upon the sanctity of elderly care (McMichael et al., 2018).

Let us now peer through the looking glass of pandemics and zoonoses, where the convergence of climate change and infectious diseases takes center stage. Climate change has the power to ignite the embers of zoonotic diseases, those that traverse the boundary between animals and humans. As temperature and precipitation patterns shift, altering the distribution and behavior of host animals, the risk of zoonoses intensifies, casting a specter of uncertainty over our future (Kolstad et al., 2020).

Consider the outbreak of SARS-CoV-2, the virus that plunged the world into the throes of the COVID-19 pandemic. In its genesis, a tangled web of connection emerged, linking the wildlife trade to the transmission of this insidious virus. Bats, those enigmatic creatures, were identified as potential reservoirs, highlighting the intricate dance between our changing climate and the emergence of novel pathogens (Wu et al., 2020). And as we witness the specter of climate change shaping the distribution and behavior of disease vectors, from mosquitoes to ticks, we must brace ourselves for the potential perils they bring, threatening to unleash a new wave of infectious diseases upon our fragile existence.

In the vast tapestry of climate change's impact on our health systems, there exists yet another thread that we must unravel – an intricate dance between humans and animals, where the delicate balance of proximity can shape the destiny of zoonotic disease transmission. Deforestation, a consequence of human activities and the relentless march of agricultural expansion, holds the key to unlocking this enigma, as it thrusts us into the realm of wildlife habitats and the fragmented eco-systems they leave in their wake (Vieira et al., 2018).

As the chainsaws gnaw through the heart of ancient forests, a domino effect is set in motion – one that brings us closer to the very creatures that inhabit these sanctuaries. In this newfound proximity, the stage is set for the transmission of zoonotic diseases, those that stealthily traverse the boundary between animals and humans, leaving destruction in their wake (Jones et al., 2008).

Consider, for a moment, the haunting echoes of the 2014 Ebola virus outbreak in West Africa – an event that shook the foundations of our understanding. It was here, in the wake of deforestation and the encroachment of humans into previously untouched forest realms, that the seeds of tragedy were sown. As we ventured into these sacred domains, we unknowingly brushed shoulders with infected animals, setting in motion a deadly chain of events (Gibbs et al., 2015). And as the forests crumbled, so too did the delicate balance that once held nature's secrets, giving way to the shifting movements of bats – those enigmatic creatures whose wings carry whispers of both mystery and danger (Daszak et al., 2020).

But let us not confine our exploration to the depths of distant jungles, for even within our bustling urban landscapes, the dance between humans and animals unfolds, revealing a precarious intermingling that bears consequences yet to be fully grasped. In areas teeming with high population densities, where sanitation and waste management infrastructure strain under the weight of human existence, a breeding ground emerges – one ripe for the emergence and transmission of zoonotic diseases.

Now, let us turn our attention to the realm of economic impacts, where the ripples of climate change spread far and wide, leaving no stone unturned. The interplay between climate and economy is a complex symphony, where each note carries the weight of fortunes won and lost. The Intergovernmental Panel on Climate Change (IPCC) offers a somber reflection, illuminating the profound implications that lie before.

Climate change, with its increasing frequency and severity of natural disasters, poses a significant threat to the global economy. The implications ripple across various sectors, including agriculture, fisheries, tourism, housing, energy, and even the insurance industry. The profound impacts of climate change on these sectors necessitate immediate action and adaptation to mitigate the potential economic losses and social instability.

When we examine the agricultural sector, it becomes apparent that changes in temperature and precipitation patterns can have dire consequences for productivity. The Intergovernmental Panel on Climate Change (2014) warns of projected decreases in yields for essential crops like wheat, maize, and rice in many regions. Such declines in agricultural productivity threaten food security and economic stability, particularly in low-income countries. The future of farming hangs in the balance as climate change disrupts the delicate balance of nature.

Similarly, climate change's influence extends to the world's oceans and the critical fisheries they support. Rising sea levels and warming ocean temperatures pose significant risks to fish stocks. The IPCC (2014) projects declines in fish populations across multiple regions, raising concerns about food security and the livelihoods of those dependent on fishing. As these valuable marine resources dwindle, the economic implications reverberate globally, affecting not only local economies but also the interconnectedness of the international market.

It is crucial to recognize that climate change's impacts extend beyond the physical environment and into financial and insurance markets. The catastrophic effects of extreme weather events, like Hurricane Katrina in 2005, highlight the vulnerability of these markets. The hurricane inflicted over $125 billion in damages and losses (IPCC, 2014), demonstrating the potential disruption to global financial markets and the availability and affordability of insurance products. As the frequency and intensity of such events increase, the stability of financial and insurance systems comes under greater strain.

The tourism industry, another critical sector, is significantly vulnerable to the impacts of climate change. Extreme weather events, such as floods, heatwaves, and hurricanes, can wreak havoc on tourism-dependent economies, resulting in substantial economic losses.

Consider the aftermath of the 2004 Indian Ocean tsunami, which devastated the tourism industry in Thailand and Sri Lanka. Thailand, heavily reliant on tourism, experienced a staggering drop of over 50% in tourist arrivals following the disaster, while Sri Lanka faced a decline of over 60% (WTTC, 2005; Balzin et al., 2014).

Climate change can also disrupt tourism through other means. The ongoing drought in Cape Town, South Africa, led to water scarcity and strict usage restrictions. Consequently, many hotels and tourist facilities had to close or reduce their operations due to the lack of water (Statistics South Africa, 2018).

Furthermore, changes in temperature and precipitation patterns can significantly impact tourism activities. The heatwave that struck Japan in 2018, causing record-breaking temperatures, resulted in a sharp decline in tourism, particularly outdoor activities like hiking and sightseeing (The Japan Times, 2023).

These examples demonstrate the substantial economic losses the tourism industry faces due to climate change. The 2004 Indian Ocean tsunami alone led to an estimated $10 billion loss for the global tourism industry, as stated by the World Travel and Tourism Council (WTTC, 2005). The ongoing drought in Cape Town has similarly resulted in an approximate $1.5 billion loss for the city's economy (Statistics South Africa, 2018).

Now Picture this: In the heart of the majestic Alps, where pristine slopes and renowned ski resorts attract winter enthusiasts from all corners of the globe, a shadow looms over this idyllic landscape. The European Environmental Agency grimly forecasts a future where the very essence of these snow-capped mountains is threatened. By the close of this century, they predict a chilling decline in the average snow depth, a reduction ranging from 30 to a staggering 70 percent (EEA, 2021).

The implications of such a profound transformation extend far beyond the boundaries of nature itself. Nowhere is this more evident than in the enchanting regions of North Italy, where the Alps play a vital role in the economy. It is here, amidst the breathtaking panoramas and adrenaline-fueled descents, that ski tourism flourishes, injecting a staggering €11 billion into the local coffers every year (Girardi, 2018). This vibrant industry serves as a lifeline for over 120,000 individuals, their livelihoods intertwined with the graceful dance between skis and snowflakes.

The German Alps, too, stand as a testament to the inseparable bond between winter sports and the local economy. With each passing year, 14 million eager skiers flock to this alpine wonderland, their presence akin to a snow-laden blessing for the surrounding communities. Yet, beneath the surface of this dazzling spectacle lies a delicate equilibrium, where the prosperity of many hangs in the balance.

Across the border, nestled in the French Pyrenees, a similar tale unfolds. The vibrant ski industry, a lifeline for approximately 18,000 individuals, forms the very fabric of the local economy. Each year, it breathes life into the region, pumping €450 million into its veins and sustaining the hopes and dreams of countless residents (Le Figaro, 2021). But alas, the future casts a foreboding shadow upon these cherished slopes.

With the disappearance of snow in these cherished alpine realms, a cataclysmic ripple effect would reverberate throughout the economic landscape. Consider the implications in North Italy, where the ski industry stands as a pillar of the region's GDP, contributing a significant 5 percent. As the vibrant tapestry of ski tourism begins to unravel, the loss of jobs and income would cast a long, cold shadow over local communities, their once-bustling streets now echoing with the sounds of uncertainty and hardship.

The German Alps, too, would face an uncertain fate. The decline in ski tourism, an industry interwoven into the very fabric of daily life, would unleash a devastating blow upon the local economy. Countless businesses, from quaint mountain lodges to bustling ski rental shops, would find themselves teetering on the edge of survival, their futures hanging by a thread of melting ice.

Yet, the ramifications stretch beyond mere economic numbers. The disappearance of snow and the subsequent decline of ski tourism would carve deep into the cultural and social soul of these Alpine havens. For generations, skiing has transcended a mere recreational pursuit; it has become a sacred tradition, etched into the hearts of the locals. The loss of this cherished activity would leave an indelible void, a missing piece in the mosaic of their lives.

As the Alps face an uncertain future, the interplay between nature's changing course and humanity's destiny reaches a critical juncture. It is a moment that calls for collective action, a rallying cry to safeguard not just the snow-capped peaks and icy slopes but also the vibrant communities that depend on their enduring allure. For in the balance lies the preservation of not only an economy and a culture but also the very essence of what it means to be an Alps resident (Schellingerhout, 2023).

Let me indulge in a bit of humor and sarcasm as we confront the grim truth that lies ahead for all of us if we stubbornly refuse to address climate change and keep embracing charlatans over science. Brace yourselves, for there's no escaping the dark reality that looms!

Picture this: no matter where we reside, we're in for a rollercoaster ride of misery if we don't put a leash on climate change. Oh, the joys that await us! As we bask in the scorching heat, trying to fry eggs on our car hoods, we'll soon realize we're not just getting a tan – we're getting poorer! The economy will suffer dearly, leaving us to ponder how our once flourishing stock portfolios have turned into dust collectors. Forget luxurious vacations, we'll be vacationing in our own backyards, pretending that the garden hose is a tropical waterfall.

But wait, there's more! As if economic turmoil wasn't enough, we'll have the pleasure of hosting a smorgasbord of new diseases. Zoonoses and pandemics will become our uninvited guests, and our daily routine will include Googling symptoms and diagnosing ourselves as exotic carriers of rare tropical viruses. "Honey, guess what? I've got, Llama Fever' today! The neighbors will be so jealous!"

And here comes the heart-wrenching part: our beloved family members will be at risk. We'll watch helplessly as our grandparents succumb to heatwaves like popsicles melting in the sun. Gone are the days of family reunions; instead, we'll gather for somber funerals with airconditioned caskets.

Oh, how could I forget the water predicament! In this not-so-rosy future, water will be as precious as gold, and its price will skyrocket like a SpaceX rocket on steroids. Forget about indulging in long, luxurious showers. We'll be reminiscing about the good old days when we could afford to soak in a bathtub filled with the tears of our enemies.

Potable water? Ha! That'll be a luxury for the elite. We'll be resorting to collecting the water we use to brush our teeth and perform some bizarre ritual of splashing ourselves with a pathetic piece of tissue. "Honey, can you pass me the spit jar? It's time for my daily shower!"

And oh, the flushing toilets! Say goodbye to that delightful "whoosh" sound. No more careless flushes! Every drop will be precious, and we'll be rationing our flushes like it's a new eco-friendly game show. "Congratulations, you've won a luxurious shower! But you'll have to pay for it with ten flushes, a week's worth of dishwashing, and a pledge to recycle everything, even your cat's hairballs."

Oh, the irony! We'll be living in a world where we have more smartphones than clean water. We'll stare longingly at our Instagram feeds, filled with pictures of pristine lakes and sparkling rivers, while we're stuck with our dusty, empty water glasses.

Could you illuminate the repercussions of climate catastrophes? They do not echo the tranquility of "The Sound of Silence" but instead resound with the thunderous vibrations akin to a South African vuvuzela during the 2010 Football World Cup.

Now, let's address those who insist on considering climate change a hoax. Oh, you jokers! The first-ever hoax that kills – quite the novelty, huh? Move over, urban legends and internet hoaxes, because this one's deadly serious. Climate change doesn't care about your skepticism; it's not concerned about your Facebook rants or your tweetstorms. It's coming for all of us like a silent, invisible predator, armed with rising temperatures and unpredictable weather.

So, let's embrace the dark humor in all this and have a good laugh at ourselves, shall we? After all, who doesn't love gallows humor when faced with an impending catastrophe? But let's not forget that this is no laughing matter. The consequences are real, and they are coming for us. So, my friends, it's time to get serious, drop the charlatans, and follow the science. Because when it comes to climate change, ignorance is not bliss –it's a recipe for disaster. And that, my friends, is definitely not funny!

III. The Role Of Climate Investments: No Capital, No Transformation, No Redemption

Throughout the annals of history, one can discern a compelling pattern: monumental leaps in human civilizations have consistently been propelled by investments. Whether in the form of financial capital, intellectual pursuits, or the sheer dedication of labor, these investments have ignited the flames of progress, reshaping the world as we know it.

From the grand expeditions of explorers like Christopher Columbus, fueled by financial backing from the Spanish monarchy, to the dazzling achievements of the Apollo moon missions, funded by substantial government investments, it becomes apparent that investments have been the bedrock upon which great achievements have been built.

Enter the world of Mansa Musa, the visionary Emperor of the Mali Empire in 14th-century West Africa. In an era when kingdoms rose and fell, Mansa Musa's strategic investments stood as a testament to his foresight and ambition, catapulting the Mali Empire into a golden age of prosperity and influence.

It was the year 1324 when Mansa Musa embarked on a pilgrimage to Mecca, known as the Hajj. Little did he know that this journey would not only deepen his faith but also unveil the staggering wealth of his empire to the world. As he made his way through the Sahara, the shimmering caravan of Mansa Musa captivated the hearts and minds of all who bore witness. His entourage was a spectacle like no other, adorned with gold and precious jewels that seemed to rival the sun itself. The mere sight of Mansa Musa's wealth was enough to leave onlookers dumbfounded, marking a pivotal moment in the Mali Empire's history.

But it wasn't just Mansa Musa's riches that set the Mali Empire apart. He understood the power of investments in shaping the destiny of nations. With his vast resources at his disposal, he embarked on a grand vision to transform the Mali Empire into a beacon of knowledge and commerce.

Timbuktu, the capital city, became the canvas for Mansa Musa's transformative investments. He poured his wealth into infrastructure projects that would leave a lasting imprint on the city's landscape. Mosques, libraries, and educational institutions sprouted like blossoms, infusing Timbuktu with an intellectual vibrancy that reverberated throughout the region. The crowning jewel was the University of Sankore, a sanctuary of learning that beckoned scholars from far and wide, drawn by the empire's patronage of education and Mansa Musa's commitment to fostering intellectual growth.

But Mansa Musa's investments extended far beyond the realm of academia. He recognized the pivotal role of trade and commerce in driving economic growth and prosperity. With a shrewd eye for opportunity, he established diplomatic ties and facilitated trade routes, particularly in the highly sought-after commodities of salt and gold. The Mali Empire's abundant gold resources, derived from the vast lands under Mansa Musa's rule, created a magnet for merchants from all corners of the known world. The empire's trading networks thrived, solidifying its position as a hub of wealth and exchange.

Through his strategic investments, Mansa Musa not only enriched the Mali Empire but also fostered a cultural and intellectual renaissance. His unwavering support for education, exemplified by his financial backing of scholars, students, and religious leaders, propelled the empire's intellectual capital to new heights. The Mali Empire became a beacon of knowledge, attracting luminaries from across Africa and the Middle East, who flocked to its centers of learning to exchange ideas and push the boundaries of human understanding.

While the Mali Empire and Mansa Musa's reign shine brightly in the annals of history, it's important to recognize that investments and development in ancient Africa stretched far beyond this era. Other African civilizations, such as the Kingdom of Kush, the Axumite Empire, and the Songhai Empire, experienced their own moments of flourishing through investments in infrastructure, agriculture, trade, and education. These investments played a crucial role in shaping the trajectory of these societies, fostering progress, and leaving a lasting legacy.

As we delve into the rich tapestry of human history, we uncover stories of visionary leaders like Mansa Musa, whose strategic investments propelled their empires to unprecedented heights. They serve as a testament to the transformative power of investments, igniting sparks of innovation, and unlocking the boundless potential of human endeavor.

Picture this: it's the late 15th century, and a bold adventurer by the name of Christopher Columbus, fueled by dreams of wealth and glory, secures the backing of Queen Isabella and King Ferdinand of Spain. Little did they know that his voyage to find a new trade route to Asia would take an unexpected turn, and instead, Columbus stumbled upon the untamed shores of America. Talk about an unintended discovery that forever altered the course of history!

Fast forward to the Age of Enlightenment and the Industrial Revolution, a time when investments in science, technology, and infrastructure set the stage for monumental change. Imagine witnessing the birth of the Liverpool and Manchester Railway in 1830, a marvel that transformed transportation as we knew it. Picture this: a trail of iron tracks stretching into the horizon, propelling steam-powered locomotives at unprecedented speeds. Suddenly, the world shrunk, trade flourished, and people marveled at the notion of traversing great distances in mere hours. All thanks to strategic investments that revolutionized the way we moved and connected.

But the power of investments didn't stop there. It reached for the stars. In the mid-20th century, the race to conquer space captivated the world. Picture a group of courageous astronauts embarking on the Apollo 11 mission in 1969, their sights set on an audacious goal: to land the first human beings on the moon. With substantial investments backing their endeavors, these intrepid explorers defied gravity and etched their names in history. Neil Armstrong's words echoed across the vastness of space as he took that one small step, leaving an indelible mark on mankind's journey of exploration.

And who could forget the sheer power of electricity? Imagine a world shrouded in darkness, where flickering candles struggled to hold back the night. But then, an investment like no other sparked a revolution. In 1879, Thomas Edison's invention of the electric light bulb transformed the darkness into a realm of possibilities. Picture the moment when the first electric lights illuminated the streets, homes, and factories, banishing the shadows and heralding a new era of progress. Investments in power plants and electrical grids laid the foundation for a world powered by this invisible force, sparking a revolution that would forever change the way we live, work, and dream.

But let's not forget the jaw-dropping moments that left us in awe. Cast your mind back to the late 20th century, a time when the dot-com bubble captivated investors with dreams of untold riches. In 1999, a social media website called theGlobe.com went public, capturing the world's attention. The sheer exuberance surrounding the internet-based companies during this period led to astonishing valuations. On its first day of trading, theGlobe.com's stock price skyrocketed from $9 to a staggering $97, leaving investors wide-eyed and gasping for breath. Oh, the highs and lows of the investment rollercoaster!

And how could we forget the enigmatic Elon Musk and his audacious ventures? Imagine witnessing his rise to prominence, from electric vehicles with Tesla to space exploration with SpaceX. Musk's investments have both amazed and amused the world, with his endeavors pushing the boundaries of innovation. Picture the exhilarating sight of a SpaceX rocket launching into the sky, defying gravity with a seemingly impossible feat. But the real jaw-dropper comes when that very rocket lands safely back on Earth, ready to be reused, potentially revolutionizing the future of space exploration and driving down costs. It's a moment that leaves us in awe, capturing the essence of what investments can achieve when fueled by ambition and vision.

But investments aren't just about the grandiose tales of discovery and technological marvels. They play a critical role in addressing humanity's most pressing challenges, including the fight against cancer. In the late 20th century, investments in research and development led to a breakthrough in cancer treatment. Gleevec, a revolutionary drug developed by Novartis, emerged as a game-changer for patients with chronic myelogenous leukemia. The investments poured into the research and development of Gleevec resulted in a highly effective treatment that offered hope and improved outcomes for countless individuals battling this devastating disease. It's a testament to the power of investments in the realm of healthcare and the potential they hold in transforming lives.

As we dive into the role of industrialization, we encounter names that have left an indelible mark on history. In the United States, the towering figures of John D. Rockefeller, Andrew Carnegie, and Cornelius Vanderbilt come to the fore. Picture Rockefeller, the oil magnate who harnessed the power of strategic investments to dominate the petroleum industry during the late 19th century. His company, Standard Oil, emerged as an empire through aggressive acquisitions and revolutionary refining techniques, forever altering the landscape of the oil industry.

Meanwhile, Andrew Carnegie, a Scottish-American industrialist, left an indelible mark in the steel industry. Picture Carnegie's investments in steel production and infrastructure, breathing life into towering skyscrapers, robust bridges, and the very sinews of a rapidly industrializing nation. His influence extended beyond mere business, as he channeled his immense wealth toward philanthropic endeavors, creating libraries, educational institutions, and supporting medical research that continue to shape society to this day.

And let's not forget Cornelius Vanderbilt, the transportation magnate who built a legacy around his investments in railways and steamships. Picture the transformative effect of his entrepreneurial ventures, connecting the vast expanse of the United States through a web of tracks and waterways. Vanderbilt's investments in transportation not only facilitated trade and travel but laid the foundation for the modern transportation networks we rely on today.

Across the Atlantic in Germany, names like Werner von Siemens, Robert Bosch, Gottlieb Daimler, and Alfred Krupp emerge as titans of industrialization. Picture Werner von Siemens, a German inventor and industrialist, whose investments in electrical engineering and telecommunications brought about a revolution. Founded in 1847, Siemens & Halske would evolve into the multinational conglomerate Siemens AG, shaping the future of electrical power distribution and propelling the widespread adoption of electricity across industries and households.

Then there's Robert Bosch, a German industrialist and engineer, whose investments in the automotive industry left an indelible mark. Picture Bosch's visionary contributions to automotive technology, fueling innovations such as fuel injection systems and advanced safety technologies. These investments paved the way for the vehicles we drive today, transforming the way we commute, travel, and experience the world.

And let's not forget Gottlieb Daimler, an engineering genius, and Alfred Krupp, an industrialist who reshaped Germany's industrial landscape. Picture Daimler's investments in automobile development, which culminated in the creation of the first high-speed internal combustion engine and the establishment of Daimler-Motoren-Gesellschaft (DMG). Meanwhile, Krupp's investments in heavy machinery and steel production propelled Germany into an industrial superpower, forging a path of progress and economic might.

Envision with me now a voyage to a bygone era steeped in antiquity, if you will, the vastness of the ancient Egyptian landscape, stretching across the desert sands like an endless tapestry of mystery and grandeur. Now, picture a pharaoh, a ruler of unparalleled ambition and vision, standing before his people, proclaiming his desire to build something so extraordinary, so awe-inspiring, that it would forever etch his name in the annals of history. That pharaoh, my dear reader, was none other than the mighty ruler of ancient Egypt, and the investment he made was nothing short of monumental: the construction of the pyramids.

Transport yourself back to a time when the pyramids stood tall, their gleaming limestone façades reaching towards the heavens. These majestic structures, built with an abundance of labor, skill, and resources, were not mere tombs for the pharaohs but rather enduring symbols of power, engineering prowess, and cultural significance. Can you fathom a modern Egypt without these architectural wonders? It is a challenge even for the most imaginative mind.

The pyramids of Giza, with their enigmatic allure, have become icons of Egypt, drawing in travelers from every corner of the globe. Imagine the countless visitors who flock to witness these ancient marvels, their hearts filled with wonder and awe. The dollars they bring, year after year, flow into Egypt's economy, creating a "BIG" return on investment that transcends the boundaries of time and inflation.

Now, let us journey from the sands of Egypt to the picturesque landscapes of my home state of Bavaria, nestled in the heart of Europe. Here, in the land of fairy tales and enchantment, a king once ruled with a madness that was, in truth, a fervent devotion to the arts and architectural splendor. His name was King Ludwig II, and his investments would forever shape the Bavarian landscape.

Close your eyes and envision the castles that grace the Bavarian countryside, like fragments of a dream brought to life. Among them, the exquisite Schloss Linderhof, a jewel of Rococo design; the Neues Schloss Herrenchiemsee, a magnificent homage to the Palace of Versailles; and, of course, the legendary Schloss Neuschwanstein, perched majestically on a hill, seemingly plucked from the pages of a fairy tale.

Now, dear reader, imagine Bavaria without these architectural wonders, without the tales of kings and their extravagant visions. It is an unimaginable sight, for these castles have become synonymous with Bavarian culture and tourism. Every year, tourists flock to witness the splendor of King Ludwig's creations, their wallets open in awe, as they contribute to a "BIG" return on investment that defies the constraints of inflation. In the words of the esteemed Denzel Washington, let us imagine these examples as if we were six years old, with eyes wide open to the world's wonders. The investments made by the Egyptian pharaohs and the Bavarian kings were not merely financial ventures; they were bold declarations of humanity's capacity to dream, to create, and to leave an indelible mark on the tapestry of history.

Governments, those at the helm of our nations, recognize the gravity of the climate crisis. To reduce greenhouse gas emissions and achieve climate targets, they embrace climate investments wholeheartedly. By directing resources towards renewable energy, energy efficiency, and sustainable infrastructure, governments can pave the way for low-carbon economies, breathing life into new job opportunities and fostering robust economic growth.

Moving into the realm of businesses, we find that climate investments offer substantial benefits. Companies keen on sustainability embark on a journey to reduce their carbon footprint. Through investments in energy efficiency, renewable energy, and sustainable supply chains, they enhance their competitiveness, mitigate climate-related risks, and safeguard their hard-earned reputation.

For the investors seeking not only financial returns but also environmental impact, climate investments present an alluring opportunity. With their sights set on renewable energy, sustainable agriculture, and clean technology, these investors stand to gain attractive financial gains while contributing to vital climate solutions.

In the midst of this symphony of climate investments, communities too reap the rewards. Targeted investments fortify communities against the impacts of climate change. Green infrastructures, such as urban parks and flood management systems, mitigate the effects of heatwaves and flooding. Investments in sustainable agriculture and water management bolster food security and ensure the availability of this precious resource.

And now, let us pause to reflect on the future generations, those who will inherit the consequences of our actions. For them, climate investments hold the key to securing a sustainable and livable planet. By investing in climate solutions today, we can mitigate the worst impacts of climate change, leaving behind a legacy of hope for generations to come.

But how do we rally all these stakeholders together and unleash the potential of climate investments? Let's journey into the world of deal structures, where governments, businesses, communities, individuals, and financial institutions weave a tapestry of support for climate solutions.

Project Finance emerges as a powerful mechanism, securing capital for specific infrastructure projects that demand significant upfront investment. By tying funding to project assets and cash flow rather than relying on traditional creditworthiness, Project Finance bestows the means to embolden high-risk endeavors in the realm of infrastructures and Real Assets. Rather than relying on the borrower's creditworthiness, Project Finance deals secure funding through the project's assets and cash flow. This approach is commonly utilized for climate solutions. Advantages for stakeholders include providing access to substantial capital for high-risk projects that might not be feasible through traditional lending. It also allows for the transfer of project risks to investors, lessening the burden on project developers. Additionally, Project Finance instances include several Greenfield Solar Power Projects presently underway in Sub-Saharan Africa with secured or anticipated funding totaling of U.S. dollars, some of which are being developed by our organization.

On the other hand, disadvantages for stakeholders encompass the complexity and time-consuming nature of structuring project finance deals, necessitating extensive due diligence and negotiation. The high level of scrutiny and due diligence can lead to increased transaction costs. Furthermore, the focus on project assets and cash flow may restrict project developers' future access to additional financing. However, within financial circles, there is an anticipation that project finance will assume a crucial role in propelling forward mitigation and adaptation projects amidst the age of climate change.

Venture Capital, on the other hand, opens doors for early-stage companies pioneering new technologies and business models in climate solutions. By providing equity investments, venture capital fuels innovation and supports high-impact projects with promising potential. The advantages for stakeholders in venture capital include access to capital for high-risk, innovative projects that may not be feasible through traditional lending. It also allows for risk transfer to investors, alleviating the burden on the company and its founders. Moreover, venture capital provides access to expertise and networks that can aid the company's growth and success. However, venture capital can be highly competitive, with numerous companies vying for limited capital. Investors may require a significant ownership stake in the company, limiting the control of the founders. Additionally, the focus on early-stage companies with high growth potential can lead to a prioritization of short-term profitability over long-term sustainability. Illustrations of companies backed by Venture Capital can be observed in the annals of publicly listed NASDAQ companies or on the webpages of renowned Venture Capital enterprises.

And let us not forget the transformative potential of public-private partnerships (PPPs). When government agencies and private sector investors join forces, they unlock a world of possibilities for funding and developing climate projects. This collaborative approach allows governments to leverage private sector funding and expertise, delivering projects that might otherwise remain beyond reach. PPPs can be employed to support various climate solutions, such as renewable energy projects, sustainable transportation systems, or energy efficiency retrofits. PPPs allow governments to leverage private sector funding and expertise to deliver projects that might not otherwise be feasible.

Advantages for stakeholders in PPPs include the sharing of risk and expertise between public and private sector partners. They also offer access to substantial capital for high-impact projects. Moreover, PPPs provide a clear framework for project development and financing, with well-defined roles and responsibilities for all involved parties. However, PPPs can be complex and time-consuming to structure, requiring extensive negotiation and coordination between partners. The focus on project profitability can sometimes overshadow long-term sustainability considerations.

Examples of PPP project is the Scaling Solar program in Africa, which aims to support the development of large-scale solar projects. The program has attracted private sector investment exceeding $5 billion and has resulted in the implementation of several significant solar projects in the region (Scaling, 2022).

But now, let us embark on an awe-inspiring journey to Africa, where the sun's golden rays and the wind's gentle breeze beckon us to witness a tapestry of innovative finance for climate solutions.

When it comes to financing climate solution projects in Africa, the game is all about innovative finance. In the global south, where traditional financing structures often fall short, imaginative approaches are being explored to bolster sustainable initiatives. And one of the brightest examples of this ingenuity shines through the remarkable tale of the Ethiopian Renaissance Dam, a project that defied all expectations with its audacious funding model (Ighobor et al., 2014).

Picture this: a grand vision unfurling in one of the world's most impoverished nations, a colossal dam worth a staggering $4.8 billion nestled in Ethiopia's Benishangul region, near the Sudanese border. Amidst the rugged construction site, some 8,500 tireless workers toil day in and day out, crafting what will soon become the awe-inspiring Grand Ethiopian

Renaissance Dam. As per the project's masterminds, upon its culmination, this hydroelectric marvel is poised to harness the immense potential of the majestic Nile River, producing an astounding 6,000 megawatts of electricity to meet both domestic demands and facilitate international exports.

At first glance, this 558-foot-tall engineering feat appears inconceivable, a seemingly implausible endeavor given Ethiopia's modest financial means. With a meager GDP per capita barely scraping $475, the audacity of this undertaking is nothing short of remarkable. Yet, back in 2011, the late Prime Minister Meles Zenawi, laying the foundation stone, boldly declared that this dam would rise without relying on any handouts from international donors. And true to his visionary words, the construction has marched forward relentlessly, buoyed by a medley of local taxes, donations, and government bonds.

A groundswell of unwavering support emerged as Ethiopians, both within the country and scattered across the globe, rallied behind this momentous project. Their contributions amounted to a remarkable $350 million, while even government employees chipped in, donating a month's salary in solidarity.

But the story doesn't end there. Ethiopia's spirit of ambition and resourcefulness attracted not just local champions but also enticed private companies and neighboring nations like Djibouti to jump aboard, investing in the dam through bond purchases. Simultaneously, the Ethiopian Electric Power Corporation, a state-owned utility, pledged its own revenue and secured loans from state-owned banks, fortifying the project even further.

Of course, as with any audacious undertaking, there were skeptics, economists among them, who cautioned that relying on private sector finance might hinder Ethiopia's future economic growth. But the government had a ready rebuttal – pointing to the long-term gains from selling electricity to the rapidly growing economies of East African countries, a brilliant vision that showcased the grander potential of this power project.

Ethiopia's ingenious financing recipe, blending bonds and taxes, has garnered the admiration of other African nations. And they have good reason to be impressed. Ethiopia has deftly deployed a sophisticated computerized system to track and collect taxes, thwarting evasion attempts at every turn. And they've gone the extra mile by consistently conducting awareness campaigns, educating the public about taxation and ensuring transparency in how the funds are allocated – with the dam serving as a shining beacon of this financial openness.

This exceptional financing approach centered around taxes and innovative funding mechanisms stands as a beacon of inspiration for other African countries seeking ways to fund their own grand-scale projects. With Africa presently collecting a mere 27% of its GDP in taxes, a figure deemed insufficient to address critical infrastructure needs for climate solutions such as roads, bridges, schools, and hospitals, there's a pressing need to explore new avenues of financing.

And in this quest, Ethiopia has blazed a trail of determination, transparency, and citizen support. Their resolute commitment demonstrates that even the most audacious projects can soar to new heights, paving the way for a sustainable future across the vast African continent. In this vibrant dance of sustainable finance, Ethiopia's approach has captivated the attention of other African nations. With a computerized tax collection system and transparent awareness campaigns, Ethiopia has illuminated the path towards innovative financing for climate solutions in the region. As Africa seeks to mobilize resources for sustainable development, Ethiopia's pioneering spirit stands as an inspiration for a brighter future across the continent.

As our journey through the world of climate investments draws to a close, let us remember that the rhyme between historical investments and the urgency of the present climate crisis is within our grasp. We possess the power to harmonize the echoes of the past with the demands of the present, embracing innovative finance and collective action. Together, we can create a symphony of climate investments that harmonizes with the rhythm of our planet, fostering resilience, sustainability, and hope for generations to come.

So, let us stand together, united in our determination to combat climate change, as we embark on a journey of discovery, driven by the vision of a greener, more sustainable world. Let us harness the power of climate investments and rewrite history, ensuring that our actions today rhyme with the needs of our planet and the dreams of tomorrow. For the future is now, and the song we sing can inspire change, making all the difference in this grand orchestration of life.

IV. Unraveling the Book's Revolutionary Tenets: A Journey through Ten Disruptive Theses

Ladies and gentlemen, buckle up for an adventure that will take us on a wild ride through the heart of our planet's most pressing challenges. As we flip through the pages of knowledge and embark on this enthralling quest, captivating narratives interwoven with meticulous research await us, ready to transport us to the very edge of discovery.

Our journey is divided into ten powerful theses, each one an electrifying revelation that will leave you with eyes wide open and minds buzzing. Prepare to be awed as we embark on this thrilling expedition, where hidden truths will be exposed, the status quo challenged, and a collective determination ignited within us all.

Thesis 1: "Climate Change: Unmasking the Undeniable Truth, as Devastation Dawns."

> In the first thesis, we confront the stark reality of climate change. As the devastating consequences loom before us, we shed light on the undeniable truth that demands our attention and action. From extreme weather events to disappearing eco-systems, we explore the urgent need for collective responsibility in addressing this global crisis.

Thesis 2: "Unsustainable Economy: A Path to Destruction, Fueling Inequalities with Ferocity."

> Our second thesis unveils the unsustainable foundations of our economy. As rampant consumerism and resource exploitation fuel inequalities, we confront the stark choice between perpetuating destruction or forging a path towards sustainable prosperity for all. It is time to question the status quo and challenge the very systems that drive environmental degradation.

Thesis 3: "Technology's Illusion: No Magic Wand for the Climate Crisis, Demanding a Revolution of Policies and Mindsets."

> In this thesis, we dispel the illusion that technology alone can save us from the climate crisis. While innovation holds promise, it cannot be a panacea. We delve into the necessity of policy changes and a collective shift in mindsets, realizing that transformative action lies in the amalgamation of human ingenuity and policy reform.

Thesis 4: "Greenwashing Beware: Unveiling Scams that Swindle with Biomass, Solar, Electric Vehicles, and Strategic Metals."

Amidst the burgeoning green movement, we expose the dark side of greenwashing. As companies capitalize on superficial eco-friendly claims, we delve into the deceptive practices that undermine genuine sustainability efforts. It is time to differentiate between true progress and hollow gestures, ensuring that our pursuit of greener alternatives remains authentic.

Thesis 5 "Shadows of Manipulation: Lobbying's Sinister Grip Hinders Climate Action's Noble Stride."

Our fifth thesis delves into the murky waters of lobbying and its impact on climate action. As vested interests wield their influence, we uncover the ways in which powerful lobbies hinder decision-making processes and prioritize the greater good above individual gain.

Thesis 6: "Investment Scams: Exploitation Unveiled, Trapping Innocence in the Web of Climate Deception."

In this thesis, we shine a light on the exploitation within climate investments. As unsuspecting individuals fall victim to deceptive schemes, we reveal the need for stringent regulations and transparent investment practices. A genuine commitment to climate action requires guarding against fraudulent endeavors that jeopardize progress.

Thesis 7: "Neocolonialism 4.0: Shackling the Global South, Amplifying Catastrophe, and Derailing Development."

Our seventh thesis examines the persistent shadows of neocolonialism. As climate impacts disproportionately affect the Global South, we confront the power dynamics that amplify catastrophe and hinder development efforts. True climate justice necessitates dismantling the remnants of colonial oppression and fostering equitable international cooperation.

Thesis 8: "Climate Justice: Who Orders, Pays."

In this thesis, we delve into the fundamental concept of climate justice, aiming to empower marginalized communities and dismantle the remnants of social imperialism respectively social feudalism. Our focus is on achieving a fair sharing of the burden of climate impacts, leaving no one behind in our collective pursuit of a sustainable future. Through an examination of the roles of different stakeholders, we seek to address the question of who orders and who pays in the context of climate justice.

Thesis 9: "Eyes of Hope: Witnessing the Promised Land, Forging a Future Steeped in Possibility."

As we reach this impact chapter that highlights our transformative "man in the moon" climate ventures, hope shines through the shadows of challenges. By acknowledging the truths, we have unveiled and embracing the potential for positive change, we set our sights on a future filled with possibility. Together, we can forge a path towards climate resilience, environmental stewardship, and a world where all beings thrive in harmony with nature.

Thesis 10: "Stewardship in Action: Ask Not What Your Land Can Do for You, But What You Can Do for Your Land."

In our final thesis, we delve into the profound concept of stewardship. Inspired by President John F. Kennedy's iconic words, we call upon individuals and communities to take responsibility for the well-being of our planet. Instead of viewing the land as a resource to be exploited, we encourage a shift in mindset to embrace the role of caretakers and guardians. This thesis challenges us to reevaluate our relationship with the Earth and recognize that we are interconnected with all living beings. By fostering a sense of reverence and respect for nature, we can embark on a journey of conscious action to heal and protect our fragile eco-systems.

As stewards of the land, we must make mindful choices in our daily lives, from reducing our carbon footprint to advocating for sustainable policies. Embracing sustainable practices in agriculture, energy consumption, and waste management becomes essential in our quest to mitigate the impacts of climate change.

Just as the land sustains us, it is our duty to reciprocate by nurturing and preserving it for future generations. By acknowledging the intrinsic value of nature and recognizing its limited capacity to endure human exploitation, we take a pivotal step towards ensuring a thriving planet for all life forms.

In conclusion, this final thesis serves as a call to action – a reminder that the fate of our planet lies in our hands. Let us not passively await salvation but instead be active agents of positive change. By embracing our role as responsible stewards, we can forge a path towards a more sustainable, equitable, and harmonious world. The journey may be arduous, but with determination and collective effort, we can indeed create a future where humanity and the land coexist in symbiotic harmony.

Prepare yourself for an immersive experience, for within these pages, captivating narratives intertwined with meticulous research shall transport you to the heart of this enthralling quest. Join me as we unveil the disruptive truths that lie hidden, challenge the status quo, and ignite a collective determination to carve a path toward a brighter and more sustainable future.

V. Artistic Echoes Of Climate Change: The Prelude Of Betrayal

CHAPTER

Three

Climate Change: Unmasking The Undeniable Truth, As Devastation Dawns

"The Lord God took the man and put him in the garden of Eden to work it and keep it."
– Genesis 2:15

I. Scientific Consensus on Climate Change

In the chronicles of human history, the catastrophe of Mount Pelée in 1902 stands as a stark reminder of the perils that await those who turn a blind eye to scientific warnings and dismiss the value of knowledge in the face of impending disaster. On the picturesque island of Martinique, nestled in the Caribbean Sea, Mount Pelée, an active volcano, had been quietly seething, its volcanic unrest revealing ominous signs of an imminent eruption. Increased seismic activity, ground deformation, and the ominous belching of gas emissions were the telltale signs that should have set alarm bells ringing in the minds of the local authorities and residents.

But alas, in the shadow of Mount Pelée, a chorus of complacency drowned out the pleas of a few wise voices. Among them was geologist Angelo Heilprin, a man of science who grasped the gravity of the situation and sought to sound the alarm. He raised his concerns, attempting to pierce the veil of ignorance that had enveloped the island. Yet, his warnings were met with a dismissive shrug from the powers that be, blinded by their belief that the volcano was nothing more than a dormant giant, lying in wait to be exploited for economic gain (Heilprin, 2011).

Amidst the idyllic beauty of Martinique, a seductive melody echoed through the air, weaving its way into the hearts and minds of the island's inhabitants. It was the siren song of economic interest, an intoxicating tune that promised untold riches and boundless prosperity. And like sailors ensnared by the mythical sirens of ancient lore, the people of Martinique found themselves ensnared in its irresistible grasp.

At the heart of this harmonious deception stood the sugarcane industry, a formidable titan with coffers overflowing with the allure of profit. Its fields stretched as far as the eye could see, the promise of sweet rewards beckoning those who dared to partake. But within this seemingly bountiful landscape, an insidious disregard for inconvenient truths took root, a disregard that would have dire consequences.

While scientists and wise voices raised alarm bells, pointing to the signs of impending doom emanating from Mount Pelée, the sugarcane industry turned a blind eye, casting aside the whispers of warning. In their pursuit of economic gain, they chose to dance with ignorance, entwined in a waltz that led them away from the stark reality unfolding before their very eyes.

The allure of profit, that great seductress, clouded their judgment and obscured their vision. It whispered promises of unending prosperity and cast a veil over the nagging doubts that tugged at the corners of their consciousness. The sugarcane industry, with its deep pockets and vested interests, had little time for the inconvenient truths that science revealed. They clung to the status quo, their desire to maintain their position of power drowning out the desperate cries for caution.

And so, the island's inhabitants, swayed by the dulcet tones of economic interest, walked willingly into the trap that had been set. They basked in the false sense of security that the sugarcane industry provided, shielded by a collective delusion that the volcano slumbered, forever dormant and docile. It was a delicate dance, where hopes and dreams intertwined with complacency and ignorance, blinding them to the imminent danger lurking just beneath the surface.

Then, on that fateful day, May 8, 1902, Mount Pelée unleashed its wrath upon the unsuspecting city of Saint-Pierre. With unimaginable force, the volcano spewed forth pyroclastic flows, a deadly torrent of hot gas, ash, and volcanic debris, hurtling down its slopes and obliterating everything in its path. The once-vibrant city, the jewel of Martinique, was reduced to a graveyard in a matter of minutes. The death toll reached staggering heights, with an estimated 30,000 lives lost, forever entombed beneath the layers of destruction.

The cataclysmic fate of Saint-Pierre stands as a poignant testament to the consequences that befall those who turn a deaf ear to the wisdom of science. Lives were needlessly sacrificed on the altar of hubris and neglect. It is a haunting reminder of the dire importance of heeding the counsel of scientific experts and taking decisive action in the face of impending calamity. For in such moments, lives hang in the balance, and the very fabric of communities can be torn asunder by the tempests of ignorance.

Yet, from the ashes of tragedy, lessons emerged. The aftermath of Mount Pelée's fury brought about a profound awakening, as the world recognized the pivotal role of scientific monitoring and understanding in the realm of volcanic activity. The disaster served as a clarion call for the advancement of volcanology as a scientific discipline, prompting a renewed focus on volcano monitoring and hazard assessment worldwide. Through this crucible of pain and loss, humankind gained invaluable insights into the workings of our volatile planet, fortifying our ability to protect and preserve lives in the face of nature's wrath.

Let the haunting memory of Mount Pelée's catastrophe forever serve as a chilling testament, a stark reminder of the perils that befall those who turn a blind eye to the resounding clarion call of science. It is in the relentless pursuit of knowledge, with science as our guiding light, that we unearth our only hope for navigating the treacherous waters of the climate crisis and forging a future that is both safe and resilient.

As we confront the immense climate challenges of our time, the urgency of embracing the truths illuminated by scientific inquiry has never been more paramount. The evidence is indisputable, echoed by experts from diverse fields, who have meticulously pieced together the puzzle of our changing climate. They weave a narrative of rising temperatures, melting ice caps, intensifying storms, and cascading ecological disruptions that reverberate across the globe.

To dismiss this scientific consensus is to wander blindly into the abyss of ignorance, where the consequences of our inaction loom with ever-increasing gravity. The consequences are manifold and far-reaching, affecting not only the delicate balance of our eco-systems but also the very fabric of human society. It is a call that demands our unwavering attention, a call that cannot be ignored without dire repercussions.

Just as Mount Pelée's eruption revealed the devastating consequences of dismissing scientific warnings, so too does the climate crisis beckon us to confront the uncomfortable truths before us. We stand at a pivotal moment, where our actions today will shape the world of tomorrow. Will we choose to embrace the knowledge that science offers, charting a course towards a sustainable and thriving future? Or will we succumb to the allure of ignorance, consigning ourselves to a path of uncertainty and regret?

Alright folks, buckle up because we're diving now deep into the science of climate change and the concepts that's been making waves lately.

The "carbon budget."

Now, we all know that climate change is a serious threat to our planet and all its inhabitants. To prevent catastrophic impacts, we need to act fast. That's where the carbon budget comes into play.

So, what exactly is the carbon budget? Well, it refers to the maximum amount of carbon dioxide (CO_2) that can be emitted into the atmosphere while keeping global warming within a specific target. This target is often defined as a certain increase in global average temperature above pre-industrial levels, like 1.5 degrees Celsius or 2 degrees Celsius (AR6, 2022).

Now, hold on to your hats because here's where it gets interesting. According to a study by Le Quéré et al. (2018), the remaining carbon budget to limit global warming to 1.5 degrees Celsius is a mere 420 gigatons (Gt) of CO_2. Sounds like a lot, right? Well, think again. When you consider that the world currently emits around 42 gigatons (Gt) of CO_2 every year, that budget would be exhausted in just 10 years. That's right, a decade.

And if you thought that was mind-blowing, get ready for more. The remaining budget to limit warming to 2 degrees Celsius is 1,170 gigatons (Gt) of CO_2, which would be gone in just 27 years at our current emissions rates (AR6, 2022). That's like the blink of an eye in the grand scheme of things (Farquhar, 2018).

To put this into perspective, let's take a trip back in time. Picture the Industrial Revolution in the late 18th century, which marked a significant increase in global emissions. It took humanity roughly 120 years to emit its first trillion tons of CO_2. But guess what? We've managed to emit an additional trillion tons in just the past 30 years (Prentice et al., 2001; Schimel et al., 1995; Watson et al., 1990; Ciais et al., 2013; Canadell et al., 2022; Le Quére et al., 2018). Yeah, let that sink in.

The urgency of action is crystal clear. We need to slash our emissions like there's no tomorrow if we want to avoid surpassing the carbon budget and keep global warming in check. It's going to take some serious changes in our energy systems, transportation methods, and land use practices. And here's the kicker: it's going to require collaboration between governments, businesses, and individuals on a global scale (Thornhill, 2021).

To sum it all up, the carbon budget concept serves as a glaring reminder of the need for immediate action. We must act swiftly to reduce emissions and prevent the catastrophic impacts that loom over our planet and all its inhabitants. Time is of the essence, my friends. Now, let's turn our attention to the overview of current emissions and future projections, as outlined in the seminal work by Peters et al. (2020). Brace yourself for the sobering reality that lies ahead.

In 2019, the world witnessed a staggering milestone in the realm of greenhouse gas (GHG) emissions. Human activities propelled global CO_2 emissions to an all-time high of 36.8 gigatons (Gt), a towering 62% increase compared to 1990 levels. As we dissect these figures, we discover that the energy sector reigns supreme, single-handedly accounting for 72% of the total CO_2 emissions in 2019.

Permit me the opportunity to elucidate this concept through the artistry of visual representation. Let me paint a vivid picture that captures the essence of the matter, a canvas where words take form in colors and shapes, weaving a tale that resonates with clarity and elegance. Step into this realm of imagery, where the intricacies of the subject unfold like brushstrokes on a masterpiece and allow me to guide you through the enchanting landscape of understanding.

Imagine you have a giant see-saw, and on one end, you have the global greenhouse gas (GHG) emissions from human activities, represented by a massive pile of colorful blocks. These blocks represent the CO_2 emissions, which are the most significant contributor to global warming.

Now, let's rewind to the year 1990. Back then, the see-saw was relatively balanced, with the pile of CO_2 emission blocks not too heavy. But as the years went by, human activities like burning fossil fuels, deforestation, and industrial processes started adding more blocks to the CO_2 side. Slowly but steadily, the see-saw began to tilt.

Fast forward to 2019, and you'll witness a jaw-dropping sight. The CO_2 side of the see-saw is now stacked to the sky, towering over the other end. It's as if someone went on a block-building frenzy, and the pile has reached an all-time high of 36.8 gigatons (Gt) of CO_2 emissions. That's a staggering 62% increase compared to the relatively balanced levels of 1990.

Now, let's zoom in on the main culprit behind this colossal pile-up: the energy sector. Picture a massive power plant with smokestacks billowing thick clouds of CO_2 into the air. This single-handedly accounts for a whopping 72% of the total CO_2 emissions in 2019. It's like the energy sector has claimed the throne, reigning supreme over the rest. But why is this significant? Well, these greenhouse gas emissions, particularly CO_2, act like a thick blanket around the Earth, trapping heat and causing the planet to warm up. This leads to climate change, with its cascade of consequences like melting ice caps, rising sea levels, extreme weather events, and disruptions to eco-systems and human societies.

So, in essence, this towering pile of CO_2 emission blocks on the see-saw represents the alarming trajectory of our planet's warming. It's a wake-up call, reminding us of the urgent need to shift the balance back in favor of sustainable practices and cleaner energy sources. The see-saw of emissions can be rebalanced, but it requires global cooperation, innovative solutions, and a shared commitment to a greener future for generations to come.

But what does the future hold? The study by Peters et al. (2021) provides us with a glimpse into the potential trajectories of emissions. In the "current policies" scenario, emissions are projected to surge by 16% between 2019 and 2030. Alas, this trajectory clashes with the goals enshrined in the Paris Agreement, which aspires to confine global warming to well below 2°C above pre-industrial levels. Even more disconcerting is the "pledges" scenario, considering the GHG reduction targets submitted by countries under the Paris Agreement. This scenario foresees a paltry decrease of merely 0.5% per year between 2019 and 2030. While marginally better than the "current policies" scenario, it still falls woefully short of the mark.

"The net-zero."

In the grand fight against climate change, there's another notion that has captured the imagination of nations and organizations worldwide – a net-zero scenario.

Picture this: a delicate equilibrium where the emissions of greenhouse gases (GHGs) into the atmosphere are perfectly counterbalanced by an equal amount of GHG removal. In this state, the net emissions of GHGs, like the notorious carbon dioxide, reach a resounding zero (UN, 2022).

Let's embark on a journey of imagination to understand this delicate equilibrium of greenhouse gases (GHGs) and their removal from the atmosphere. Imagine a grand scale that represents the Earth's atmosphere, and on one side of this scale, we have the emissions of GHGs being poured in. These emissions include carbon dioxide, methane, and other gases that trap heat and contribute to global warming.

Now, on the other side of the scale, picture a series of magical filters and natural processes that remove an equal amount of GHGs from the atmosphere. These filters act like nature's guardians, diligently capturing and neutralizing the harmful gases that we release. They include lush forests absorbing carbon dioxide through photosynthesis, vast oceans absorbing excess heat and carbon, and even some human-made technologies capturing emissions from industrial sources.

As we gaze upon this marvelous equilibrium, the scale remains perfectly balanced. For every GHG molecule that rises into the atmosphere, another is swiftly captured and removed, leaving the net emissions at an awe-inspiring zero. It's like a cosmic dance of give and take, where nature's equilibrium ensures that our actions don't disrupt the delicate balance of the planet.

In this harmonious state, the Earth breathes a sigh of relief. The planet remains in a state of climate harmony, free from the dangerous effects of unmitigated GHG emissions. The skies are clear, and the air is fresh, as if nature herself is rejoicing in the balanced symphony of carbon exchange.

But how do we get there? It demands nothing short of an all-out assault on emissions. Countries and organizations must unleash a torrent of aggressive actions, employing an arsenal of strategies. They're embracing renewable energy sources, overhauling their energy efficiency, adopting sustainable practices in sectors ranging from transportation to agriculture, and even delving into the realm of carbon capture and storage technologies. The concept of net-zero has taken center stage in the fight against climate change. The Paris Agreement, a global treaty signed back in 2015, stands as a testament to this growing urgency. Its lofty ambition? To limit the increase in global average temperature to well below 2 degrees Celsius above pre-industrial levels, with a feverish pursuit of capping it at 1.5 degrees Celsius. Achieving such audacious goals necessitates the attainment of a global net-zero emissions state – a state of equilibrium – by the mid-century, typically hovering around the year 2050 (UN, 2022).

But net-zero scenarios aren't merely about reining in emissions; they also delve into the delicate realm of emission removal. It's a realm where carbon sinks hold sway, natural processes like afforestation and reforestation serving as sanctuaries to soak up those lingering emissions. And there's a touch of technological wizardry as well, with the likes of direct air capture and carbon capture and storage (CCS) throwing their hats into the ring.

Now, it's crucial to grasp that achieving net-zero doesn't imply that GHGs are banished from existence altogether. Some sectors, like aviation and certain industrial processes, may continue to belch out those infamous GHGs. But here's the catch: they're obliged to offset their emissions by an equivalent number of removals from the atmosphere. It's a precarious balancing act, a dance on the tightrope of equilibrium.

All in all, the net-zero scenario represents the historic "holy grail" in the epic battle against climate change. Like the mythical quest for the holy grail, achieving net-zero emissions has been the elusive pursuit of generations, a noble endeavor steeped in legend and mystery. It symbolizes the ultimate prize, a state where the emissions of greenhouse gases (GHGs) are perfectly countered by their removal, akin to the sacred chalice that grants eternal life. Just as knights of old embarked on daring quests to find the holy grail, humanity now stands united in the quest for a net-zero world, a transformative journey that will define our future and safeguard the very essence of our planet. The quest may be arduous, but the rewards are immeasurable – a sustainable future where the balance of nature is restored, and the threat of climate change is vanquished. It's a rallying cry, a beacon guiding humanity towards a sustainable, low carbon future. But make no mistake – it demands the collective resolve of governments, businesses, and individuals alike. We must summon the courage to slash emissions, invest in the dazzling array of clean technologies, and weave sustainable practices into the very fabric of every sector of our economy. The stakes couldn't be higher, and the time to act is now.

"The tipping point."

In the vast tapestry of the world of sciences, there exists a fascinating concept known as tipping point theory. This theory suggests that certain systems, whether natural or societal, possess critical thresholds that, once surpassed, trigger profound and often irreversible changes. Tipping points are akin to delicate balances, delicately poised on the edge of a precipice, waiting for a nudge that could send them hurtling into an entirely new state (Lenton et al., 2019).

Indulge me in the artistry of crafting an intricate scenario within the boundless realm of your imagination, where I shall shed light on the concept of tipping points, disentangled from the intricacies of climate science. Let us venture beyond the conventional confines and explore the universal essence of these critical thresholds that transcend disciplines.

Imagine a giant Jenga tower, where each wooden block represents a different system in the world - the weather, eco-systems, economies, and even social dynamics. These blocks are carefully stacked, forming a delicate balance that holds everything together.

Now, picture these blocks as the tipping points in the world of sciences. Each one has a critical threshold, a point at which it becomes unstable and vulnerable to change. It's like when you remove a Jenga block, and the tower teeters on the edge of collapse, waiting for that one little nudge that could send it tumbling.

In the weather system, for instance, think of a snow-capped mountain. As temperatures rise due to climate change, the snow begins to melt. There comes a point where the melting snow exposes more rock, which absorbs more heat, causing even more snow to melt. It creates a chain reaction, pushing the system past its tipping point, and suddenly, the mountain becomes snow-free, altering the landscape and affecting the surrounding eco-system.

In the social system, consider a community facing rising tensions and inequalities. There's a tipping point where the pressure becomes unbearable, and even a small spark can trigger significant social unrest or change in political dynamics, leading to unforeseen consequences.

These tipping points are like dominoes, with one affecting the next, setting off a series of cascading events. Once they are triggered, they can lead to profound and often irreversible changes – much like when the Jenga tower collapses, and it's almost impossible to recreate the exact same structure.

Understanding tipping points is crucial because they can have far-reaching effects on our world. It's like a game of balance, where we must be mindful of how we interact with our environment and society. By recognizing these critical thresholds, we can work to prevent or manage tipping points, ensuring that we don't push our systems past the edge into unknown territory.

Within the Earth's delicate climate system, there exist precarious tipping points that, if crossed, hold the power to unleash profound and irreversible changes upon our planet. Picture a world where Arctic Sea ice vanishes, the Amazon rainforest withers away, permafrost thaws, and the colossal West Antarctic Ice Sheet crumbles. The consequences of these tipping points are far-reaching and demand our immediate attention.

Let us embark on a journey into the first scenario: the loss of Arctic Sea ice.

The Arctic, a region teetering on the edge of rapid warming, bears witness to an alarming phenomenon – the melting of its iconic sea ice. If this disheartening trend persists, we may find ourselves facing an ice-free Arctic during the summer months in just a few short decades. Brace yourself, for this eventuality carries staggering implications for global weather patterns, sea levels, and the vibrant tapestry of biodiversity.

The loss of Arctic Sea ice sets off a perilous chain reaction, a treacherous feedback loop. Once the shimmering, reflective ice cover succumbs to its liquid fate, a dark ocean surface emerges, greedily absorbing an abundance of solar radiation. This triggers a merciless cycle of further warming and accelerated ice melt. This insidious positive feedback amplifies the warming process within the Arctic, casting its long, ominous shadow upon the delicate intricacies of global climate dynamics (NSIDC, 2020).

Moreover, the melting of sea ice introduces copious amounts of freshwater into the Arctic Ocean, disrupting the established flow of ocean currents. This upheaval bears consequences for the climate of the northern hemisphere, manifesting as altered rainfall patterns and capricious temperature extremes. Its tendrils reach far beyond the Arctic, orchestrating a symphony of atmospheric disruptions that reverberate across the globe's diverse weather systems (Screen et al., 2018).

But the repercussions of Arctic Sea ice loss extend even further – they extend to the extraordinary wildlife that calls this frigid expanse their home. Majestic polar bears, resilient walruses, and elusive seals find their very existence inextricably linked to the presence of sea ice. With diminishing ice cover, their habitats and precious food sources teeter on the precipice of peril, threatening their populations and destabilizing the delicate Arctic eco-system (AMAP, 2017).

And there, amidst the icy panorama, lie the vulnerable coastal communities of the Arctic. The ice, acting as a natural bulwark, stands guard against the onslaught of storms and the relentless erosion of fragile coastlines. Yet, stripped of this elemental defense, these communities face an ever-growing vulnerability, requiring resilience and adaptation efforts to ensure their safety and survival in the face of an uncertain future.

Now, let us shift our gaze to the second scenario that haunts the dreams of conservationists – the Amazon and African rainforests dieback.

To gain a clearer understanding of this situation, let your mind's eye envision the vast and lush expanse of the Amazon Forest, stretching like a verdant tapestry across the horizon. Now, envision Africa's tropical forests, with their rich biodiversity and emerald canopy. Together, these majestic woodlands are often referred to as the "lungs of the Earth."

Imagine these forests as colossal breathing organs, much like our own lungs. Just as we inhale and exhale, absorbing oxygen and releasing carbon dioxide, these forests play a vital role in the Earth's respiratory system. They inhale carbon dioxide, a greenhouse gas, and convert it into life-giving oxygen through the process of photosynthesis.

As the sunbathes the forest canopies with its warm embrace, an orchestra of greenery springs to life. Trees stand tall like guardians of the Earth, their leaves reaching out to absorb sunlight. Through this magical alchemy, they transform carbon dioxide into the very air we breathe, exhaling oxygen to sustain life on our planet.

But here's where the experimental scenario takes a thought-provoking twist. Imagine a world where these forest lungs are under threat. Human activities, like deforestation and forest degradation, start to diminish these vital breathing organs. Just as we feel the strain when our own lungs struggle, the Earth faces a similar challenge.

As these forests shrink, the balance of our planet's respiratory system falters. The exchange of carbon dioxide and oxygen becomes imbalanced, leading to an accumulation of greenhouse gases (GHGs) in the atmosphere. The result? A warmer world, with rising temperatures and more extreme weather events.

In this scenario, we witness the profound impact of preserving versus depleting these forest lungs. When we protect and restore these natural treasures, we safeguard the delicate balance of our planet's climate. We maintain a harmonious symphony of life, where these forest lungs continue to cleanse the air and provide a sanctuary for countless species of plants and animals.

By understanding the significance of the Amazon Forest and Africa's tropical forests as the lungs of the Earth, we come to appreciate the crucial role they play in sustaining life on our blue planet. Just as we cherish and care for our own lungs, let us extend the same reverence and protection to these magnificent woodlands, ensuring they continue to breathe life into our world for generations to come.

The destruction of the Amazon rainforest (Mongabay, 2023) strikes at the heart of our planet's carbon regulation. With its demise, its ability to absorb carbon dioxide falters, leaving the atmosphere burdened with greater levels of this potent greenhouse gas. The result? An intensification of global warming, a catalyst for more frequent and extreme weather events such as crippling droughts, searing heatwaves, and devastating floods. Moreover, the loss of this irreplaceable rainforest threatens to plunge countless plant and animal species into the dark abyss of extinction, forever erasing their unique place within Earth's grand tapestry of life (Nepstad et al., 2014).

Yet, the perils of the Amazon rainforest dieback extend beyond the realm of climate impacts alone. Millions of souls call this verdant expanse their home, drawing sustenance and livelihood from its bountiful embrace. The ruination of the rainforest heralds an era of displacement, poverty, and fierce resource conflicts. What's more, the economic value of coastal areas imperiled by rising sea levels – closely tied to the rainforest's fate – amounts to a staggering sum, reaching into the trillions of dollars (Malhi at al., 2009).

Now, let us cast our gaze upon the third scenario that looms on the horizon – permafrost thaw.

Across vast stretches of the Northern Hemisphere, frozen soil, and rock – permafrost – reigns supreme. Yet, as temperatures soar, this ancient icy tapestry begins to unravel, unleashing a potent concoction of carbon dioxide and methane into the waiting embrace of the atmosphere. The implications of permafrost thaw extend far beyond the confines of climate change, reaching into the very fabric of our planet's delicate equilibrium.

Let's embark on an experimental scenario to grasp the enigmatic concept of "permafrost" that everyone can easily relate to.

Imagine you are on a winter adventure in a snowy wonderland, where the ground is frozen solid. The surface layer is covered in a thick blanket of snow, while beneath lies a hidden world of ice. This frozen ground is what scientists call "permafrost."

Now, picture yourself in a cozy cabin surrounded by frosty landscapes. The cabin's floor is chilly to the touch, as the permafrost underneath keeps the ground frozen year-round. It's like having a built-in refrigerator that preserves the icy conditions regardless of the season.

But here's where the experiment takes an intriguing turn. As the cabin's heater warms the air inside, it begins to affect the permafrost beneath the floor. Slowly, the ground starts to thaw, and you notice a few cracks forming on the icy surface. This is what happens in the real world when permafrost is subjected to rising temperatures due to climate change.

As the thawing continues, you observe that the ground becomes softer and more unstable. The once-frozen surface now becomes muddy and unstable, similar to walking on a partially melted ice rink. This thawing of permafrost can lead to problems for buildings, roads, and infrastructure in the Arctic regions, as the ground becomes less reliable for supporting structures.

Now, let's take this experiment even further. The thawing permafrost releases ancient organic matter, such as dead plants and animals, that has been preserved in the frozen ground for thousands of years. As the organic matter decomposes, it produces greenhouse gases (GHGs) like methane and carbon dioxide. These gases are released into the atmosphere, contributing to global warming and further exacerbating climate change.

This experimental scenario helps us understand the significance of permafrost beyond its icy exterior. It acts as a critical player in the Earth's climate system, storing vast amounts of carbon and influencing the global climate.

As you leave the cabin and step back into the snowy landscape, take with you the understanding that permafrost is more than just frozen ground – it is an integral part of the delicate balance of our planet's climate. Our actions and choices today can impact the fate of permafrost and the broader implications for our environment. By addressing climate change, we can work towards preserving this frozen world and safeguarding the Earth's delicate equilibrium for future generations (Alley et al., 2005; DeConto et al., 2016; Bintanja et al., 2013).

Allow me to expound upon this matter with clarity.

The release of greenhouse gases (GHGs) from the thawing permafrost sets in motion a disconcerting dance of intensified global warming, further fueling the fires of our collective predicament. A treacherous feedback loop takes hold, amplifying the impacts of climate change, creating a harrowing cycle that proves challenging to halt or reverse. Our race against time grows ever more desperate as we witness the unraveling of this ancient icy kingdom.

In the vast expanse of scientific consensus on climate change, a clear narrative emerges, interweaving the concepts of carbon budget, net-zero and tipping points. Scientists warn us that we have a limited carbon budget – the amount of greenhouse gases (GHGs) we can emit before crossing dangerous thresholds. To avoid catastrophe, we must strive for net-zero emissions, where any remaining emissions are balanced by removals. However, tipping points loom ominously, with the loss of Arctic Sea ice, Amazon rainforest dieback, and permafrost thaw presenting grave risks. Projections paint a stark picture, urging immediate action to curtail emissions, transition to sustainable practices, and protect delicate balances. The time to act is now, for the fate of our planet hangs in the balance.

To close this section, let our Jenga tower be a reminder that the world is interconnected, and every action has consequences. Just like carefully playing the game, we must handle our planet's delicate balances with care, knowing that one wrong move could set off a chain of events we might not be able to reverse. By safeguarding these tipping points, we can build a more stable and sustainable future for all.

II. Common Myths And Misconceptions

In the annals of human history, there lies a profound lesson to be learned from the interplay of myths and misconceptions. Across civilizations and epochs, we have witnessed the power, and the potential pitfalls of distorted truths.

The lesson echoes through time, reminding us of the importance of discernment and critical thinking. Myths, with their captivating narratives, have shaped cultures and ideologies, sparking imagination and forging identities. Yet, they also taught us the significance of questioning, of seeking deeper truths beyond the surface.

In the shadows of myths, misconceptions often lurked, luring us into the embrace of half-truths and misunderstandings. These misconceptions served as cautionary tales, illustrating the dangers of blind belief and the need for constant inquiry.

But as the story of mankind unfolds, a brighter revelation emerges. We have learned that the pursuit of knowledge is not a linear path but an evolving journey. We have discovered that the acknowledgment of misconceptions is a catalyst for growth, leading us to challenge preconceived notions and broaden our understanding.

In the present moment and with this perspective in mind, let us first set sail on a captivating journey, immersing ourselves in an enthralling tale that unfolds in my homeland of Germany. Before we delve into the myths and misconceptions surrounding the climate change crisis, this captivating narrative awaits, ready to enchant and enlighten us.

In September 2022, the German electorate spoke through a closed election, ushering in a coalition government comprising the Green Party, the Liberal Democrats, and the Social Democrats. This progressive government, led by a self-proclaimed "Climate Chancellor," seemed poised to orchestrate a remarkable green revolution, enacting legislative and administrative transformations toward a net-zero economy. While their intentions were undoubtedly noble, good intentions alone do not guarantee success.

To grasp the current sentiment that surrounds my story, I turned to the vox populi, listening attentively to the voices within my own circles – friends and acquaintances who are not climate deniers but individuals willing to make personal sacrifices for the greater good. Yet, something had gone awry in Germany, causing these well-meaning individuals to turn their backs on the Climate Chancellor and the Progressive Pro Climate Government. It was the arrogance of those who believed themselves to be the righteous fighters, dismissing people's anxieties as mere myths and misconceptions. But even within these narratives, lies an intriguing thesis that deserves attention, for neglecting it may allow it to fester like malignant cells, endangering democracy, the rule of law, and potentially paving the way for fascism.

One such thesis, not necessarily propagated by climate skeptics but rather by the exhausted and powerless masses, revolves around their immediate concerns – feeding their families, preserving their homes, and securing a decent retirement. They question the heavy burden imposed upon them to save the planet while the rest of the world fails to follow suit. They fear that Germany alone will endure misery while countries like China, the USA, and India prosper. And indeed, there is a grain of truth in their observations, with giant industrial companies such as BASF already announcing their departure from Germany in pursuit of cheaper energy (NZZ, 2023). Does this fear hold the semblance of a mere myth, or is it an urgent challenge that necessitates intelligent solutions for our triumph in the battle against the climate crisis? In the labyrinth of uncertainties, one must not falter, for the failure to discern between baseless charlatans' myths and genuine concerns that warrant attention may pave the way to misfortune, disaster, or even the abyss of something far more sinister – the chilling specter of fascism. With vigilant discernment and clarity, we must navigate this perilous terrain, steering towards intelligent solutions that safeguard our shared future and preserve the essence of humanity's noble aspirations.

Allow me to illustrate further.

Again, step back in time to the advent of the newly elected German government in the year 2022, brimming with ambitious visions of fostering environmentally friendly housing solutions. Embracing the "Gebäudeenergiegesetz" legislative package, their transformative quest aimed to infuse newly installed heating systems in both new and existing buildings with a minimum of 65 percent renewable energy, heralding a brighter, greener future from the dawn of January 1st, 2024 (Traufetter et al., 2023).

Yet, amid these noble endeavors, a disconcerting consequence came to light – homeowners with modest incomes felt burdened by the state's mandate, fearing it could lead to financial strain and even the dreaded specter of homelessness. In the wake of such discontent, a far-right faction, the AfD (AfD), surged in the polls, gaining prominence as the largest party in Eastern Germany. While the region's complex history, transitioning from the shadows of Hitler's rule to Communist governance and embracing Western democracy, might have played a part, it wasn't the sole root cause of this unexpected development.

The correlation between the noble aim of achieving net-zero in the housing sector and the hurried, ill-advised implementation by the Ministry in charge of climate protection raises concerns of a potential catastrophe. Like an undeniable piece of evidence – smoking gun – this situation exemplifies how misguided policies can unwittingly fuel the rise of ideologies that threaten the very foundations of democracy.

In this enchanting yet cautionary tale, we are reminded that even the most well-meaning initiatives require the gentle touch of meticulous consideration, sensitivity to the diverse needs of citizens, and a dash of charming foresight. The pursuit of environmental stewardship must be coupled with empathy, for it is in the delicate balance of thoughtful policy implementation that we chart the course towards a prosperous, harmonious, and democratic future for all. As we traverse this delicate path, let us remember that a touch of charm and thoughtful wisdom can guide us towards a world where sustainability and humanity thrive in harmony.

My cherished friends, it is quite likely that you, too, have discerned this sagacious observation within the vast realm of the climate change debate. In our quest for the most viable path forward, we often witness critical voices being unjustly vilified rather than constructively engaged with.

In contemplating this matter, one cannot help but share the sentiment that instead of resorting to disparagement, it behooves us to redirect our efforts towards proposing inclusive solutions that resonate with a wider audience. Let us cast our net wider, embracing the hearts and minds of diverse communities, rather than solely catering to a select group of environmentally conscious individuals, amidst their champagne-laden gatherings. In fostering a culture of open dialogue and collaboration, we cultivate an environment where innovative ideas can bloom, transcending the barriers of difference and skepticism. Together, hand in hand, we shall forge a brighter future, where collective efforts lead us towards sustainable solutions that embrace the needs and aspirations of all.

Specifically, let us delve into a scenario of paramount importance, which necessitates careful examination. It is perplexing to witness how those advocating for green tariffs on goods with questionable carbon footprints in their manufacturing process, as well as proponents of product traceability and a fair sharing of the financial burden between the carbon-intensive affluent and the masses, are unjustly labeled as communists.

Allow me to provide a comprehensive elucidation on the intricate workings of green tariffs and their profound impact in the global fight against climate change. These tariffs serve as a powerful tool to incentivize sustainable practices in international trade. By attaching monetary value to the carbon footprint of goods, green tariffs prompt manufacturers and producers to embrace environmentally responsible methods.

As the world grapples with the urgent need to reduce greenhouse gas emissions, green tariffs present an opportunity to create a level playing field. Manufacturers who choose eco-friendly practices are rewarded, gaining a competitive edge in the global market. Conversely, those who continue with unsustainable processes bear the cost of their carbon-intensive choices, fostering a transition towards more sustainable alternatives.

The benefits of green tariffs extend far beyond economic competitiveness. By encouraging businesses to reduce their carbon footprints, these tariffs contribute significantly to mitigating climate change on a global scale. With each sustainable decision, the cumulative impact resonates across industries and borders, leaving a lasting impression on the health of our planet.

Moreover, green tariffs complement the efforts of nations striving to meet their climate commitments. They empower governments to proactively address climate concerns while fostering a culture of environmental stewardship within their borders.

In the face of political complexities, where vested interests may sometimes overshadow the will of the people, green tariffs provide a tangible means for true representation of the collective desire for a sustainable future. By embracing this progressive approach, politicians can demonstrate their dedication to safeguarding the environment and enhancing the well-being of their constituents.

In the nutshell, green tariffs are not a manifestation of ideology but a pragmatic and impactful solution in the battle against climate change. They embody the essence of cooperation, shared responsibility, and a commitment to preserving our precious planet for generations to come. Embracing green tariffs is not merely a choice, but a compelling imperative in shaping a world where sustainable practices thrive, and the fight against climate change becomes a unified global endeavor. On occasion, instead of demeaning those who express critical views on climate change, we can channel our efforts more productively by proposing solutions that garner widespread support. Our focus must transcend the realms of eco-enthusiasts who revel in extravagant gatherings, reaching out to the hearts and minds of the broader public.

Contemplating this situation, I am left wondering why advocates of green tariffs on goods sourced from high-carbon manufacturing processes, those advocating for product traceability, and proponents of equitable financial responsibility are consistently branded as adherents of a particular ideology. Where are the diverse perspectives that could enrich the conversation?

Everywhere I turn, I observe politicians attentively heeding the voices of their puppet masters, neglecting the genuine interests of the vox populi. In this labyrinth of complexity, the call for visionary leadership and open dialogue becomes ever more imperative.

It is time to bridge the gaps and seek common ground, cultivating solutions that resonate with a broader audience. By embracing inclusive approaches, we can transcend ideological barriers and unite in a collective pursuit of a sustainable future.

In our quest for enlightenment and understanding, let us navigate away from the treacherous waters of divisive rhetoric, and instead, embrace a serene environment where genuine dialogue and constructive proposals find fertile ground to flourish. By doing so, we liberate ourselves from the confinements of labels and vested interests, embarking on a transformative journey towards a future where the well-being of our planet and all its inhabitants resides at the core of our collective aspirations.

As we embark on this voyage, our compass points towards a tour of the celebrities of climate myths and misconceptions. Guided by the wisdom of esteemed scientists and experts, we shall navigate through the murky waters of misinformation, skillfully debunking each myth that seeks to obscure the undeniable truths of climate science.

In the company of renowned climate scholars, we shall uncover the main fallacies that have taken root in the public discourse, untangling the web of misconceptions that clouds our path towards meaningful action. Armed with knowledge and evidence, we shall pierce through the shadows of doubt, illuminating the way forward with the brilliance of truth (Maslin, 2023; Maslin, 2014).

Together, we shall celebrate the triumph of facts over fiction, and the victory of reason over denial. Let us embark on this illuminating journey, where the light of knowledge guides us towards a future of unity, stewardship, and a sustainable world for generations to come.

Our first stop on this expedition brings us face to face with a persistent fallacy: the notion that climate change is merely a part of the natural cycle. Ah, but fear not, for the study of palaeoclimatology, which examines past climates, reveals a different story altogether. The changes witnessed in the last 150 years, since the advent of the industrial revolution, have been truly exceptional, defying the boundaries of natural variability. Model projections even hint at the possibility of unprecedented future warming, unlike anything experienced in the past five million years.

To further debunk this myth, let us challenge the tale of the Earth's climate recovering from the cooler temperatures of the Little Ice Age, or the resemblance of today's temperatures to the Medieval Warm Period. Alas, both climatic events were regional in nature, affecting specific areas such as north-west Europe, eastern America, Greenland, and Iceland. The true magnitude of change can only be witnessed in the last 150 years when a staggering 98% of the Earth's surface has undergone warming simultaneously.

Prepare to embark on a celestial odyssey, where we unravel the enigmatic dance between sunspots and galactic cosmic rays, an interplay that wields a subtle yet profound impact on Earth's climate. Sunspots, those mesmerizing tempests that ebb and flow upon the sun's fiery canvas, hold the key to a cosmic spectacle that captivates both scientists and stargazers alike.

In their dynamic presence, sunspots possess a unique ability to influence the climate of our beloved blue planet. Their magnetic might create a dance of radiation, gently weaving a complex tapestry of energy that touches our very atmosphere. For centuries, inquisitive minds have sought to understand the extent of their influence, delving into the very heart of celestial phenomena. Yet, let us journey further, beyond the realm of folklore and fable, to a realm where empirical data illuminates the way. Satellite observations, a modern marvel of scientific ingenuity, have diligently surveyed the cosmic stage since 1978. And what have they revealed? A captivating revelation, indeed! For in the annals of satellite records, there lies no discernible upward trend in the sun's radiant energy reaching our precious planet.

Thus, while sunspots exude their magnetic charm and orchestrate the cosmic ballet, they cannot, with certainty, be attributed as the sole cause of the recent global warming we grapple with today. Instead, a symphony of forces – both terrestrial and celestial - play their part in shaping the climatic narrative.

Now, prepare for a captivating clash between the realms of common sense and scientific revelation. Some skeptics assert that carbon dioxide's minute presence in the atmosphere renders it incapable of exerting any significant warming effect. Yet, let us embark on a journey through the corridors of scientific history and uncover the pioneering work of the remarkable American scientist, Eunice Newton Foote.

Long before the modern discourse on climate change emerged, in the year 1856, Eunice Newton Foote graced the stage of scientific inquiry with her brilliance. A visionary woman ahead of her time, she was not only an inventor and an advocate for women's rights but an amateur scientist who dared to explore the interplay of the sun's rays with various gases.

Through her meticulous experiments, she postulated that higher levels of carbon dioxide would usher in a warmer planet. Foote ingeniously designed an experiment, employing an air pump, glass cylinders, and thermometers, to showcase the greenhouse effect of carbon dioxide. When exposed to sunlight, the cylinder filled with carbon dioxide retained more heat and maintained its warmth for a remarkably longer duration than the one containing normal air.

Her groundbreaking discoveries continue to resonate through time. While countless subsequent scientific experiments persistently validate the greenhouse effect of carbon dioxide, a handful of skeptics, including the renowned Roy Spencer, tenaciously adhere to the idea that, Foote's pioneering experiments may not fully apply to the vastness of our atmosphere. They bolster their stance as a matter of "common sense," arguing that carbon dioxide's minute presence in the atmosphere renders it impotent to induce a substantial warming effect.

So, let us quell this skepticism with an illuminating analogy borrowed from another public source (Nuccitelli, 2016). Consider this: a mere 0.1 grams of cyanide, a fraction of 0.0001% of an adult's body weight, has the chilling power to bring about death. Similarly, carbon dioxide, presently comprising 0.04% of the atmosphere, wields a potent greenhouse effect. It is a force to be reckoned with, defying the notion that its small presence could be inconsequential. In contrast, nitrogen, constituting a staggering 78% of the atmosphere, remains relatively inert despite its overwhelming abundance.

As we venture through this intersection of science and sensibility, we are reminded that appearances can deceive. The seemingly minute concentration of carbon dioxide belies its immense influence on our planet's climate.

As a pristine landscape unveils itself before us, a new realm of exploration emerges when we confront an alternative perspective put forth by climate change skeptics. Their assertion suggests that scientists covertly manipulate data to craft a narrative of a warming climate, leading us away from the actual truth. However, it's essential to note that this claim, put forth by climate skeptics, lacks substantial evidence and has been widely discredited through rigorous scientific and comprehensive investigations.

In this territory of skepticism, we navigate through the intricate web of evidence and reason by connecting the dots in one confrontation, seeking to unveil the undeniable truths that lie beneath the surface.

Let us embrace the pursuit of this intellectual confrontation now.

Numerous studies and independent analyses have confirmed the robustness and integrity of climate data. For example, a comprehensive study published in the journal Environmental Research Letters (Sillmann, 2014) examined temperature data from multiple sources, including satellites, weather balloons, and surface stations. The study found that all datasets consistently showed a warming trend, aligning with the conclusions drawn by leading climate scientists (NASA, 2023; AAAS, 2014; ACS, 2019; AGU, 2019).

Furthermore, investigations into alleged data manipulation have been conducted by independent organizations and found no evidence of systematic or deliberate misrepresentation. For instance, the Berkeley Earth Surface Temperature project, led by physicist Richard Muller, was specifically initiated to examine concerns of data manipulation. After conducting an extensive analysis, the project confirmed the warming trend and attested to the credibility of the existing temperature records (Hausfather, 2023).

The Intergovernmental Panel on Climate Change (IPCC), a reputable international body of climate scientists, has undergone rigorous peer-review processes for its reports, ensuring transparency and accuracy in climate assessments.

These references, along with numerous other scientific studies and assessments, underscore the integrity and reliability of climate data and dispel claims of data manipulation. It is essential to rely on reputable scientific sources and peer-reviewed studies to gain a comprehensive understanding of climate change and its impacts.

Before concluding this exploration of myths and misconceptions surrounding climate change, it is imperative to address a particularly perilous and misguided anecdote. In an article penned by Simon Usborne and published in The Guardian (2022), under the title "More people is the last thing this planet needs: the men getting vasectomies to save the world," a concerning trend is brought to light. Usborne reports that an increasing number of young, childless men are opting for sterilization as a drastic measure driven by environmental concerns in the face of the escalating climate crisis.

However, it is important to challenge the premise underlying this narrative. While it is true that the climate crisis demands urgent attention and action, the focus should not be fixated solely on individuals' reproductive choices. The crux of the matter lies in the carbon footprint we leave behind, not in the footprint of our reproductive organs and sperms. As a famous U.S. politician once quipped, "It is the carbon footprint, stupid." Our efforts must be directed towards reducing greenhouse gas emissions, transitioning to renewable energy sources, and implementing sustainable practices across industries, rather than solely scrutinizing individual decisions regarding parenthood.

In the realm of climate change, it is paramount to distinguish between the systemic changes required to combat carbon emissions and the personal choices individuals make. By refocusing the conversation on addressing the root causes of environmental degradation, we can make significant strides toward safeguarding the future of our planet, transcending the realm of anecdotal, and potentially misleading, distractions.

III. Urgency Of The Situation

In the face of an urgent and pressing global challenge, it is all too easy to become entangled in the complexities of theoretical studies and statistical analyses. While these endeavors certainly have their place in understanding the gravity of the situation, there is something inherently powerful in connecting with the human experience, in envisioning the potential scenarios that could unfold based on our unique circumstances. It is with this spirit of imagination and projection that we embark on a journey through the following chapter.

Rather than relying solely on dry data and academic discourse, let us explore a tapestry of possibilities. Let us dare to envision a world where our income, health, age, and country of residence intertwine with the daunting realities of climate change. Through a touch of fantasy and the ability to connect the dots, we can paint vivid pictures of what lies ahead if we do not act with urgency.

In the pages that follow, we will delve into compelling scenarios, each one grounded in the intricacies of personal circumstances and global challenges. These scenarios will serve as catalysts for action, igniting a flame of motivation within us to sound the alarm bell, reach out to our representatives, mobilize our communities, and perhaps even embark on our own projects alongside our peers. The time for complacency has long passed; now is the moment to rise and do something.

As we embark on this exploration, let us remember that the urgency of the situation knows no boundaries. It does not discriminate based on wealth or power. It affects us all, albeit in different ways. By embracing our capacity to imagine and project, we can uncover the reasons why sleep should evade us, why wakefulness should become our ally in the pursuit of change.

Together, let us uncover the threads that bind us, the shared responsibility we bear, and the transformative actions we can take. As we turn the page, be prepared to confront the reality of our world, the consequences of inaction, and the possibilities that lie within our grasp. The time for apathy has passed; the time for action is now.

In the face of climate change, a sobering reality emerges – its consequences are far from equitable. The burden of its impact disproportionately falls on marginalized communities, already burdened by disadvantages and lacking the resources to adapt or mitigate the fallout. It is within these communities that the true cost of climate change becomes starkly apparent, as access to fundamental necessities such as food, water, and healthcare becomes increasingly precarious. To truly comprehend the magnitude of this issue, we must delve into the experiences of marginalized communities across developed nations, the Global South, and in particular the African continent.

Within developed countries, the faces of marginalized communities are diverse, encompassing low-income individuals, people of color, indigenous populations, and those residing in vulnerable areas like floodplains, coastal regions, or industrial zones. These communities face compounded challenges due to poor air quality, placing them at heightened risk of respiratory conditions. Moreover, they bear the brunt of extreme weather events, such as heatwaves, floods, and hurricanes, which result in displacement, property, and infrastructure damage, and tragically, loss of life. The intricate details of the unbridled destruction, giving rise to scenes of tumultuous uproar and chaos, will be meticulously examined in the forthcoming sections.

Venturing into the Global South, marginalized communities often comprise rural inhabitants, indigenous groups, and those residing in informal settlements within urban areas. Their livelihoods heavily rely on farming, fishing, or natural resource extraction, linking their existence intimately to fragile eco-systems ravaged by climate change. Droughts, floods, and erratic weather patterns threaten their ability to produce food, access clean water, and sustain their means of survival. For instance, Bangladesh grapples with rising sea levels and intensified storms, leading to the displacement of millions, many of whom were already trapped in the grip of poverty (IPCC, 2014; Roy et al., 2022).

In Africa, marginalized communities often include rural populations, indigenous groups, and pastoralist communities intricately intertwined with natural resources for sustenance, such as agriculture, fisheries, and livestock farming. Climate change exacerbates their existing hardships of poverty, food insecurity, and conflict, directly jeopardizing their way of life. In the Sahel region, recurring droughts and desertification accelerate soil degradation and the loss of fertile land, further intensifying food insecurity and displacement.

What compounds the hardships faced by these marginalized communities is the disconcerting reality that well-off, wealthy, and well-connected populations may even benefit from climate change. Coastal property owners, for instance, witness their holdings appreciate as sea levels rise.

Likewise, disaster response companies witness a surge in profits in the aftermath of extreme weather events. This dichotomy only serves to widen the gap between the privileged and the marginalized, exacerbating existing inequalities and fostering a climate change-induced social injustice.

Step into the haunting aftermath of Hurricane Katrina and let the emotions of the victims envelop you. In the wake of this catastrophic force that besieged the United States in 2005, the truth of its disproportionate impacts lies in the harrowing tales of specific events. Among the broken remnants, vivid reminders emerge, etching the pain and despair faced by the most vulnerable.

Amidst the chaos, the city of New Orleans bore the brunt of devastation, and within its heart, low-income African American neighborhoods endured the harshest blows. The weight of the tragedy bore down heavily, shattering the lives of the marginalized communities. Visualize the once thriving streets now submerged under floodwaters, with 300,000 homes left damaged or completely obliterated. In the predominantly African American Lower Ninth Ward, where life already clung to a thread, a staggering 4,500 out of 5,500 homes were reduced to rubble.

As the floodwaters receded, the haunting silence echoed the loss of livelihoods and shattered dreams. The toll of job losses surged as businesses succumbed to the unforgiving tempest, leaving families adrift in a sea of uncertainty. The pillars of healthcare, already fragile in these neighborhoods, were further eroded, leaving the vulnerable without the care they so desperately needed.

Education, the beacon of hope for a better future, crumbled like ancient ruins. The education systems, already strained by inequalities, could no longer offer solace to their students. Innocent children found themselves uprooted, their sense of security ripped away. In this dark aftermath, the weight of pre-existing inequalities only intensified, leaving a scar that would take generations to heal.

Within the wreckage, one cannot help but feel the anguish and despair of those left grappling with the aftermath. Lives upturned, dreams shattered, and futures uncertain – Hurricane Katrina became a haunting embodiment of the deep-seated disparities that persist in our society.

These vivid tales from the aftermath of Hurricane Katrina remind us of the painful truth – that when nature unleashes its fury, it is often those already burdened by the weight of inequality who bear the heaviest load.

We must heed these stories and strive to build a world where the most vulnerable are not left defenseless against the storms of life.

Picture now stepping into the heart-wrenching aftermath of the recurring California wildfires, and feel the emotions that engulf the victims. These infernos, unleashed with merciless fury, scorch vast expanses of land in the Western United States, leaving in their wake a trail of devastation that knows no bounds. Amidst the charred remains, the poignant reality surfaces – the flames show no mercy, and the burdens fall disproportionately on the shoulders of the vulnerable.

Visualize the scenes unfolding in the low-income communities, where families struggle to eke out a living, living in mobile homes or remote rural areas. The fire's merciless grasp takes hold, leaving nothing but ashes and sorrow in its wake. In 2018, the Camp Fire roared through these communities, leaving a haunting legacy of destruction. Over 18,000 structures turned to rubble, and among them, 13,000 homes were reduced to mere memories. Families, already burdened by financial constraints, are now left grappling with the harsh reality of not having the means to rebuild or relocate ("Camp Fire", 2018).

In the midst of this inferno, the most heart-rending tragedies unfold - the loss of numerous lives, each a story of pain and tragedy. The victims, predominantly elderly or disabled individuals, lacked the swift means to evacuate from the path of the encroaching blaze. Their vulnerability transformed the wildfire into a relentless adversary, with lives tragically lost in its unyielding grip.

Amidst the smoky haze, the grief and sorrow are palpable. Families left homeless, dreams reduced to ashes, and lives forever altered - the California wildfires spare no one, but their wrath falls most heavily upon those already burdened by the weight of poverty.

In the aftermath of these calamities, the heartache lingers, a testament to the urgent need for change. The recurring wildfires serve as a stark reminder that climate change does not discriminate – it takes aim at all, but it is the marginalized who suffer the most.

Amidst the relentless blaze of a scorching summer, the heartache of heatwaves intensified by the relentless force of climate change unfolds in harrowing detail. Picture the humble neighborhoods of low-income communities, where the absence of air conditioning or cooling mechanisms transforms homes into veritable ovens, trapping families in an unyielding embrace of suffocating heat. In this unforgiving landscape, vulnerability takes root, and the struggle for survival becomes a grueling ordeal.

In the Pacific Northwest, the summer of 2021 bore witness to an unprecedented heatwave that stormed through the region, leaving a trail of devastation and sorrow in its wake. As temperatures soared to unimaginable heights, the vulnerable among us found themselves besieged, unable to escape the unyielding grip of the merciless heat. Tragedy struck the elderly and the homeless with particular ferocity, as their meager means left them ill-equipped to combat the sweltering temperatures (Perkins-K., 2020).

The elderly, once pillars of wisdom and resilience, now stood defenseless against the relentless onslaught of the sun. Their fragile bodies, weathered by the passage of time, trembled under the oppressive weight of the heat, seeking refuge that seemed forever elusive. The homeless population, already burdened by the harsh realities of life on the streets, found no respite from the searing temperatures that rendered their makeshift shelters unbearable.

As the sweltering days turned to restless nights, the air was thick with grief and loss. Families mourned the lives cut short, and communities grappled with the void left behind by the departed. The echoes of anguish reverberated through the streets, a haunting lament for those claimed by the unforgiving heatwave.

Now, as the July 2023 heatwave grips Southern Europe, the implications for relatively poorer regions loom ominously. The elderly, with their small pockets and limited resources, face an uphill battle against the relentless heat. The sweltering temperatures weigh heavily on these vulnerable souls, as the specter of heat-related illnesses and even death casts a shadow over their lives.

In this landscape of hardship and adversity, the urgency of addressing climate change becomes ever more pronounced. The July 2023 heatwave serves as an unyielding reminder of the stark inequalities that define our world. It calls us to action, urging us to champion sustainable practices and advocate for justice in the face of climate-induced hardships.

As the sun sets on another day of scorching heat, will a glimmer of hope emerge?

The heart-wrenching tale of the recent flooding in Ahrtal, Germany, strikes a personal chord within me, for it is a place where cherished family and dear friends reside. In the aftermath of the devastating floods, I found myself standing shoulder to shoulder with my friends, volunteering to help the victims rebuild their shattered lives. As I witnessed the resilience of the inhabitants of this idyllic and romantic region, Ahrtal, Germany, my admiration for their courage and strength soared.

The images of the July 2021 flooding are etched into my mind like haunting canvases of despair. Nature's fury unleashed with unbridled force, turning serene landscapes into scenes of destruction and loss. In the quiet villages nestled along the riverbanks, the floodwaters swept away not only homes but also the dreams and aspirations of the people who called this place home.

In the face of the catastrophe, rural, low-income communities bore the brunt of the devastation. The floodwaters showed no mercy, sparing neither livelihoods nor hope. For farmers and small business owners, the impact was especially profound, as their lives revolved around the land and the businesses they had nurtured for generations. Their crops, their livestock, their homes – all swallowed by the relentless deluge.

As the floodwaters receded, they left behind a landscape of desolation and heartache. Lives upended, dreams shattered, and the weight of uncertainty hung heavy in the air. The financial toll, soaring into the billions of euros, laid bare the vulnerability of these communities, revealing the urgent need for comprehensive support and solidarity.

The harsh reality of climate change looms large in the aftermath of the flooding. As extreme weather events become more frequent and intense, the suffering of marginalized communities is exacerbated, making it a clarion call for action. Ahrtal, Germany, stands as a poignant reminder that climate change does not discriminate; it seeks out the most vulnerable and leaves behind a trail of devastation.

In the face of such adversity, the resilience and strength displayed by the inhabitants of Ahrtal, Germany, inspire awe. Their indomitable spirit, forged in the crucible of hardship, propels them forward in their journey of rebuilding and healing. Journey with me to the sun-scorched lands of the Global South, where the beauty of nature clashes with the harsh realities of climate change. Here, amid the sprawling landscapes of Mozambique and Madagascar, we bear witness to the profound struggles faced by marginalized communities, their very existence in the throes of an unforgiving climate.

As we approach Mozambique in 2019, we are met with a sight that wrenches the heart – Cyclone Idai, a tempest of destruction, unleashing torrents of rain and winds upon the land. In its wake, a haunting aftermath unfolds. Homes lie in ruins, possessions scattered like broken dreams, and lives lost in the relentless fury of nature's wrath. We see them, the rural and indigenous communities, grappling with despair as they search for loved ones amidst the debris, their cries echoing into the void.

In the labyrinth of makeshift settlements, desperation hangs heavy in the air. The lack of rescue services leaves them feeling abandoned, like castaways adrift in a sea of calamity. We stand beside them, witnessing the desperation etched upon their faces as they climb the trees, begging for their lives, clinging to hope in the face of impending doom.

Venturing further, we arrive in drought-stricken Madagascar. The earth, once teeming with life, now lies parched and barren, the cries of the suffering echoing across the vast plains. We see them, the rural farmers and indigenous groups, toiling tirelessly in the blistering sun, their sweat mingling with tears as they fight a losing battle for sustenance.

In 2021, the drought reaches a breaking point, pushing over a million people to the brink of desperation. Hunger gnaws at them, and the lack of water is a relentless torment. Parents retire to bed at night, their hearts heavy with the burden of providing for their children, uncertain if there will be food to fill their bellies come morning.

As we delve deeper, we witness the devastation wrought by Cyclone Enawo in 2017, another cruel twist of fate for the people of Madagascar. The cyclone's fury tears through their homes and crops, leaving in its wake a trail of destruction. We see them, the rural farmers and indigenous groups, left to pick up the pieces of their shattered lives, grappling with the harsh reality of loss and displacement.

In the face of such adversity, the resilience of these communities is awe-inspiring. Their determination to survive in the face of unfathomable hardship is a testament to the strength of the human spirit. Yet, it is a struggle that should not be theirs alone to bear.

The challenges posed by climate change have laid bare the inequalities that persist in our world. As we bear witness to the devastation and hardship endured by the marginalized, we are called to action. The plight of Mozambique and Madagascar is a call to break the chains of indifference and empower those most vulnerable among us.

In all instances pictured in this section, marginalized communities bear the disproportionate weight of climate change despite contributing minimally to its causes. Shockingly, a mere 10% of emissions are attributed to the poorest half of the global population, while the richest 10% accounts for over 50%. This staggering inequality underscores the urgent need for climate solutions that prioritize equity, justice, and the inclusion of marginalized communities (IPCC, 2014; UNICEF, 2015).

Climate change is not solely an environmental issue; it is an issue of social justice. Its impact deepens existing inequalities, amplifying the struggles of marginalized communities. As we forge a path forward, it is imperative that solutions to the climate crisis prioritize the principles of fairness and inclusion, recognizing and addressing the unique needs and perspectives of those most vulnerable to its effects (Chancel et al., 2023).

In the realm of contemplating the urgency of action, one might be inclined to dismiss these discussions as mere theoretical exercises, akin to what the French refer to as "hypothèse d'école." However, allow me to take you on a trip to the future, say, the year 2050, to offer a glimpse into what life might look like in your city – assuming there is still a city to speak of – and how these factors may impact what remains. While my predictions draw upon various sources, including The Future We Don't Want project, a collaborative effort involving C40 Cities (C40, 2018), Global Covenant of Mayors (Mayors, 2023), and the Urban Climate Change Research Network (UCCRN, 2023), as well as insights from esteemed scholars and literature, I must acknowledge that some inspiration has been derived from mainstream media sources (Nova, 2023; Williams, 2023; de Trenqualye, 2023).

Our expedition to the year 2050 is a random endeavor, allowing us to employ common sense to locate a city analogous to yours, for the purpose of understanding the potential consequences of the climate crisis. Regardless of your location in 2050, your life will undoubtedly be influenced – whether positively or negatively – by a range of factors, including extreme heat, the intersection of extreme heat and poverty, water availability, food security, sea level rise, and the vulnerability of nearby power plants to rising sea levels.

Now, step aboard my future climate ship as we embark on our exploration of 2050. Presently, approximately 350 cities on Earth already experience extreme heat conditions, where the 3-month average maximum temperatures reach at least 35°C (95°F). Surpassing 200 million people reside in these cities, representing 14 percent of the global urban population currently subjected to high heat conditions. However, by the 2050s, over 970 cities will regularly encounter these scorching temperatures, exposing more than 1.6 billion individuals to extreme summer heat. Astonishingly,

45 percent of the global urban population will be residing in cities with high summer temperatures, signifying a 700 percent increase in the number of people affected compared to today.

Step into the haunting realm of a future shaped by the relentless force of climate change, where cities once bustling with life and prosperity now teeter on the precipice of catastrophe. As we delve into the report by Moody's Analytics (Kamins, 2023), a chilling reality emerges, painting a vivid picture of the life-altering impacts that await us.

Picture yourself in the vibrant city of San Francisco, a thriving metropolis nestled between the rolling hills and the sparkling sea. Yet, hidden beneath its beauty lies a lurking threat. Rising sea levels threaten to engulf this iconic city, transforming it from a beacon of innovation into a sinking symbol of climate despair. The streets, once bustling with energy, may soon be submerged, leaving the cherished homes and businesses buried beneath the waves.

As we journey further, the bustling streets of New York City come into view. The heartbeat of America, a testament to human ingenuity and ambition, now faces an existential struggle against the forces of nature. Extreme heat grips the city, turning its concrete jungle into a furnace of unforgiving temperatures. The air, once filled with the promise of opportunities, now hangs heavy with the stifling weight of climate-induced hardships. As businesses falter, and health struggles, the very fabric of society strains under the relentless pressure.

In the sun-scorched landscape of Phoenix, Arizona, we confront a vision of life unbearably transformed by extreme heat and water stress. The once-thriving desert oasis now battles for survival, as water becomes a scarce and precious commodity. The struggle for sustenance and shelter weighs heavily on its residents, amplifying the inequalities that have persisted for generations.

Across the Atlantic, the city of Miami, Florida, stands as a haunting symbol of our planet's fragility. The uttered sentiments resonate within the corridors of our thoughts, foretelling a grim fate. If global temperatures rise by a mere 4°C, Miami risks becoming a ghost town, its glittering skyline lost beneath the unforgiving waves. The very essence of this coastal paradise hangs in the balance, as rising sea levels inch ever closer, threatening to erase the city from the map.

The consequences of climate change extend far beyond city limits, rippling across borders and triggering waves of migration. Faced with the encroaching tides, communities are forced to leave their homes, seeking refuge in new lands, displacing lives, and reshaping societies. The challenges that lie ahead are monumental, demanding bold and urgent action.

The sun beats down relentlessly on the vast expanse of Australia, its scorching rays penetrating even the bustling cities, leaving no soul untouched by the oppressive heat. Now, Australia calls us to bear witness to its struggle with extreme heat, a battle that reverberates far beyond its remote regions, seeping into the heart of major cities.

Imagine yourself in Western Sydney, a region that stands as a testament to the magnitude of the crisis. The Australian Institute paints a stark picture, revealing that this very area is one of the most severely affected in the entire country. In the not-so-distant past, the number of days per year with temperatures soaring above 35°C (95°F) stood at a challenging average of 9.5 days. However, with the unforgiving passage of time, that number has surged to a staggering 15.4 days in the past decade alone. Yet, the worst is yet to come, as we face an alarming prophecy for the year 2090 - a haunting vision of up to 46 scorching days annually, and certain suburbs enduring an inconceivable 58 days of blistering heat.

The implications of such relentless heat go far beyond discomfort. The lucky ones may endure painful muscle cramps or debilitating exhaustion, while others may suffer severe heatstroke, their bodies unable to withstand the relentless assault. Tragically, in the most heart-wrenching cases, the unforgiving heat claims lives here.

Life, as once known in this city, now teeters on the edge of survival. The once-vibrant streets, filled with laughter and bustling commerce, now bear witness to struggles for survival. Houses stand as sanctuaries against the heat, yet even they offer little respite from the relentless assault of the sun. The very fabric of society is reshaped, as businesses falter and health systems strain under the weight of an unprecedented crisis.

Migration, too, becomes a defining theme in this unfolding drama. As the mercury climbs, communities are forced to grapple with the idea of leaving their homes in search of cooler havens. The struggle for resources intensifies, as competition for water and shelter takes a toll on the most vulnerable.

As we step into the heart of Jakarta, Indonesia, a city teeming with life and culture, we are met with a scene that reveals the harsh reality of its vulnerability. The air hangs heavy with pollution, a noxious cloud that lingers, threatening the health and well-being of its inhabitants. As we navigate its bustling streets, we are confronted with the magnitude of the challenges that plague the capital.

A report by Verisk Maplecroft (2021) lays bare the grim truth – Jakarta is deemed the most vulnerable city on the planet. A convergence of environmental threats looms over this vibrant metropolis, creating an inescapable vortex of peril. The city rests upon a precarious foundation, facing both the tremors of seismic activity and the ominous specter of flooding. But perhaps the most haunting revelation is that Jakarta has earned the unnerving title of "the world's fastest-sinking city."

The city's battle with environmental adversity is further compounded by a dire water crisis. With limited access to clean water, its more than ten million citizens are forced to extract it from the very ground they stand on. This relentless extraction causes the land to subside, exacerbating the sinking phenomenon that threatens to engulf Jakarta. The water they depend on, a source of life and sustenance, turns against them, claiming the very land they call home.

In the face of such a dire situation, the struggle for survival becomes a way of life. Housing, once a refuge from the chaos of the world, now stands as a fragile fortress against a relentless onslaught. The health of the people is compromised, as pollution fills their lungs, and the looming threats of seismic activity and flooding haunt their every step. Businesses grapple with the reality of an uncertain future, as they navigate the treacherous terrain of environmental instability.

The plight of migration looms large, as the crisis forces difficult decisions upon the city's residents. The question of whether to stay and brave the storm or seek refuge in more stable lands hangs heavily in the air. The struggle for resources intensifies, as the search for water and shelter becomes a matter of survival.

Amidst this uncertain landscape, we are left to ponder the fate of Bali and Java Island, the cherished havens of natural beauty and enchantment. The very essence of these paradises is entwined with the delicate balance of the environment. Will they remain the idyllic destinations we have come to cherish, or will the relentless march of climate change and environmental challenges alter their beauty forever?

The audacious plan to build a new city in Kalimantan, on the island of Borneo, emerges as a daring vision of hope amid the shadows of adversity. President Joko Widodo's ambitious endeavor holds the promise of a new beginning, envisioning a future where the nation's capital can thrive on more stable ground. But the path ahead is fraught with challenges and uncertainties, as Jakarta's ten million citizens grapple with the weight of an unprecedented crisis.

As we venture into the bustling metropolis of Delhi, India, we are met with a city that stands at the forefront of the climate crisis. Verisk Maplecroft's risk report paints a stark picture of the challenges that confront this vibrant urban center, ranking it as the second most affected city by climate change. The gravity of the situation is underscored by the fact that India is home to thirteen of the top twenty highest-risk urban areas, highlighting the widespread vulnerability of its urban populations.

The air hangs heavy with pollution, casting a pall over the city and its inhabitants. Delhi grapples with severe contamination, where the very air they breathe becomes a silent assailant on their health. The haunting truth emerges – "noxious air causes almost one in five deaths in India." Each breath carries the burden of danger, as respiratory illnesses and cardiovascular diseases become all too common among the city's residents.

Amidst this toxic environment, life becomes a precarious dance with danger. Housing, once a sanctuary of solace, now offers little respite from the relentless onslaught of pollution. Businesses struggle to survive, navigating the treacherous waters of economic uncertainty, as the health of their workforce is compromised.

The struggle for health becomes an unyielding battle, as the population grapples with the dire consequences of alarming pollution levels. The toll on public health is immense, with the report revealing that water pollution alone claims approximately 400,000 lives annually. The lack of access to clean water, coupled with the insidious presence of pollutants, exacts a heavy price on the well-being of the people.

The plight of the most vulnerable becomes even more pronounced, as certain classes of society bear the brunt of the crisis. Poverty weaves its way into the fabric of life, exacerbating the challenges faced by those already marginalized. Struggles for survival become a daily reality, as the battle against pollution intertwines with the struggle to escape the clutches of poverty.

Migration becomes an arduous journey, as some seek refuge in less polluted regions, hoping to find a breath of fresh air in new surroundings. The search for a healthier environment becomes a matter of life and death, as families seek solace and security for their future.

As we journey further, we arrive in Lima, Peru, a city teetering on the brink of environmental vulnerability. Like its Indian counterparts, Lima suffers from the scourge of air pollution, primarily stemming from vehicle emissions.

The environmental vulnerability of Lima has a profound impact on the people, business, and health of the city. The heavy air pollution presents a significant threat to the well-being of the population. The constant exposure to pollutants, such as particulate matter and toxic gases, can lead to a wide range of health issues, including respiratory problems, cardiovascular diseases, and even premature death. Children, the elderly, and those with pre-existing health conditions are particularly susceptible to the harmful effects of pollution.

From a business perspective, the pervasive air pollution can hinder economic growth and productivity. Reduced air quality may result in increased sick leave among the workforce, leading to reduced productivity and increased healthcare costs for businesses. Additionally, the poor air quality may discourage investment and tourism, which could have long-term negative effects on the local economy.

Startling research from the University of Chicago suggests that adhering to the World Health Organization's pollution guidelines could extend the population's average lifespan by 4.7 years (AQLI, 2023).

Yet, Lima's challenges do not end there. Its inhabitants also grapple with the poor quality of housing and infrastructural deficiencies, compounding their vulnerability. Apart from air pollution, Lima's residents face additional challenges stemming from inadequate housing and infrastructure. Substandard housing conditions, such as overcrowding and lack of proper sanitation, can increase the risk of various diseases and reduce overall living standards. Moreover, infrastructural deficiencies can hinder access to essential services such as clean water, healthcare facilities, and education, further exacerbating the vulnerability of the population.

These challenges have a cascading effect on various aspects of life in Lima. People's health is compromised by living in subpar conditions, hindering their ability to pursue education and economic opportunities. Businesses face difficulties in finding a healthy and productive workforce, and the lack of proper infrastructure may hamper their operations and growth potential.

Furthermore, this cycle of environmental vulnerability and poor living conditions may perpetuate social inequalities, as marginalized communities often bear the brunt of these issues disproportionately. This creates a need for a comprehensive approach that addresses not only environmental concerns but also social and economic disparities.

As our voyage continues, our ship sets its course for Lagos, Nigeria, a vibrant and populous city in Africa. However, beneath its lively surface, the specter of unlivability looms large. For years, annual floods have been a customary challenge in Nigeria, but the recent past has brought forth extreme rainy seasons that wreak havoc on the city's economic activity, exacting a staggering cost of more than $4 billion each year.

The Climate Change Vulnerability Index by Verisk Maplecroft (2021) delivers a stark message, highlighting the disproportionately severe impact of environmental threats on African cities. Lagos, with its limited resources to combat these challenges, finds itself facing an uphill battle against the unforgiving forces of nature.

The impact on people is palpable, as the city's swelling population strains already limited infrastructure, leaving many struggling for adequate housing, clean water, and sanitation. The floods displace families and communities, deepening social disparities and putting mental well-being at risk.

Businesses, entangled in Lagos' thriving commercial landscape, bear the brunt of the relentless rainy seasons. Supply chains falter, transportation grinds to a halt, and financial losses ripple through the city's economy, affecting local enterprises and global trade alike.

Yet, it is the health of Lagos' residents that perhaps suffers the most. Contaminated water sources and waterborne diseases become prevalent, leaving vulnerable populations, especially the young and elderly, at the mercy of the floods' aftermath. Access to healthcare becomes uncertain, leading to increased mortality rates and a diminishing average life expectancy.

In this city of contrasts, even the prospect of an exclusive island-based luxury gated community emerges. Yet, as rising waters threaten Lagos' existence, the question arises – can such a sanctuary endure the relentless forces of nature?

Venturing eastward, our journey brings us to Muscat, Oman, where extreme temperatures above 41.3°C (106.3°F) defy expectations. This scorching heat poses significant challenges to the city's inhabitants, impacting their health, livelihoods, and daily lives. The oppressive heatwave can lead to heat-related illnesses, affecting the vulnerable population, such as the elderly and children, the most. Moreover, businesses and industries face disruptions, with potential energy shortages and reduced productivity under such extreme conditions.

Powerful cyclones batter Muscat, serving as a stark reminder of how climate change is intensifying the frequency and intensity of natural disasters. These cyclones' impact on infrastructure, businesses, and homes can leave lasting scars on the city's economy and people's well-being, leading to financial losses, property damage, and displacement.

Continuing our odyssey, we reach Manila, Philippines, where the inexorable rise of sea levels threatens coastal communities. The alarming rate at which the waters of Manila Bay are rising, over four times faster than the global average, puts lives and livelihoods in peril. The consequences of sea-level rise extend beyond mere inundation, as it exacerbates the risk of storm surges and flooding during typhoons and heavy rainfall events. This poses a dire threat to public health, displacing people from their homes, and increasing the likelihood of waterborne diseases and injuries.

The rapid urbanization and overpopulation in Manila further compound the city's vulnerability. The strain on infrastructure and resources heightens the challenges of disaster response and recovery. At the end of 2020, a staggering 145,000 people found themselves injured or displaced by disasters, highlighting the urgent need for resilient urban planning and disaster preparedness measures.

Now, we arrive in Shanghai, China, a bustling metropolis teeming with twenty million souls. As a coastal city, Shanghai stands on the frontlines of climate change-induced flooding. Projections by Climate Central paint a concerning picture – if global temperatures rise by 3°C, a staggering 17.5 million people could face displacement due to rising waters. The city's expansive waterfront makes it susceptible to storm surges and sea-level rise, threatening lives, homes, and businesses.

Though flood prevention walls have been erected along the waterfront, the city remains perched on the precipice of an uncertain future. The potential for catastrophic flooding looms large, necessitating continuous efforts to strengthen infrastructure, improve early warning systems, and engage in climate adaptation strategies.

As Europe beckons, a daunting question looms - what will life be like in the next 50 years amidst the unyielding grip of the climate crisis? A prophecy awaits us, revealing the destined fate of European capital cities, their coastal lines to be redrawn by the relentless forces of nature.

An ominous study paints a vivid picture, showcasing the capitals that shall bear the brunt of intensified droughts. Among them, Athens, Lisbon, Madrid, Nicosia, Sofia, and Valleta stand as parched domains, sweltering under soaring temperatures. Central Europe, akin to a fiery furnace, will endure the harshest heat waves, with scorching spikes reaching 2°C to 7°C in the low-impact scenario and a blistering surge of 8°C to 14°C in the high-impact scenario. Brace yourselves, for the likes of Athens, Nicosia, Prague, Rome, Sofia, Stockholm, Valleta, and Vienna shall confront the most severe heat wave severity and frequency.

Yet, that is not the extent of Europe's trials. A relentless torrent of floodwaters is destined to surge, especially in the United Kingdom, where 85% of its cities lying along flowing rivers, including the grandeur of London, shall suffer the battering floods. Under the high-impact scenario, specific cities like Cork, Glasgow, and Wrexham shall bear witness to a dramatic escalation in flood severity, with staggering increases ranging from 77% to 115%. Across Europe, capitals such as Dublin, Helsinki, Riga, Vilnius, and Zagreb are poised to encounter the greatest increase in flooding severity and frequency.

Returning to the scorching heat waves that shall pervade all 571 cities in the European Union's official database, the furnace of central Europe will endure the most blistering blow. Heat wave days and their maximum temperatures will rise across the continent, but central Europe shall bear the brunt, experiencing fiery surges ranging from 2°C to 7°C in the low-impact scenario and a searing blast of 8°C to 14°C in the high-impact scenario. Among these European capitals, Athens, Nicosia, Prague, Rome, Sofia, Stockholm, Valleta, and Vienna shall face the most severe and frequent heat waves.

With the profound lyricism of the iconic Bob Marley, who once resonated, "Everywhere is war, me say war," let us underscore the all-encompassing battle against the ravages of climate change, a fight that transcends national boundaries, continents, ethnicities, and diverse cultures. As our exploration ship ventures into the depths of the unknown, we bear witness to a distressing reality: the relentless destruction of our planet, a planet betrayed. There is no escape from the consequences of our actions, at least not within the confines of this home we call Earth. Without a shred of doubt, the relentless force of climate change shall sow a trail of devastation, leaving behind a wreckage of shattered lives for all of humanity to bear witness to. Unless, of course, we muster the courage to embark on an audacious expedition to another celestial body.

Is it truly possible? Can we cast aside our apathy and wake up from the collective slumber that has allowed us to wreak havoc on our environment? The notion of venturing to another planet might seem like a far-fetched dream, a fanciful escape from our self-inflicted predicament. But perhaps, just perhaps, it is a wake-up call we desperately need.

The signs of our planet's suffering are omnipresent, a somber symphony of destruction playing out before our very eyes. Droughts scorch once-fertile lands, leaving parched soil in their wake. Floodwaters surge through city streets, a grim reminder of our vulnerability. Rising temperatures turn heat waves into infernos, pushing our limits and testing our resilience. And the rising sea levels, inch by inch, encroach upon coastal cities, threatening the very existence of communities.

Bob Marley's words echo through the corridors of time, a poignant reminder of the battles we face on all fronts. The war against climate change and environmental degradation demands our unwavering commitment and united efforts. It is a war that requires us to confront our own shortcomings, to challenge the status quo, and to embrace the radical changes necessary to secure a sustainable future.

So, let us heed the call to action. Let us wake up from our complacency and dare to imagine a world where we are stewards of the planet, rather than its conquerors. Together, we can forge a new path, one that leads us towards a future where we cherish and protect the fragile beauty of our only home. The journey may be arduous, but the destination is worth fighting for.

IV. Artistic Echoes Of Climate Change: Warming Whispers

CHAPTER

Four

Unsustainable Economy: A Path To Destruction, Fueling Inequalities With Ferocity

"Is it not enough for you to feed on the good pasture, that you must tread down with your feet the rest of your pasture; and to drink of clear water, that you must muddy the rest of the water with your feet? And must my sheep eat what you have trodden with your feet, and drink what you have muddied with your feet?"

– Ezekiel 34:18-19

I. The Concept Of The Unsustainable Economy

Within the realm of venture capitalism, the art of crafting a compelling equity story, conveyed in simple terms, stands as a rewarding yet arduous endeavor. In this domain, where disruptive innovations and complex concepts intertwine, I seek solace and draw inspiration from unconventional sources. As I navigate the daunting task of elucidating groundbreaking ideas to my investment committee, some of whom lack a background in science or technology, I have unearthed two extraordinary realms that wield the potential to captivate and enlighten: the dynamic world of football and the timeless wisdom enshrined within the ancient scriptures of the Bible.

In these seemingly disparate arenas, I find a reservoir of narratives, metaphors, and analogies that transcend mere words and awaken profound understanding. In the grace and finesse of football, I uncover tales of strategy, teamwork, and resilience, mirroring the essence of successful investment ventures. The elegant choreography on the field symbolizes the interplay of various stakeholders in a business, where each player's unique strengths converge, giving rise to a harmonious symphony of growth and prosperity.

Similarly, the treasured scriptures of the Bible offer allegorical insights that resonate with the intricacies of venture capitalism. The timeless parables and teachings hold profound lessons of risk and reward, resilience, and foresight. Just as a visionary leader navigates through trials and tribulations, ancient wisdom imparts invaluable guidance, drawing striking parallels between the trials faced by pioneering entrepreneurs and the trials of legendary biblical figures.

With these diverse narratives at my disposal, I embark on the compelling odyssey of storytelling, weaving together the language of finance with the spirit of football and the wisdom of ancient texts. As I breathe life into the esoteric world of investments, I recognize that it is not merely about presenting data and projections but about crafting a tapestry of experiences and emotions that resonate with every listener.

Talking of the Bible, let us embark on a poignant journey through the annals of time, tracing back to an epoch around 1445-1405 BCE, where the ancient Israelites forged a path of liberation during the Exodus from the clutches of slavery in Egypt, destined for the Promised Land.

Within the sacred book of Leviticus, where laws, rituals, and moral codes intertwine, a vivid tapestry of their life conditions unfolds. To grasp their profound struggles, we must venture into the heart of their experience, where unyielding hardships tested their faith, resilience, and reliance on divine guidance.

Picture a multitude of weary souls, weary from generations of enslavement, daring to march forth into the unforgiving wilderness, leaving the security of familiar lands behind. The scorching sun bore down on their brows, its relentless gaze mirroring the fiery trials that lay ahead. With each footstep, their journey echoed the rhythm of a symphony of deprivation and yearning for freedom.

As they wandered through the vast expanse of the desert, a harsh and inhospitable terrain, thirst clung to their parched throats like a relentless tormentor. The scarcity of water mirrored the scarcity of hope, and they found themselves drawing on reservoirs of faith to sustain them through the darkest hours.

Their nomadic existence lacked permanence, for they carried their homes upon their backs like the wings of a wandering bird. Makeshift tents offered meager shelter against the unforgiving elements, and yet, within these humble abodes, a strong sense of community thrived. Together, they sought solace and support, finding strength in unity as they ventured toward an uncertain destiny.

The brutal elements, however, were not their sole adversaries. A myriad of dangers lurked in the shadows of the wilderness. Hunger gnawed at their bellies, mirroring the hunger for a better life, for a land flowing with milk and honey. The arduous task of procuring sustenance from the wilderness was a test of survival, a test that called upon their ingenuity and resourcefulness to conquer each day's trials.

Yet, in the midst of their trials, the ancient Israelites were propelled forward by the alluring promise of a "land flowing with milk and honey." This symbolic phrase transcended its linguistic confines to embody a vision of unparalleled richness and prosperity, a vision etched within their hearts and minds. "Milk," a captivating symbol of abundance, embodied the thriving livelihoods they yearned for in the Promised Land.

Imagine the vast herds of cattle dotting the lush landscape, a testament to the land's fertility and favor. The gentle lowing of the cattle echoed the harmonious rhythm of nature's bounty, as their milk and dairy products became a source of sustenance, nourishing the people and nurturing their dreams of a better life.

Within this bounteous land, they found themselves immersed in the art of animal husbandry, their lives intertwined with the grazing beasts. Their hands skillfully tended to the cattle, cultivating a bond of mutual dependence. The flourishing pastures afforded ample resources for their herds, signifying a realm where prosperity and abundance reigned.

And then, "honey" – a tantalizing representation of sweetness and copiousness – unfolded like a golden tapestry across the land. Fruits, crops, and agricultural produce adorned the landscape in a vivid display of nature's generosity. The land's fertility seemed boundless, offering a plenitude of resources to nourish their bodies and souls.

As they wandered through the Promised Land, they delighted in the sweet taste of the fruits, a testament to the fertile soil that effortlessly yielded its gifts. Orchards, heavy with the weight of ripened fruits, became a tangible reminder of the land's abundant blessings. From vineyards to fields of golden grain, the earth seemed to yield its treasures willingly, nurturing a profound sense of gratitude and awe.

The promise of milk and honey was not merely a fanciful reverie but a tangible vision visible to anyone who stepped foot upon that blessed soil. It epitomized a land of plenty, a land where their toils would yield boundless rewards, and their struggles would find solace in the lap of prosperity.

As they carried this vision within their hearts, the ancient Israelites continued their arduous journey, their spirits aflame with hope and determination. The symbolism of milk and honey served as a guiding light, illuminating their path through the wilderness of uncertainty and trials. It was a beacon that reminded them that beyond the struggles lay a destiny of fulfillment, a destiny carved in the very essence of the Promised Land.

While this vision of a land ripe for infinite growth may sound enticing, the biblical narrative reveals a different perspective. According to the Scriptures in the book of Leviticus 25: 2–7, upon their entry into the Promised Land, the Israelites were instructed to observe a sabbatical year every seventh year. During this year, the land would lie fallow, devoid of any sowing or pruning. Not even the grains that had fallen during the previous harvest were to be harvested. Instead, the produce that grew on its own would be left for the owner, slaves, hired laborers, foreign residents, and the poor. Even domestic animals and wild beasts would partake of it.

Moreover, every fiftieth year, known as the Jubilee, the land was to remain in complete rest, mirroring the principles of the Sabbatical year. The same regulations applied to the produce of the soil. This system required careful planning, as the harvest of the forty-eighth year had to be abundant enough to sustain the nation not only during the Sabbath year but also through the following year and the months until the harvest of the fifty-first year, which followed the Jubilee.

In this biblical context, we are presented with a different ethos than that espoused by Gordon Gekko. The infamous character from 'Wall Street' proclaimed that "greed is good" and urged a relentless pursuit of profit. However, the biblical account teaches us that we cannot operate solely on the principle of unbridled greed. The lessons from Leviticus demonstrate the importance of balance, stewardship, and the recognition that natural resources must be preserved and shared equitably for the greater good.

Now let us embark on this discussion about growth, infinite growth, greed, and the sustainability of economic systems. While some may find it more appealing to approach these topics from an economic standpoint, I would like to introduce interdisciplinary concepts that I learned during my days as a young Electrical Engineer at the prestigious Technical University of Munich. These concepts may be disregarded or dismissed by some economists as nerdy, but as one of my Venture Capital Partners once said, mathematics and physics do not lie. Let us therefore set forth on an expedition through the realm of science, guided by the timeless wisdom of the First Law of Thermodynamics. Within this sacred domain, we encounter a profound truth - the total energy within a closed system remains constant, undergoing transformation from one form to another. As we venture into the world of modern innovation, we grapple with the profound implications of this fundamental law.

Amidst the dynamic landscape of progress, innovation stands as the lifeblood of growth, pushing the boundaries of human achievement. From revolutionary technologies to awe-inspiring breakthroughs, the engine of innovation roars with boundless potential. However, concealed within these transformative endeavors lies an insatiable appetite for energy – a vital force that propels the wheels of progress.

Consider the vast expanse of cloud computing, a digital universe where enormous amounts of data are processed, analyzed, and stored. The power-hungry behemoth of cloud computing consumes energy voraciously, necessitating a constant supply to fuel its relentless pursuit of knowledge and insight. Like energy vampires, these data-driven technologies draw upon the lifeblood of electricity, leaving their mark on our power grids and resources.

In a world fascinated by the idea of infinite growth, the concept of perpetual motion captures our imagination. The vision of boundless expansion, where the economy knows no limits, beckons like an enchanting melody. However, we must pause and contemplate – does this vision align with the laws of physics? Can perpetual motion also known as Perpetuum Mobile truly coexist with the immutable truths revealed by the First Law of Thermodynamics?

Imagine a Perpetuum Mobile – a marvelous contraption that, once set in motion, never falters, spinning tirelessly and elegantly with no need for an external energy supply. Its allure lies in the promise of boundless motion, an eternal dance of gears and wheels defying the wear of time. Yet, as enchanting as this vision may be, the immutable laws of thermodynamics remind us that such a creation is but a beautiful mirage, forever eluding our grasp.

At the heart of this tale is the essence of energy – a finite and precious resource. Picture it as a golden stream flowing through the universe, moving from one form to another in a grand cycle of transformation. While this energy can take on different guises – from fossil fuels forged over millions of years to the sun's radiant embrace – it cannot arise from nothingness nor last indefinitely. Just as a river flows within its banks, energy remains bound by the laws of nature, unyielding in its conservation.

Consider the very earth beneath our feet, where ancient forests once thrived. Over eons, they transformed into the fossil fuels that now power our modern world. Yet, as we harness this potent energy to fuel our ambitions, we must confront the consequences of our consumption. The once-dormant energy finds release in the form of emissions and pollution, casting a shadow upon the delicate balance of nature.

The law of energy conservation, like a wise sentinel, cautions us that energy conversions are never perfect. When we seek to extract work from energy, a portion slips through our fingers like sand, irretrievable and lost. In the pursuit of infinite growth, we face a quandary – the more we rely solely on increased energy consumption, the more we encounter diminishing returns. Like the flickering light of a candle nearing its end, our desire for boundless expansion encounters the sobering limits imposed by inefficiencies.

Moreover, as we extract energy to power our ambitions, we must bear in mind the environmental toll of such endeavors. The great symphony of energy conversion and consumption reverberates across the globe, leaving a trail of consequences. Pollution stains the air, waters turn turbid with waste, and delicate eco-systems crumble under the relentless pressure. The pursuit of infinite growth becomes not just an economic endeavor but a profound moral test, challenging us to balance our desires with the preservation of our planet's majesty.

In this intricate dance of energy and progress, we are left with a profound truth - the notion of infinite growth, like the mythical Perpetuum Mobile, dances at the edge of possibility. The mirage of boundless expansion beckons, but its elusive promise is tempered by the realities of finite resources, inefficiencies, and environmental consequences.

As you can imagine dear friends, the law of energy conservation truly has a way of poking fun at those devout followers of the infinite growth theory, doesn't it? Those folks who confidently march down the path of dissociating growth from resources, blissfully believing that innovation will magically solve all issues of scarcity. But lo and behold, this law comes along and reminds us, ever so sarcastically, that energy is indeed finite and cannot simply be conjured out of thin air. Oh, the audacity of it! How inconvenient for those who believe that endless consumption will lead us to everlasting prosperity! The law of energy conservation, with its sly smile, reminds us that there are pesky environmental considerations to be made. Yes, my dear infinite growth enthusiasts, your dreams of boundless expansion must grapple with the reality of environmental implications.

Oh, I hear you, dear scholars of economics, protesting my audacious use of the law of energy conservation. "Wait a minute," you say, "that doesn't directly disprove our beloved theory of decoupling growth from resource limitations through technological advancements." Ah, the antithesis of my argument, how charming! Let us bask in the glory of differing perspectives, for there may not be an absolute right or wrong, but only a kaleidoscope of interpretations.

But fret not, my esteemed interlocutors, for I shall graciously address your objections and embark upon a captivating exploration of the intricate relationship between the venerable law of energy conservation and the alluring theory of decoupling growth from resource limitations. Brace yourselves, for this intellectual voyage shall undoubtedly enlighten and delight!

Let us first immerse ourselves in the realm of energy input, where the wonders of technological advancements hold sway. Ah, how they mesmerize with their ability to enhance energy efficiency and diminish the resources required per unit of economic output. These marvels offer tantalizing prospects of achieving the mythical decoupling of growth from resource consumption. Picture the enchantment of improved energy conversion processes, the artistry of optimized industrial practices, and the captivating allure of energy-saving technologies. Through their conjurations, we may bear witness to the magical attainment of greater economic output while sipping sparingly from the cup of energy. Oh, what a delightful prospect indeed!

Yet, amidst this enchantment, let us not forget the solemn admonition of the energy conservation law. Energy, my esteemed companions, is a capricious mistress – it cannot be created nor destroyed. It merely transforms from one form to another, leaving behind traces of energy losses as it traverses the realms of conversion. Even as we bask in the radiance of improved energy efficiency, we must humbly acknowledge that these conversion processes forever bear the burden of such losses. Alas, the complete decoupling of economic growth from the clutches of energy consumption remains an elusive dream, a mirage on the horizon of our aspirations.

Ah, the mischievous rebound effects, those capricious rascals! With their cunning ways, technological advancements that enhance energy efficiency may usher in unintended consequences. These effects materialize when the gains in efficiency, driven by their devilish charms, entice us towards heightened consumption and expanded economic activities. Oh, the treachery! How these rebound effects cunningly offset the very resource savings we believed to have achieved! They serve as a gentle reminder, a whisper in our ears, that the extent of decoupling growth from the voracious jaws of resource consumption may be circumscribed, restrained by forces beyond our control.

And lo, let us now venture into the labyrinthine depths of system complexity. Our theory of decoupling growth from the confines of resource limitations sings hymns of praise to the wonders of technological advancements. But let us not be beguiled, my wise comrades, by the sirens of progress alone. For as these magnificent contrivances are birthed into existence, their very production and eventual disposal contribute to the depletion of resources and the degradation of our cherished environment. How delightfully ironic! We mustn't disregard the holistic evaluation of technologies, tracing their impact throughout the entirety of their life cycle. Only then shall we grasp the true breadth and depth of their consequences.

And now, dear souls, let us bask in the radiant glow of the energy source transition, the beacon of hope that ignites our fervent dreams. Ah, the shift towards renewable energy sources, those heralded saviors! To decouple growth from the clutches of resource limitations, we must bid farewell to the finite and pernicious energy sources of yore – those wicked fossil fuels that have ensnared us for far too long. Yes, our technological advancements may extend their helping hand, but let us not be swayed by their beguiling promises. This grand transition demands monumental investments, the development of intricate infrastructures, and sweeping systemic changes. Oh, and let us not overlook the practical limitations imposed by the fickle availability and scalability of those whimsical renewable energy sources.

Now picture, if you will, an intricate web of detrimental impacts on the environment, society, and the fates of future generations. The notion known as an unsustainable economy encompasses an economic model that recklessly depletes precious natural resources, engenders copious amounts of waste and pollution, fans the flames of social inequalities, and callously dismisses the far-reaching repercussions of its actions.

Allow me to illuminate this concept through a series of vivid examples that illustrate the perils of an unsustainable economy, leaving an indelible imprint on your consciousness.

First, behold the harrowing tale of overconsumption and resource depletion. In the realms of an unsustainable economy, rampant consumption reigns supreme, heedless of the finite nature of resources. Imagine a nation feverishly guzzling fossil fuels for energy production and transportation, oblivious to the urgent need for investment in renewable energy sources. Such a heedless reliance inexorably drains the well of non-renewable resources, all while hastening the ominous specter of climate change.

Next, let us traverse the desolate landscapes scarred by environmental degradation – a dire consequence of an unsustainable economy. Witness the wanton destruction of majestic forests, the erosion of once-fertile soils, and the insidious pollution that permeates the air we breathe. Industries, driven by their insatiable quest for profit, unleash toxic chemicals into our pristine water bodies, while the unchecked emissions of copious amounts of greenhouse gases (GHGs) conspire to ravage delicate eco-systems and jeopardize human health.

Ah, the shadow of social inequity looms large over the landscape of an unsustainable economy. In this realm, the mighty few amass immeasurable wealth and power, their dominion unchallenged, while the needs and rights of marginalized communities languish in neglect. A chasm widens, rife with injustice and inequality, threatening to rupture the very fabric of society. Unrest simmers beneath the surface, poised to erupt into a tempest of conflict, eroding the precious bonds of social cohesion.

Cast your gaze upon the mountains of waste that tower ominously in an unsustainable economy. Excessive, prodigious, and often unnecessary, this relentless production of waste pollutes our surroundings, imperiling both the environment and our own well-being. Single use plastics abound, despoiling land and sea alike, while the improper disposal of electronic waste further compounds the cycle of pollution and wanton resource squandering.

And let us not forget the tragic consequences of a myopic focus fixated solely on short-term gains and economic growth – a hallmark of an unsustainable economy. As the relentless pursuit of immediate prosperity unfolds, the toll on our finite resources mounts, while opportunities for investment in sustainable technologies and practices wither away, neglected and forgotten.

In conclusion, dear readers, the concept of an unsustainable economy reveals the stark realities we must confront. Its clutches threaten to strangle the delicate balance of our world, endangering the environment, perpetuating social inequities, and undermining the well-being of future generations. Let us heed this cautionary tale, for only by embracing sustainable alternatives can we hope to forge a path towards a brighter, more resilient future.

II. The Unsustainable Economy: A Path To Destruction

Ladies and gentlemen, buckle up as we embark on a riveting journey across time and continents, exploring the devastating consequences of unsustainable practices that have wreaked havoc on our beloved planet. Prepare to witness stories of tragedy and destruction, unveiling the true cost of our relentless pursuit of growth.

Our first destination takes us to Guinea, a West African nation blessed with the world's largest bauxite reserves. In the pursuit of coveted metals essential for electric vehicles (EVs), Guinea finds itself on the frontier of a modern-day "gold rush." But this quest for growth comes at an immense price. The booming demand for aluminum, driven by the rising popularity of EVs, has transformed Guinea's northwestern region of Boké into a battleground of destruction.

As the trucks and trains laden with bauxite rumble through the newly constructed roads and tracks, the landscape of Boké undergoes a drastic metamorphosis. Behind the scenes, however, lies a tale of misery. We hear the voices of the villagers, their lives forever altered by the greed-driven frenzy. Through countless interviews with residents, non-profit groups, and industry experts, a chilling truth emerges.

Mining companies, propelled by their insatiable hunger for profit, acquire vast swaths of land once teeming with life and fertile farmland. The villagers, custodians of the land for generations, are left displaced and dispossessed, their pleas for fair compensation falling on deaf ears. As we dig deeper, we discover that the devastation extends far beyond the loss of homes and livelihoods. A government study paints a haunting picture, revealing that over the next two decades, an area nearly the size of Delaware will be ravaged by bauxite mining, erasing not just farmland but also precious natural habitats (Chason et al., 2023).

Let's now embark on another eye-opening expedition, this time to the war-torn landscapes of Iraq. In our relentless pursuit of understanding, we delve with the eyes of The Conflict and Environment Observatory (CEOBS, 2016), into the unrecorded chapters of environmental destruction caused by the harrowing oil fires that rage unchecked amidst a dire humanitarian crisis.

Imagine the scenes in Qayyarah, a town in northern Iraq, where the sky fills with toxic plumes billowing from burning oil wells. As we immerse ourselves in the story, a troubling fact emerges amidst the chaos and devastation, no one seems to be taking on the vital task of environmental monitoring. While humanitarian agencies are present, their focus is fixated solely on the immediate needs of the displaced, leaving the lasting impact on our planet largely unrecorded.

Let us peer through the lens of time and witness the birth of these infernos. The fires were ignited by retreating forces of the Islamic State, engulfing wells, storage tanks, and even a refinery. The conflagration that began years ago continues to cast its suffocating shadow over an already war-torn region. Yet, as the flames dance defiantly, there is a disheartening absence of dedicated experts scrutinizing the ground-level consequences.

Qayyarah, once a bustling town, has become a haven for those fleeing the conflict in Mosul and its surroundings. Many residents, despite the dangers, choose to remain in their homes rather than seek refuge in displacement camps. It is amidst this backdrop that we hear the anguished cries of the people, as the smoke and pollution from the fires take a toll on their health. Reports emerge of respiratory illnesses plaguing the community, a direct consequence of the environmental catastrophe unfolding before their eyes.

Attempts to extinguish the fires have been met with limited success, hampered by the presence of mines and improvised explosive devices scattered across the scorched terrain.

Oxfam, the first aid agency to raise its voice, implores the Iraqi government to take immediate action. The stakes are high, for with every passing day, the fires continue to consume the land, polluting the air and threatening the delicate balance of eco-systems. Satellite imaging, our window into the afflicted region, reveals the vast extent of the plumes' reach. Over 256 square kilometers have been blanketed by the ominous shroud of smoke for more than 21 days, enveloping towns in its suffocating grip. The urgency for on-the-ground data collection and rigorous monitoring cannot be overstated. While satellite images and models aid in mapping areas of concern, they fall short of providing the comprehensive understanding that only field research can offer.

To comprehend the true extent of the danger posed by the raging infernos, we must explore the hazardous substances they release. Burning heavy sour crude, the oil field near Qayyarah emits a toxic concoction comprising volatile organic compounds, carbonyls, polycyclic aromatic hydrocarbons, dioxins, furans, and a cocktail of gaseous pollutants. These noxious elements permeate the air, posing grave risks to human health and leaving an indelible mark on the environment.

The behavior of the smoke plumes holds the key to understanding the potential consequences. In previous instances, such as the Kuwaiti oil fires during the Gulf War in 1991 or the Buncefield oil depot fire in the UK in 2005, atmospheric conditions allowed the plumes to rise to high altitudes, minimizing their immediate impact on the ground. However, Qayyarah presents a different scenario altogether, with the smoke settling in close proximity to the fires and contaminating the surrounding areas. With the arrival of winter rains, the region faces the additional peril of widespread soil and water contamination, further exacerbating the ecological crisis.

The challenging circumstances on the ground make the task of research and assistance even more arduous. Contaminated water sources, agricultural land, and the daily threat of air pollution necessitate immediate action. Establishing a responsive health registry and providing ongoing assistance to those affected present formidable challenges. While UN agencies stand ready to support the Iraqi government, the ball lies firmly in their court. Only with their consent and collaboration can we begin to address the intertwined humanitarian and environmental risks that plague Qayyarah and its surrounding regions. And above all, the fires must be extinguished, bringing an end to this ecological nightmare.

Our next travel stop takes us to the heart of Brazil, where a troubling report (BBC, 2020) paints a bleak picture of the Amazon rainforest. It accuses the Bolsonaro government of launching a full-scale assault on this invaluable eco-system. Strap yourselves in as we delve into the chilling details.

First, let's confront the grim reality of deforestation rates. The Amazon rainforest, a natural wonder teeming with life, is being decimated at an alarming pace. Brazil's National Institute for Space Research (INPE, 2023) reveals that in 2020 alone, a staggering 11,088 square kilometers of forest was lost. To put it in perspective, this represents the highest deforestation rate witnessed in the past 12 years.

But deforestation is far more than just numbers on a page. It signifies a catastrophic loss of habitat for countless species, plants, animals, and insects. The Amazon rainforest, with its mind-boggling biodiversity, is home to an estimated 400 billion individual trees and millions of unique species. Sadly, their existence is now hanging in the balance, their homes torn asunder by the relentless destruction (WWF, 2023).

The consequences extend far beyond the borders of the Amazon itself. This rainforest plays a pivotal role in regulating global climate patterns. It acts as a natural carbon sink, absorbing and storing massive amounts of carbon dioxide, thus mitigating the impact of greenhouse gas emissions. However, deforestation disrupts this delicate balance, unleashing stored carbon into the atmosphere and contributing to climate change. Startlingly, between 2000 and 2012, deforestation in the Brazilian Amazon alone accounted for approximately 8% of global CO_2 emissions. The implications for our planet are dire (Fraser, 2014).

But let us not forget the human tragedy unfolding amidst the trees. Indigenous communities, deeply intertwined with the Amazon rainforest, suffer the brunt of this environmental assault. Approximately 400 indigenous tribes call this lush paradise their home, their lives inextricably linked to the forest's bounty. Yet, as deforestation encroaches upon their ancestral territories, these communities face displacement and disruption. Their traditional knowledge and cultural practices, honed over generations, are being irretrievably lost. It is a heartbreaking blow to both their livelihoods and their cultural identity (Survival, May 2023).

As we peel back the layers of this environmental crisis, a sobering truth emerges. The ongoing deforestation in the Amazon rainforest stands as a stark reminder of the dangerous dance between an unsustainable economy and the insatiable quest for infinite growth. Multiple factors contribute to this destructive trajectory.

Economic exploitation lies at the heart of the matter. Logging, agriculture, and mining are the driving forces behind deforestation in the Amazon. These activities, often fueled by a thirst for short-term profits, extract resources without regard for the long-term consequences. The insidious pursuit of infinite growth and unbridled economic expansion wreak havoc on the environment and the local communities that depend on it.

Unsustainable agricultural and livestock practices further accelerate the destruction. Large scale operations, hungry to meet the growing global demand for soybeans and beef, carve out vast expanses of forest using slash-and-burn methods. The land, once rich with life, is reduced to ashes. This rapacious model of constant growth and consumption fuels the flames of an unsustainable economy.

But the devastation does not stop there. The Amazon rainforest is a vital cog in the intricate machinery of our planet's eco-system. Its services are immeasurable. From carbon sequestration and climate regulation to the preservation of biodiversity, the rainforest performs indispensable tasks. Deforestation disrupts these delicate balances, plunging us further into the abyss of climate change, greenhouse gas emissions, and the loss of precious habitat .

And what of the social and economic inequities that entwine themselves with this environmental crisis? The consequences of deforestation in the Amazon fall disproportionately on the shoulders of local communities, particularly indigenous peoples. Their ancestral lands are grabbed, their homes destroyed, and their resources plundered. In this relentless pursuit of infinite growth, the rights and well-being of marginalized communities are cast aside. The resulting social injustice perpetuates an unsustainable economic system that prizes profit above all else.

To break free from this destructive cycle, we must embrace a more sustainable path. It requires valuing the preservation of eco-systems, respecting human rights, and promoting equitable economic development. Only through a fundamental shift in our approach can we hope to salvage what remains of the Amazon rainforest and forge a more harmonious relationship between our planet and ourselves.

However, if we direct our attention to President Jair Bolsonaro's policies and their impact on gold mining in the Amazon rainforest, a more specific narrative unfolds. Here we witness a microcosm of unsustainable economic practices and the relentless pursuit of infinite growth.

Under the Bolsonaro administration, concerns have mounted regarding increased deforestation and environmental degradation in the Amazon. Within this troubling landscape, illegal gold mining activities have emerged as a grave concern. The pursuit of this precious metal entails wanton destruction. Vast areas of forest are cleared, rivers and water sources polluted with mercury, and habitats ravaged. The consequences for biodiversity and local communities are severe (BBC, 2020).

It is important to acknowledge that the issues at hand are multifaceted and extend beyond the actions of any single administration. Nonetheless, the policies and rhetoric of the Bolsonaro government have drawn criticism for potentially enabling and encouraging activities like illegal gold mining, thus exacerbating the already grave environmental and social challenges facing the Amazon.

India, 1984.

In eloquent accounts, as exemplified by distinguished publications such as "The Bhopal Syndrome: Pesticides, Environment and Health"

(Weir, 1998), "The Bhopal disaster and its aftermath: a review" (Broughton, 2005), and "Environmental Ethics: Overcoming the Curse of Bhopal" (Lappé, A., 2014) we find a stark warning to all. This tragedy evokes a haunting chapter in the annals of industrialization, a chilling testament to the dark underbelly of progress. In the heart of Bhopal, India, this catastrophe unfolded with devastating consequences. Toxic gas escaped from the Union Carbide India Limited (UCIL) pesticide plant, engulfing the unsuspecting population and the delicate eco-system that surrounded them. The aftermath reverberated through the lives of countless individuals, leaving an indelible scar upon the fabric of society (Weir, 2022).

The figures that emerge from the chaos and despair paint a grim picture of the magnitude of the catastrophe. Lives were shattered, extinguished in an instant. The official death toll, a solemn statistic provided by the Indian government, hovers around 3,800 souls. But whispers of dissent suggest a far more chilling reality. Estimates spiral upwards, indicating that the toll may have climbed to an unfathomable range of 15,000 to 25,000 lives lost. The enormity of such devastation defies comprehension, as grief echoes through the corridors of time.

Beyond the veil of death, a sea of survivors emerged, bearing the physical and emotional scars of a calamity that forever altered their existence. Toxic gas, a silent predator, inflicted injuries of varying severity upon thousands. The exact number lost to the haze of uncertainty remains elusive, obscured by the chaos of the aftermath. Yet, estimates weave a tapestry of suffering, suggesting tens of thousands, even hundreds of thousands, left to grapple with the enduring legacy of toxic exposure. Respiratory disorders, vision impairments, neurological afflictions, and reproductive complications persist as harbingers of a tragedy that lingers, unrelenting in its grip.

The tendrils of this tragedy extended far beyond the physical and emotional realms. The web of economic repercussions entangled the affected region, ensnaring livelihoods and strangling prosperity. Disrupted and shattered, the delicate threads of income generation withered, leaving a scarred landscape of financial instability. The burden of increased healthcare costs, a constant reminder of the lingering agony, added insult to injury. The financial compensation and the exorbitant costs of the cleanup efforts etched a lasting imprint upon the economic trajectory, ensuring that the scars of Bhopal would be felt for generations to come.

In the aftermath of devastation, a glimmer of hope emerged in the form of a legal settlement. Union Carbide Corporation (UCC), the entity responsible for this catastrophe, faced the weight of accountability. A settlement, a sum offered in recompense for the suffering endured, was reached. Yet, as the echoes of justice reverberated, a collective outcry rose. $470 million, a staggering figure by most accounts, paled in comparison to the scale of tragedy and the vast sea of affected souls. The inadequacy of this settlement became a haunting testament to the limitations of justice in the face of such immense sorrow.

The Bhopal Gas Tragedy stands as an enduring testament to the inherent perils embedded within an unsustainable pursuit of growth and profit. It reveals a confluence of factors that perpetuated the catastrophe. Cost-cutting and the lack of safety measures became the norm, as the pursuit of profit overshadowed the sanctity of human life. Regulatory bodies entrusted with safeguarding the well-being of the populace faltered, revealing a systemic deficiency in their ability to enforce safety regulations. The tragedy also showcased the chilling truth of externalizing costs, burdening the affected population and the environment while those responsible sought to evade accountability. It whispered a tale of a community forsaken and an environment ravaged, as the pursuit of economic growth trampled upon their lives and the eco-system they called home.

The Bhopal Gas Tragedy serves as a clarion call, a somber reminder of the perils that await those who prioritize profit and growth over the well-being of people and the environment. It urges us to forge a future rooted in sustainable and responsible economic practices, one that values the sanctity of life, embraces safety and environmental stewardship, and upholds the fundamental rights of all. As we gaze upon the lessons etched upon the fabric of history, let us strive to build a more equitable, compassionate, and sustainable world.

The Magnitogorsk Industrial Complex, nestled in the heart of Magnitogorsk, Russia, serves as a poignant example of the consequences of infinite growth theory within the realm of Soviet industrialization. This colossal complex, boasting an expansive steel plant and related industries, epitomized the Soviet Union's unyielding pursuit of rapid economic expansion and industrial might (Hosan, 2021).

The theory of unlimited economic growth, also known as the infinite growth theory, forms the bedrock upon which the complex was built. This theory, closely aligned with the overarching philosophy of Marxism-Leninism, held that economies could perpetually expand without encountering inherent limits or negative repercussions. Within this ideological framework, the Magnitogorsk Industrial Complex emerged as a tangible manifestation of boundless growth and industrial prowess.

The genesis of this behemoth dates back to the 1930s, when Joseph Stalin's grand vision for the Soviet Union materialized into ambitious plans for industrial transformation. The Magnitogorsk Industrial Complex was a key player in this transformative agenda, with its sights set on maximizing steel production and driving rapid economic growth, particularly in heavy industry. The complex was designed to meet the burgeoning demands of industrialization, facilitate modernization, and fortify the military-industrial complex.

Centralized planning formed the backbone of the complex's operations, with the state dictating production targets and priorities. The driving force behind this centralized approach was the relentless pursuit of high output levels, often at the expense of environmental considerations and long-term sustainability. The consequences of this myopic focus were manifold and far-reaching.

Environmental concerns were frequently disregarded or pushed to the periphery in the relentless quest for swift industrial growth. The Magnitogorsk Industrial Complex cast a dark shadow upon the local environment, releasing copious amounts of air pollutants, contaminating water sources, and ravaging nearby forests. Eco-systems suffered, with the Ural River succumbing to pollution and the once-thriving forests facing destruction.

The infinite growth theory, embodied by the Magnitogorsk Industrial Complex, turned a blind eye to the finite nature of resources and the potential environmental fallout stemming from unbridled industrial expansion. Immediate economic gains and production targets took precedence, often at the expense of long-term sustainability and ecological equilibrium.

While the complex played a pivotal role in the Soviet Union's industrialization and economic ascent, its legacy serves as a somber reminder of the environmental and social costs entwined with the pursuit of limitless economic expansion. In recent years, efforts have been made to address the environmental ramifications and promote more sustainable practices. However, the formative years of the Magnitogorsk Industrial Complex and its alignment with the infinite growth theory serve as potent reminders of the challenges and consequences accompanying unchecked economic expansion.

Though precise figures documenting the extent of destruction caused by the Magnitogorsk Iron and Steel Works (MMK) may prove elusive, historical records and studies offer glimpses into the environmental consequences wrought by this industrial colossus.

Firstly, air pollution emerged as a pressing concern attributable to the MMK. During its initial operational years, the plant spewed significant quantities of sulfur dioxide (SO_2), nitrogen oxides (NO_x), and particulate matter, contributing to poor air quality and adverse health effects among the local populace. A study published in the journal Environmental Pollution (2020) underscored the MMK's responsibility for releasing millions of tons of pollutants into the atmosphere over several decades, inflicting substantial environmental damage.

Secondly, water pollution emerged as another adverse consequence of the MMK's operations, particularly impacting the Ural River. The plant's discharge of untreated or inadequately treated wastewater, laden with heavy metals, oil, and other contaminants, led to water pollution and the degradation of aquatic eco-systems. Although quantifying the exact quantities of pollutants released into the river proves challenging, the environmental impact has been undeniably significant.

Lastly, the establishment and expansion of the MMK necessitated substantial land clearance, resulting in widespread deforestation in the surrounding areas. The construction of the plant, infrastructure development, and the demand for fuel and raw materials all contributed to the loss of forested land and the destruction of delicate eco-systems.

While exact numerical data on the magnitude of destruction caused by the MMK may be scarce, several sources provide valuable insights into the environmental impacts.

One notable resource is the study titled "Environmental pollution by heavy metals around the Magnitogorsk Iron and Steel Works (Russia)" published in the journal Environmental Pollution (2020). This study meticulously examines soil, water, and air contamination in the vicinity of the MMK, shedding light on the significant environmental pollution unleashed by the plant.

Another informative piece is the research article titled "The environmental degradation of the Ural River and its potential for bioremediation," featured in the journal Environmental Science and Pollution Research (2016). This article delves into the contamination of the Ural River resulting from industrial activities, including those of the MMK. It also explores potential strategies for remediation and restoration.

The Magnitogorsk Industrial Complex, with its alignment with the infinite growth theory and its ecological fallout, stands as a poignant testament to the perils of unchecked economic expansion. It serves as a stark reminder that sustainable practices and a consideration of the environment are imperative, irrespective of the economic system in place. The lessons of Magnitogorsk reverberate beyond its Soviet context, reminding us of the delicate balance we must strike between growth and the preservation of our natural world.

Now we reach our final stop, the heart of one of the most catastrophic environmental disasters in American history: the Deepwater Horizon oil spill of 2010. This devastating incident, well illuminated in The National Commission on the BP Deepwater Horizon Oil Spill and Offshore Drilling's comprehensive report, "Deep Water: The Gulf Oil Disaster and the Future of Offshore Drilling" (2011) and etched in the annals of marine oil spills, serves as a stark reminder of the intertwined relationship between the relentless pursuit of growth and its far-reaching consequences.

The Deepwater Horizon disaster, connected to the concept of infinite growth, unfolds on the backdrop of an industry driven by the pursuit of profits. The ill-fated drilling rig sought to tap into the oil-rich Macondo Prospect, nestled deep beneath the ocean floor. As global energy demands soared, the pressure to extract and sell more oil intensified, leading to riskier and more challenging drilling operations.

The expansion of offshore drilling, propelled by the belief in continuous growth in oil production, set the stage for this calamity. Companies ventured into uncharted territory, striving to access untapped reserves in remote and treacherous locations, including deepwater areas. This expansion came at a cost, amplifying the risks associated with drilling and inflicting environmental impacts on sensitive eco-systems.

The Deepwater Horizon disaster laid bare the oil industry's disregard for safety and environmental protection. In their relentless pursuit of growth and fatter profit margins, corners were cut, safety protocols overlooked, and regulations loosely enforced. The absence of precautionary measures and inadequate response plans contributed to the severity of the spill and the subsequent environmental devastation.

The environmental consequences of the Deepwater Horizon oil spill were profound. Marine and coastal eco-systems bore the brunt, witnessing the loss of wildlife, destruction of habitats, and contamination of water resources. The repercussions extended beyond local boundaries, casting a dark shadow on the delicate balance of eco-systems and biodiversity, both in the Gulf of Mexico and beyond.

This catastrophe serves as a poignant reminder of the risks entailed in an economic model fixated on perpetual growth, without due consideration for environmental sustainability, safety measures, and the long-term impacts on eco-systems and communities.

The consequences of the Deepwater Horizon oil spill reverberated across the environment, marine life, and the economy. The spill's sheer magnitude cannot be understated, with an estimated 4.9 million barrels (210 million gallons) of oil released into the Gulf of Mexico over 87 agonizing days. It stands as one of the largest marine oil spills in history, leaving an indelible mark on the region.

The environmental impact was catastrophic, as the spilled oil blanketed shorelines, marshes, and beaches, wreaking havoc on delicate habitats, mangroves, and sensitive wetlands. Marine species, ranging from fish to birds, sea turtles, and dolphins, fell victim to this ecological tragedy. The long-term consequences for these eco-systems continue to be studied, their full extent yet to be comprehended.

Beyond the environmental toll, the spill exacted a heavy economic price. The fishing, tourism, and recreational industries along the Gulf Coast suffered significant setbacks. Fishing grounds were forced to close, and the negative perception surrounding oil-affected beaches dealt a severe blow to businesses and individuals, leading to job losses and financial hardships.

Moreover, the health and well-being of those involved in cleanup efforts and local communities were put at risk. Exposure to toxic chemicals and oil, along with the dispersants used in the cleanup, gave rise to a host of health concerns. Respiratory issues, skin problems, and mental health effects became all too common among those directly affected by the spill.

As we traverse the globe, we encounter these cautionary tales of unchecked greed and shortsighted decisions, each carrying a heavy toll on the environment, eco-systems, and the very fabric of our societies. The true cost of unsustainable practices becomes painfully evident as we witness the irreversible damage caused by our insatiable hunger for more.

In this journey through time and space, we are faced with a stark realization. Our pursuit of growth, driven by an unsustainable economic model, exacts a heavy toll on our planet and its inhabitants. It is a call to action, urging us to reevaluate our priorities, embrace sustainability, and chart a new course towards a future where progress and preservation go hand in hand.

So, let us heed the lessons of these stories, for they serve as cautionary tales in our collective quest for a more sustainable and harmonious world.

The quest for economic growth exacts a heavy toll on our environment, leaving a trail of devastation in its wake. As nations strive to expand their economies, the thirst for natural resources becomes insatiable. They extract, consume, and exploit with reckless abandon, heedless of the consequences. The consequences? Climate change, deforestation, and the depletion of irreplaceable resources. Picture the extraction and combustion of fossil fuels, fueling industries and transportation, and spewing forth greenhouse gases (GHGs) that accelerate the warming of our planet. Witness the unrelenting demand for water, pushing regions towards scarcity. Behold the mountains of waste, choking the air, poisoning the water, and suffocating the soil. These environmental calamities not only imperil our fragile eco-system but also exact a heavy toll on our economies and societies.

Alas, the pursuit of economic growth creates an environment conducive to breeding social inequalities. It bestows upon a privileged few the spoils of success, while relegating others to the periphery of society. Wealth and income disparities widen, with a select few amassing unprecedented power and influence. These powerful titans exploit natural resources, trampling over the rights of indigenous communities, leaving in their wake a trail of displacement and marginalization. The cycle repeats as goods and services are produced in developing countries under exploitative conditions for the consumption of the privileged few in developed nations. The chasm between the haves and the have-nots widens, leading to a fracturing of social cohesion and an erosion of trust.

But the consequences of unbridled economic growth extend far beyond environmental destruction and social inequalities. Its dark underbelly reveals a stark reality – public health hangs in the balance. The very air we breathe becomes polluted, tainted by the emissions of industries and transportation. Respiratory ailments and cardiovascular diseases ensue, plaguing those who bear the brunt of the noxious fumes. As forests fall to the insatiable appetite for land, the natural habitats of countless species are destroyed, setting the stage for the emergence of zoonotic diseases, such as the formidable COVID-19. And it is the marginalized communities, often confined to areas with degraded environments and limited access to healthcare, who bear the greatest burden of these health crises.

III. Fueling Inequalities With Ferocity

As I sit here on the 18th of June 2023, contemplating my final thoughts and musings on this section, I can't help but reflect on Juneteenth. It beckons us to honor and commemorate Juneteenth – the very Juneteenth that has etched its place in history as Juneteenth National Independence Day, a federal holiday in the United States. It is a day that stands as a poignant reminder of the emancipation of enslaved African Americans.

The name itself carries weight, a fusion of "June" and "nineteenth," a nod to the historic order that echoed through the land on June 19, 1865. It was Major General Gordon Granger who, two and a half years after the Emancipation Proclamation, declared freedom for the slaves in Texas. Originating in Galveston, Juneteenth has since been commemorated annually in various corners of the United States, often serving as a vibrant celebration of African-American culture.

But it was only in 2021, after the tireless efforts of Lula Briggs Galloway, Opal Lee, and countless others, that Juneteenth was officially recognized as a federal holiday. It was President Joe Biden who affixed his signature to the Juneteenth National Independence Day Act, ushering in this newfound status. And here we are, on this Juneteenth, as I grapple with the weighty subject at hand – the unsustainability of our economy and the destructive path it treads, fueling inequalities with unfettered ferocity.

It becomes glaringly apparent that any honest exploration of "Unsustainable Economy: A Path to Destruction, Fueling Inequalities with Ferocity" must delve into the harrowing legacies of slavery and colonialism. These two glaring examples serve as vivid reminders of the dire consequences that come hand in hand with unsustainable growth.

Slavery, a dark chapter in human history, holds a mirror to our collective past. It was a system built upon the exploitation of fellow human beings; a monstrous enterprise driven by economic motives. Enslaved individuals were stripped of their humanity, their labor mercilessly extracted to fuel the relentless pursuit of profit. The echoes of this abominable institution continue to reverberate through time, manifesting as enduring inequalities that plague our society and hinder any notion of true progress.

And then there is colonialism, a force that swept across continents, leaving in its wake a trail of devastation and inequality. The colonizers, driven by a hunger for power and wealth, imposed their will upon colonized lands and peoples. Resources were pillaged, indigenous cultures suppressed, and economic systems skewed to favor the colonizers. The scars of colonialism remain etched upon the landscapes and societies it touched, perpetuating disparities and sowing the seeds of discontent that persist to this day.

It is through the lens of slavery and colonialism that we witness the dire consequences of an unsustainable economy. Their tales serve as cautionary reminders of the perils that await when unchecked growth and inequality are left unchecked. They lay bare the truth that our economic systems must be built upon the pillars of sustainability, justice, and equality if we are to escape the destructive path, we find ourselves on.

As I progress further within this section, on this Juneteenth, let us honor the resilience and triumphs of those who fought for freedom and justice. Let us commit ourselves to crafting a future that uproots the inequities of the past and fosters a sustainable economy – one that paves the way for a more just and inclusive society. Only then can we truly break free from the shackles of an unsustainable path, charting a course towards a brighter and more equitable future for all.

A comprehensive study on the Trans-Atlantic Slave Trade published by Cambridge University Press (Eltis et al., 2013) sheds light on the economic enrichment of European powers through slavery. This compelling study reveals a staggering figure of approximately 12.5 million African individuals forcibly transported to the Americas, primarily to serve as labor in European colonies. This influx of enslaved labor had a profound impact on the productivity of agricultural plantations, particularly those cultivating lucrative crops like sugar, tobacco, and cotton. The profits generated from these plantations flowed into the coffers of European slave traders, merchants, and plantation owners, fueling their economic prosperity.

Further reinforcing the link between slavery and wealth accumulation, studies, and seminal works (Fogel et al., 1985) have highlighted the central role of slavery in the early economic development of the United States. The institution of slavery formed the bedrock of the agricultural economies in the southern states, paving the way for substantial wealth accumulation among plantation owners. Astonishingly, by the mid-19th century, enslaved African Americans represented the largest financial asset in the United States, surpassing the combined value of all U.S. manufacturing and railroads. This economic reliance on slavery created immense disparities in wealth and power, perpetuating social inequalities within both European and American societies.

The ecological ramifications stemming from the transatlantic slave trade demand meticulous consideration, esteemed scholars (Silkenat, 2022) extensively explore this facet. The expansion of plantation agriculture, driven by the transatlantic slave trade, exacted a heavy toll on the environment, particularly in the Americas. Extensive land clearing for large-scale cultivation of crops like sugar and tobacco led to deforestation, soil erosion, and the loss of biodiversity. The ecological aftermath of this intensive agricultural system still reverberates in regions that were once dominated by plantation economies, underscoring the long-lasting environmental degradation caused by the slave trade.

The transatlantic slave trade also left a lasting legacy of widening wealth gaps and social inequalities. Illustrating this facet, "Empire of Cotton: A Global History" (Beckert, 2014) serves as a poignant case study. It vividly showcases how the vast profits stemming from the slave trade and plantation economies prolonged and exacerbated socio-economic inequalities across both the continents of Europe and America. The concentration of wealth among a privileged merchant class further entrenched the power dynamics, leaving marginalized communities, particularly people of African descent, excluded from economic opportunities. Systemic discrimination and exploitation became deeply ingrained, exacerbating social divisions, and fueling ongoing inequality.

Turning our attention to Africa, it is evident that the transatlantic slave trade played a pivotal role in the continent's underdevelopment and economic dependency. Prof. J. E. Inikori (2004) argues this perspective in "Africans and the Industrial Revolution in England: A Study in International Trade and Economic Development." The extraction of human resources and natural wealth through the slave trade had devastating consequences for African societies. The loss of millions of able-bodied individuals and the disruption of local economies created a vicious cycle of poverty and economic dependency that continues to hinder sustainable development efforts today.

These examples, supported by robust studies and historical data, provide a stark portrayal of the profound impact of the transatlantic slave trade. From economic enrichment and environmental degradation to widening inequalities and the underdevelopment of entire regions, the legacy of this dark chapter in human history continues to reverberate. Acknowledging these historical realities is crucial for forging a path towards a more equitable and sustainable future.

The frustration over mankind's failure to learn from history is palpable, especially in the context of modern slavery and the relentless pursuit of economic growth. Even though one might expect humanity to have evolved beyond the horrors of slavery, the lyrics of Bob Dylan's "Blowin' in the Wind" serve as a poignant reminder that such progress is far from certain. I quote:

"How many years can some people exist.

Before they're allowed to be free?

How many times can a man turn his head.

And pretend that he just doesn't see?"

The persistence of injustice and oppression throughout history is evident in the modern-day scourge of human rights abuses, particularly in relation to cheap food exports. Italy and Spain provide sobering examples of such exploitation. An investigative report by The Guardian titled "Are your tinned tomatoes picked by slave labor? How the Italian mafia makes millions by exploiting migrants" (Jones et al., 2019) sheds light on the dire situation of migrant workers in Italy. The report reveals a form of exploitation that surpasses historical slavery, devoid of even the basic concern for the health and well-being of the workers.

In Italy, rather than acknowledging and addressing this exploitation, the country's interior minister, Matteo Salvini, has strategically labeled immigrants as the "new slaves." This narrative of victimizing the victims serves political purposes, perpetuating the marginalization of immigrants and fueling far-right rhetoric. Salvini and his allies twist the logic, portraying the Italian people as the victims and immigrants as the criminals, further entrenching the exploitative system.

Reports and investigations have exposed the gravity of the situation. The Global Slavery Index estimates that there are 50,000 enslaved agricultural workers in Italy, with a total of 145,000 people enslaved across various sectors (2023). The United Nations' special rapporteur on slavery warns that almost 100,000 agricultural workers in Italy endure inhumane conditions. Similarly, Spain has witnessed labor rights violations in its agriculture sector, as highlighted in The New York Times article "In Spain, Workers in Strawberry Fields Speak Out on Abuse." Migrant workers, particularly those from North Africa, face low wages, excessive working hours, and substandard living conditions (Alami, 2019). These violations include precarious employment contracts, wage theft, and limited access to social security benefits.

It is important to note that these examples represent specific cases, and human rights abuses in relation to cheap food exports can extend beyond Italy and Spain. Efforts to combat these abuses involve enhancing labor protections, strengthening regulations, promoting responsible sourcing practices, and supporting transparency initiatives across food supply chains. The aim is to ensure the respect of human rights and fair treatment of workers throughout the agricultural and food production sectors.

The echoes of history are unmistakable, revealing mankind's continued failure to learn from the past. The lyrics of Dylan's "Blowin' in the Wind" serve as a haunting reminder that unless we actively address these issues and confront the uncomfortable truths, history will repeat itself.

In the dark annals of colonial history, the economic exploitation of colonized peoples in the relentless pursuit of infinite growth stands as a grotesque testament to human greed. Countless studies have delved into this painful chapter, revealing the scars it has left on African families across generations. The German colonization of Kamerun serves as a striking example, where treaties with coastal kings were mere formalities disregarded in their quest for domination. Under the auspices of the now-defunct "League of Nations," Kamerun was reduced to a mere Potemkin protectorate façade, granting the Germans license to orchestrate a campaign of ruthless deportation, and forced labor.

The German colonizers displayed unparalleled mastery in uprooting tribes from their ancestral lands and conscripting them into grueling labor for their rubber plantations and the construction of railways leading to the port cities. These railways, a conduit for the stolen goods and commodities destined for the German homeland, came at a staggering cost to the lives and well-being of countless African men and women. The harrowing tales passed down from my ancestors bear witness to the horrors they endured – endless hunger, merciless beatings, and unspeakable acts of castration etched deep into their bodies, leaving indelible scars that they shared with their descendants.

My own grandfather, a keeper of our family's painful history, recounted time and again the plight of the young men from the Eton Tribes in Cameroon, who were forcibly deported to the tribal Bassa regions surrounding Edea, Eseka, and beyond, to toil under the oppressive weight of German colonial ambitions. Their suffering, etched into the very fabric of their beings, serves as a haunting reminder of the depths to which colonial powers stooped in their insatiable hunger for wealth and control.

Numerous studies have dissected the intricate web of exploitation woven through the vehicle of colonialism, laying bare the devastation inflicted upon Africa's people, environment, and economy. While a wealth of research exists, I will highlight two pivotal works that, in my view, provide a comprehensive understanding of this multifaceted tragedy. These studies, like interlocking puzzle pieces, connect the dots and expose the true magnitude of the crimes committed against Africa and its people.

Enter for example the realm of Adam Hochschild's monumental work (2020), "King Leopold's Ghost: A Story of Greed, Terror, and Heroism in Colonial Africa." This book stands as a profound testament to the horrors inflicted upon the Congolese people during the merciless colonization by King Leopold II. Now, let us embark on this journey together, for the Congo, more than any other place in Africa, epitomizes the adage, "Africa is not a poor place, it is a place where poor people live." I offer my own translation: "Africa is a rich place where impoverished souls reside."

Hochschild's findings resonate across multiple dimensions, each shedding light on the depths of this historical tragedy. Firstly, in terms of quantifying the magnitude of suffering, Hochschild uncovers a staggering toll inflicted upon the Congolese people under Leopold's rule. The population of the Congo Free State, under Leopold's iron grip, dwindled by millions due to the horrors of forced labor, violence, disease, and starvation. It is a chilling estimate that as many as 10 million Congolese lost their lives during this dark period.

But it is not merely the numbers that reveal the horrors of Leopold's reign. Hochschild's qualitative analysis delves into the heart of the matter, exposing the brutal methods employed by Leopold's agents in their ruthless pursuit of profit from the Congo's abundant natural resources, particularly rubber. The pages of Hochschild's work unmask a disturbing reality of widespread forced labor, the unspeakable mutilation and torture inflicted upon Congolese workers, and the relentless suppression of any resistance or dissent. Moreover, Hochschild unravels the intricate web of complicity woven by European companies and individuals who aided and profited from the merciless exploitation of the Congo.

Yet, Hochschild's narrative does not stop there. It pierces through the veil of deceit to illuminate the international complicity and moral bankruptcy that enabled these crimes to persist. With deft precision, he unveils the calculated manipulations employed by Leopold, who cunningly cloaked his actions under the guise of humanitarianism while perpetrating abhorrent acts against innocent lives. Hochschild meticulously examines the collusion of European governments, businesses, and individuals who, fully aware of the atrocities, chose to avert their gaze or, even more reprehensibly, reveled in the spoils of the Congo's ravaged resources.

Through "King Leopold's Ghost," Hochschild crafts a powerful narrative that exposes the darkest corners of history. He unearths the depths of human suffering endured by the Congolese people and unveils the shadowy alliances that perpetuated this harrowing chapter. It is a stark reminder of the immense capacity for greed and the dire consequences of unchecked power. May we heed the lessons of the past and strive for a future where such atrocities are forever banished from the annals of human history.

Now, as I introduce you to the captivating world of Walter Rodney's seminal work (2018), "How Europe Underdeveloped Africa," be prepared to witness the profound impact of colonialism on African societies, a story that echoes through the corridors of time.

Rodney's words become a symphony of knowledge, orchestrating a vivid picture of Africa's struggle against the forces of exploitation and oppression. As we delve into his profound findings, let us embark on a journey that will awaken your senses and open your eyes to the untold chapters of history.

Listen closely as the echoes of systematic plundering reverberate through the ages. The colonial powers, driven by insatiable greed, descended upon Africa's lands like voracious predators. They stripped the continent of its precious resources, leaving behind a trail of devastation. Minerals, once embedded in the heart of the earth, were wrested away, while the fertile soils of Africa yielded their harvests to feed foreign lands. Africa's wealth flowed like a river, but it was a river that never returned to quench the thirst of its own people.

Feel the weight of unequal trade relations bearing down upon Africa's shoulders. Like a mighty beast of burden, the continent was saddled with the burden of supplying raw materials to fuel the industrial machine of the colonial powers. In return, Africa received manufactured goods at exorbitant prices, a mockery of fair trade. The scales of justice tilted precariously, and Africa's economic aspirations were smothered beneath the weight of exploitation.

Observe the shattered remnants of indigenous economic systems, once vibrant and thriving. They were the lifeblood that coursed through Africa's veins, connecting communities and fostering prosperity. But colonialism's iron fist shattered this delicate eco-system, replacing it with foreign economic structures that bled Africa dry. Trade networks crumbled, traditional practices were discarded like forgotten treasures, and Africa's self-sufficiency became a distant memory.

Let the poignant voices of cultural erosion resonate within your soul. Africa's tapestry of diverse cultures, woven with threads of resilience and heritage, faced an onslaught from the juggernaut of colonialism. European values, education systems, and governance structures cast their shadows over African lands, suffocating indigenous traditions. The flames of cultural preservation flickered, threatened by the winds of change, leaving behind scars of loss and displacement.

Bear witness to the grotesque dance of socio-economic inequalities perpetuated by colonial rule. Elites conspired with colonial authorities, forging unholy alliances that sealed the fate of millions. Wealth was extracted with callous disregard, cementing the chains of poverty for generations to come. The chasm between the haves and the have-nots widened, creating a haunting legacy that still haunts the African continent today.

While Rodney's work grazes upon the environmental consequences of colonialism, it is the economic and social dimensions that dominate his narrative. Yet, we catch a glimpse of the scars inflicted upon Africa's natural environment. The extraction of minerals, the pillaging of timber, and the clearing of vast swathes of land for agricultural purposes left a scarred landscape in their wake. Environmental degradation, deforestation, soil erosion, and the loss of precious wildlife became the silent casualties of colonialism's insatiable hunger for profit (Harvey, 2020; Harvey, 2012).

However, to uncover a more detailed and comprehensive understanding of the environmental and climate impacts of colonialism in Africa, we must venture beyond Rodney's chronicles. Other realms, such as environmental history or postcolonial studies, harbor tales that breathe life into the intricate connections between humanity and nature. There, we can uncover a tapestry of narratives that illuminate the environmental footprints left by colonial powers, ensuring that these chapters of history are never forgotten.

Ladies and gentlemen, I can feel some of you itching to shout at me, "Hey, Mister! Most African countries have been independent for over sixty years now. Stop complaining and do better!" It's a revisionist song echoing in my ears, as 50 Cent would say in his lyrics of "Ghetto Qu'ran

(Forgive Me)." Allow me to quote a few lines that resonate with the sentiment:

"...I recall memories, filled with sin
Over and over again, and again (And again and again)..."

Indeed, my dear companions, it is an undeniable fact that the shadows of colonialism have endured even in the wake of gaining independence. However, allow me to spare you the tedium of a didactic discourse. Instead, I shall enthrall you with a narrative, one meticulously chronicled within the esteemed annals of The Washington Post (Koven, 1981), under the evocative title "Bokassa's Version of Giscard Link: Portrait of Paternal Relationship."

This article delves into the intriguing relationship between French President Valery Giscard d'Estaing and Jean-Bedel Bokassa, the former dictator of the Central African Republic. It portrays a complex bond steeped in paternalism, where lavish gifts, favors, and business arrangements became the currency of their connection.

Then in exile in the Ivory Coast, Bokassa blames Giscard for his downfall and the loss of his wealth and power. He accuses Giscard of exploiting their relationship for personal gain, claiming that if he was involved in corruption and diamond theft, then Giscard should bear the weight of accountability as well.

Bokassa's bitterness and desire for revenge against Giscard are palpable, particularly during the period when Giscard faced a presidential election runoff against Socialist Francois Mitterrand. Bokassa sought to depose Giscard by exposing the alleged misdeeds and manipulations of the French president.

While some of Bokassa's allegations were vicious, comical, or salacious such as the legendary "For 10 years, I was with Giscard. If I was a cannibal, he was a cannibal. For 10 years, I was with Giscard. If I stole diamonds, he should be punished, too, because he got his diamonds." and have been previously reported in the French media (Delpey, 1985) the Giscard government has never officially responded to these accusations. Giscard himself dismisses Bokassa's claims as nothing more than the vengeful ramblings of a fallen leader.

What emerges from this article is a glimpse into Giscard's fascination with Africa, his passion for big-game hunting, and the perceived arrogance that ultimately contributed to his electoral defeat. Accepting diamonds and other lavish gifts, while maintaining a sprawling hunting preserve during his time in office, painted a picture of the Central African Republic as a mere extension of France – an object of patronizing manipulation by its former colonial power.

These diamonds, symbolic of Giscard's privileges and questionable connections with Bokassa, become the focal point of scrutiny. Bokassa alleges that over eight years, he bestowed numerous diamond gifts upon Giscard and his family, including some of significant value. He cynically comments on the financial needs of the Giscard family, leaving us pondering the details left unsaid.

In essence, this article offers us more than a glimpse into the dynamics of power, corruption, and neocolonialism in Africa. It presents a tapestry of consequences with implications for poverty, the environment, and the welfare of African nations.

Now, my friends, let us explore some specific examples that reveal the economic exploitation and neocolonial practices perpetuating poverty and hindering equitable distribution of wealth and resources in African nations.

One such example lies in the extraction of Africa's abundant natural resources by former colonial powers and multinational corporations. These entities, driven by their insatiable appetites, have plundered Africa's minerals, oil, and gas, leaving behind meager benefits for local communities while reaping enormous profits that often flow out of the continent.

Furthermore, unfair trade practices imposed by developed countries favor their own economic interests, stifling African nations' ability to compete in global markets. Tariffs, subsidies, and non-tariff barriers create an imbalanced playing field, where African nations export raw materials at low prices and import finished products at exorbitant costs, perpetuating poverty and hindering local industrial development.

Let us not forget the exploitative labor practices that plague Africa, with multinational corporations taking advantage of low wages, poor working conditions, and the absence of workers' rights. This unscrupulous extraction of cheap labor further widens the gap between the haves and the have-nots, perpetuating poverty and inequality.

Land grabbing, another neocolonial practice, sees foreign investors swooping in to acquire vast tracts of land, often with the collaboration of local elites and governments. This land grab displaces communities, disrupts livelihoods, and exacerbates food insecurity and poverty, as commercial agriculture and resource extraction take precedence over local needs.

Lastly, the burden of debt cannot be overlooked. African countries find themselves shackled by debt incurred from loans provided by former colonial powers and international financial institutions. These loans often come with stringent conditions that drain resources, diverting them from poverty alleviation, healthcare, education, and environmental conservation.

The examples I've presented shed light on the culture of corruption and wealth inequality inherent in the relationship between Giscard and Bokassa. Accepting diamonds as gifts signifies a corrosive system where political elites enrich themselves while the wider population suffers. Public resources are misused or diverted for personal gain, depriving communities of much-needed development opportunities.

The lack of accountability surrounding these allegations further erodes trust in government institutions. Patronage networks form, where leaders cultivate relationships with autocratic rulers, solidifying their influence and benefiting from ill-gotten wealth and resources. Meanwhile, ordinary citizens are left marginalized and impoverished, with little access to the resources that could uplift their lives.

But let us not overlook the environmental consequences that can arise from unregulated hunting activities. Hunting, when not properly controlled, can disrupt eco-systems, deplete wildlife populations, and degrade habitats. The consequences ripple through the delicate balance of nature, affecting not only the flora and fauna but also the livelihoods of local communities dependent on natural resources.

While the specific impact of Giscard's hunting preserve remains elusive, numerous global examples demonstrate the detrimental effects of unregulated hunting on ecosystems and local communities. From the poaching of elephants for ivory to the illegal trade in bushmeat, overhunting has resulted in significant declines in wildlife populations, ecological imbalances, and the erosion of traditional practices tied to the natural environment.

Ladies and gentlemen, these examples serve as a stark reminder of the complex web woven by power, corruption, and neocolonialism in Africa and beyond. They illustrate the urgent need for equitable and transparent relationships between former colonial powers and African countries or between the "haves" and the "haves not". Only by addressing these systemic issues can we foster sustainable development, alleviate poverty, and improve the welfare of African populations.

Let us strive for a world where the echoes of neocolonialism fade away, replaced by a symphony of collaboration, empowerment, and shared sustainable prosperity.

IV. Fueling The Climate Crisis With Ferocity

In the vast expanse of our oceans, a quiet battle rages beneath the surface – a battle against an invisible enemy that threatens the delicate balance of marine eco-systems and casts a shadow over the future of our planet. This enemy is plastic waste, a growing global concern with far-reaching environmental implications. As we delve into the depths of this issue, we uncover a world where insufficient waste management, inadequate recycling, and widespread pollution converge to wreak havoc on our climate and the creatures that call the oceans home.

To fully grasp the magnitude of this problem, we must first understand the sheer scale of plastic production. Each year, over 400 million tons of plastic are manufactured for a multitude of applications across the globe. A staggering amount, to say the least. However, what is even more alarming is the fate that awaits a significant portion of this plastic once its purpose is served. Approximately 14 million tons find their way into the ocean every year, comprising a staggering 80% of all marine debris, from the surface waters to the deepest recesses of the seabed (Statista, 2023; IUCN, 2023).

As this plastic invades the marine realm, it leaves a trail of destruction in its wake. Marine species, unsuspecting victims of this human-made plague, suffer injuries and fatalities due to ingestion or entanglement in the debris. From majestic whales to tiny plankton, no creature is immune to the perils of plastic pollution. But the consequences don't stop there. Plastic waste seeps into the very fabric of our environment, tainting our food sources and jeopardizing our health. It disrupts the delicate balance of eco-systems, threatening biodiversity and diminishing the resilience of marine life.

I hear you saying that the overall impact of plastics on ocean temperatures is relatively small compared to the broader climate change factors.

You raise an astute point, my friend. While the direct impact of plastics on ocean temperatures may seem relatively small in the grand scheme of climate change, it is the potential for feedback loops and tipping points that warrants our attention and concern. Science has indeed shed light on the broader ecological and environmental considerations that can amplify the effects of plastic pollution in the seas, ultimately contributing to larger-scale changes with far-reaching consequences.

Let us delve into these aspects, for they unveil a deeper understanding of the complex relationship between plastics and the climate.

First, we encounter the albedo effect – an often overlooked but significant factor. Plastics, especially those that float upon the ocean's surface, possess the power to alter the reflective properties of the water, reducing its albedo. This reduction means that more sunlight is absorbed by the ocean, leading to increased water temperatures. This seemingly minor change can have cascading effects, potentially accelerating the melting of ice caps and releasing additional greenhouse gases (GHGs), thereby intensifying global warming. It is a chain reaction that we must interrupt if we are to preserve the delicate balance of our planet's climate.

Moreover, the ecological disruptions caused by plastic pollution cannot be underestimated. The intricate tapestry of marine eco-systems relies on a delicate balance of interconnected relationships. When plastic infiltrates these systems, it disrupts food chains, alters habitats, and impairs the reproductive capabilities of marine organisms. Such disruptions reverberate throughout the eco-system, diminishing its ability to sequester carbon and mitigate climate change. We find ourselves caught in a precarious positive feedback loop, where the impacts of climate change are exacerbated by compromised eco-systems – a cycle we must strive to break.

Carbon sequestration, a vital process for stabilizing atmospheric carbon dioxide levels, is also impacted by the presence of plastics in the seas. As plastics release chemicals and toxins into the water, they can interfere with the ability of marine organisms, such as phytoplankton, to perform photosynthesis and capture carbon. When the carbon sequestration capacity of marine eco-systems is compromised, atmospheric carbon dioxide levels rise, further contributing to climate change. It becomes clear that our fight against plastic pollution is intertwined with our battle against global warming.

Lastly, we confront the specter of ocean acidification – a consequence of increased carbon dioxide absorption. Plastic debris, through its interaction with seawater, releases certain chemicals that contribute to the acidification of our oceans. This phenomenon poses a grave threat to marine life, particularly organisms that rely on calcium carbonate structures like coral reefs and shell-forming creatures. As these eco-systems succumb to acidification, the delicate balance of carbon dioxide absorption and release is disrupted, potentially exacerbating the effects of climate change.

Yet, the implications of plastic waste reach far beyond the boundaries of our oceans. They extend to the realms of climate change and the global fight against carbon emissions. Plastics, intrinsically linked to the fossil fuel industry, derive 99% of their raw materials from fossil fuel feedstocks. As the petrochemical industry churns out plastic, it simultaneously fuels global oil demand growth – a connection often overlooked as we focus our efforts on reducing fossil fuel usage in other sectors. The extraction, refining, and manufacturing processes involved in plastic production generate substantial carbon emissions, to the tune of 1.96 gigatons (Gt) of carbon dioxide equivalent in 2015 alone, accompanied by a staggering price tag of approximately $341 billion annually (Landrigan et al., 2023).

The story doesn't end there. Improper disposal methods, such as incineration and open burning, release significant greenhouse gases (GHGs), further exacerbating the plastic-related carbon footprint. In fact, if left unchecked, plastic waste is projected to contribute up to 13% of the remaining carbon budget by 2050 – a haunting prospect for our climate's future. But the impact of plastics on climate change goes beyond emissions. I reiterate, these pervasive materials have the power to disrupt the ocean's natural ability to absorb and sequester carbon dioxide. By interfering with the carbon sequestration process, plastics impede the vital role of marine eco-systems as natural carbon sinks, potentially accelerating the pace of climate change (Joshi, 2021).

The intricate interconnectedness of Earth's systems demands our comprehensive efforts to reduce plastic waste, mitigate climate change, and safeguard marine eco-systems. It is a complex web we must navigate with caution and diligence if we are to secure a sustainable future for our planet.

So, my friend, let us not be complacent in the face of this challenge. Instead, let us embrace the urgency that arises from the interplay of plastics, climate change, and the oceans. Through collective action, relentless research, and a commitment to stewardship, we have the power to turn the tide. Let us stand united, driven by the understanding that our fate and the fate of our planet are inextricably linked.

As we dive deeper into this abyss of environmental concern, it becomes evident that urgent action is necessary. Binding agreements must be explored and implemented to address the relentless march of marine plastic pollution. Awareness must be raised, and practices must be changed. We must reduce plastic production, improve waste management systems, and promote recycling and circular economy practices. The battle against plastic waste requires a comprehensive, multifaceted approach – one that considers the interconnectedness of our planet's systems.

Amidst the saga of sea plastic and its far-reaching implications on the climate crisis, yet another profound anecdote emerges to captivate our attention.

In the steel-gray waters of the North Sea, a remarkable sight captures the attention of onlookers in the port of Wilhelmshaven. It is the Höegh Esperanza, an imposing tanker that stretches across the expanse of three football fields. But this is no ordinary vessel. The German government has commissioned it as a solution to the nation's energy crisis – a Floating Storage and Regasification Unit (FSRU).

The purpose of the Höegh Esperanza is to convert liquefied natural gas (LNG) back into its gaseous form. This converted gas can then be injected into natural gas pipelines, providing an alternative energy source for Germany. The need for such a solution arose when Russia's invasion of Ukraine led to the cutoff of Russian pipeline gas, leaving Germany in search of new options.

Germany has turned subsequently to LNG imports from countries like the United States to meet its energy needs. However, this decision has not been without controversy. Some climate advocates argue that investing in fossil fuel infrastructure, such as LNG terminals, contradicts Germany's goal of transitioning to a carbon-free future. Nevertheless, the German government sees it as a crucial step toward energy security.

But here's the twist: Germany, which has banned shale gas fracking domestically since 2017, is now importing with that platform LNG primarily sourced from fracked natural gas in the United States. This apparent contradiction raises questions about the government's stance on fracking and its commitment to environmental sustainability. According to a report from the German Ministry of Energy, fracking is a contentious method used to extract natural gas, oil, and geothermal energy from unconventional sources. Unconventional natural gas deposits, such as shale gas and coal bed methane, require hydraulic fracturing (fracking) to access these trapped resources. The process involves injecting a high-pressure fluid mixture of water, sand, and chemicals into the rock formations to create or widen fractures, allowing the extraction of natural gas (BMWK, 2023).

Esteemed comrades, it is incumbent upon me to apprise you of the prevailing legal framework in Germany, which bears tremendous significance. Within this context, it is imperative to note that the extraction of gas from shale, clay, marl, and coal bed formations, denoted as unconventional fracking projects, is predominantly prohibited by law. This decision is rooted in profound apprehensions regarding environmental risks and an inherent dearth of comprehensive knowledge and experience in handling such endeavors. Only a limited number of pilot projects in these formations are allowed for scientific purposes, subject to strict conditions. Conventional fracking in other rock formations is also subject to stringent regulations.

As attested by the same esteemed German Ministry, currently engaged in importing fracking gas, it becomes abundantly clear that fracking engenders a myriad of environmental perils. These encompass the perilous contamination of groundwater and drinking water, the insidious dissemination of air and noise pollutants, the extensive imposition on land resources, and the voracious consumption of water. The historical deployment of hazardous fracking fluids within Germany has elicited deep-seated apprehensions concerning potential threats to human health and ecological balance. As of now, prevailing regulations restrict the use of only non-hazardous or minimally hazardous fracking fluids, yet it remains incumbent upon us to subject the potential impact of water withdrawals on groundwater reservoirs and surface water bodies to scrupulous scrutiny under the purview of water authorities.

My cherished companions, we find ourselves pondering the abrupt shift in the stance of the Minister for Economic Affairs and Climate Action, now embracing LNG imports derived from fracked natural gas. The answer to this enigma lies in a state of undeniable apprehension, which I label as the 3Ps: Panic, Panic, Panic. Germany currently confronts an acute energy crisis, grappling with soaring energy costs and the imminent prospect of companies relocating their operations overseas. The German economy, heavily dependent on affordable gas supplies from Russia, now faces peril, compelling the government to prioritize the pursuit of economic growth above all else. Indeed, the climate champions from the Green Party, entrusted with the leadership of that esteemed Ministry, have regrettably succumbed to hypocrisy. Their grandstanding ideals have been exposed as mere façade, akin to the fable of "the emperor has no clothes."

To discern the whereabouts of the emperor's attire, let us conjure vivid images where German newspapers incessantly paint a somber picture with headlines bearing witness to companies departing the nation and the specter of an energy crisis looming over their very existence. Fear, an unwelcome companion, has precipitated a sense of impending deindustrialization, prompting the government to prioritize immediate growth over steadfast environmental considerations. As urgency grips the corridors of power to secure affordable energy sources and preserve industrial vitality, the once profound analyses and studies on fracking and its ecological repercussions have been relegated to the shadows.

Caught in the throes of this palpable panic, the German government has chosen to disregard its own published reports highlighting the perils of fracking. The pursuit of growth, even at the expense of sustainable practices, has emerged as the driving force behind their decisions. The allure of inexpensive gas and the promise of economic stability have cast a veil over their vision, preventing a discerning gaze upon the potential consequences of escalating methane missions and the exacerbation of climate change.

Alas, the German government finds itself treading a precarious path, its choices seemingly at odds with professed climate ambitions. The delicate equilibrium between short-term economic stability and the pursuit of long-term environmental sustainability remains unresolved. The future of Germany's energy landscape hangs precariously in the balance, symbolized by the Höegh Esperanza's presence in the North Sea – a vessel embodying both hope for energy security and the intricate challenges entwined with forging a sustainable trajectory forward. My esteemed friends, the poignant scene before us serves as a compelling testament that not all manifestations of climate destruction, stemming from the consequences of unsustainable growth, can be attributed to an Ex Machina root cause. Instead, it underscores the very real and intricate web of human actions and decisions that bear the weight of responsibility in shaping our environmental destiny.

In the vast expanse of our ever-changing world, where passions and pleasures intertwine, there exists a curious nexus between music, spirits, and the savory delights of a perfectly cooked steak. An unlikely trio, yet bound by an invisible thread that weaves its way through our desires and indulgences.

Confession is a powerful thing, for it unveils the hidden facets of our being. And so, I confess to you now, my dear reader: I have fallen under the enchantment of country music. Its twangy melodies and heartfelt lyrics have seeped into the deepest recesses of my soul. But that's not all. Another confession demands my lips to part once more, revealing a love for whiskey – a golden elixir that warms the spirit and ignites the senses.

As I pondered the intersection of these two passions, a revelation struck like a bolt of lightning on a darkened night. It came in the form of Chris Stapleton's soul-stirring ballad, "Tennessee Whiskey."

In its verses:

"You're as smooth as Tennessee whiskey,
You're as sweet as strawberry wine,
You're as warm as a glass of brandy,
And honey, I stay stoned on your love all the time."

I found a harmonious fusion of the smoothness of whiskey and the sweetness of strawberry wine. But what intrigued me most was the refrain, an ode to the intoxicating power of love – a love that has the ability to leave one perpetually entranced.

Yet, as my mind wandered to the sizzle of a well-marbled steak on a scorching grill, I couldn't help but imagine a remix of this refrain – one tailored to the delight of my taste buds and the carnal pleasures of devouring a perfectly cooked piece of meat. And so, my version came to life:

"You're as smooth as Filet Mignon,
You're as sweet as Ribeye,
You're as warm as T-Bone,
And honey, I stay stoned on your love all the time."

But let us now delve into weightier matters, for even as we revel in our passions, we cannot ignore the looming specter of the climate crisis that engulfs our planet. In this age of awakening, where the consequences of our actions demand introspection, it becomes imperative to consider the impact of our indulgences on our fragile environment.

For in the world of steaks lies a hidden truth, a truth that cannot be ignored nor taken lightly. The livestock industry, with its insatiable appetite for resources, has been identified as a significant contributor to greenhouse gas emissions. The production of meat, particularly beef, requires vast amounts of land, water, and feed. Methane, a potent greenhouse gas, is released by cattle during their digestion process, further exacerbating the environmental strain.

Scientists have confirmed that livestock farming, a key player in the arena of agriculture, contributes a hefty 32% to the methane tally, while in comparison paddy rice cultivation chips in with around 8%. This methane, a potent greenhouse gas, holds an 80-fold greater warming potential than carbon dioxide over two decades, and it fuels the creation of hazardous ground-level ozone (UNEP, 2021; Smith, 2021; NASA, 2022).

Behold the livestock sector, responsible for over 40% of its own climate footprint through methane emissions, primarily from beef and dairy cattle. Tackling this methane predicament stands as the paramount opportunity to put the brakes on global heating by 2040 (FAO, 2023).

In contemplating the grand tapestry of climate data, a disquieting revelation comes to light: Methane emissions surge forth at an unprecedented pace, surpassing even the deceleration witnessed in carbon dioxide emissions during the pandemic-induced lockdowns of 2020. This fervent ascent serves as an unequivocal clarion call, underscoring the urgent imperative to address methane emissions as a pivotal linchpin of climate change mitigation.

Among the culprits of this worrisome trend, livestock, encompassing the likes of cows, sheep, and goats, shoulders a significant burden, contributing approximately 14% of humanity's climate emissions. The methane exhaled during their digestive processes, be it through burps or the release of manure, presents both a grave concern and a golden opportunity in our collective struggle against global warming (Paddison, 2023).

Integral to our endeavors towards the ambitious objectives laid out in the United Nations Climate Change Conference (ccacoalition, 2021) and the Intergovernmental Panel on Climate Change (AR4, 2007) is the imperative to curtail methane emissions from the agricultural domain. Resolute experts assert that the reduction of methane holds the utmost potential to temper the relentless onslaught of warming, manifesting as the most consequential opportunity to shape a more sustainable and resilient future between now and the horizon of 2040 (UNEP, 2021).

Indeed, my cherished friend, I wholeheartedly recognize the pursuit of curbing methane emissions as a noble quest, a swift and potent strategy against the tempestuous impact of climate change. In this noble endeavor, the exploration of alternative feeds for cattle, the concerted effort to reduce food loss and waste, and the thoughtful moderation of meat and dairy production all assume pivotal roles, converging harmoniously to foster a sustainable and resilient future. And yet, despite the urgency of addressing methane emissions, many countries falter in setting specific targets or commitments for the agricultural sector. A few, like New Zealand and Denmark, venture forward, but progress remains painfully inadequate. For example, the lack of political will and precise policies to bolster sustainable agriculture and usher in dietary shifts prove formidable obstacles on the path to curbing methane emissions. An imperative shift away from large-scale meat and dairy industries, particularly in high-income countries, becomes essential. The goal: forge healthier and more sustainable food systems.

Against this backdrop, the European Union envisions a remarkable reduction of emissions by at least 55% by the year 2030. Achieving this ambitious feat necessitates profound changes, including cutbacks in livestock and meat consumption. Remarkably, despite the absence of climate-specific targets for livestock farming within Europe, the urgent need for comprehensive policies in this vital sector echoes loudly.

However, a glimmer of hope emerges from countries like the Netherlands, where visionary plans are taking shape to cull livestock numbers in response to pressing environmental challenges such as ammonia pollution. These decisive steps signal the potential for stringent regulations to firmly establish themselves in the realms of meat and dairy, forging a pathway towards a more sustainable and environmentally conscious future.

While some countries, like the United States and the European Union, pledge to trim methane emissions, the farming sectors lack specific targets. This void calls for fortified and coordinated efforts to tackle methane emissions from agriculture on a global scale. Brazil and Argentina, commanding roles as major beef producers and cultivators of animal feed crops, resist entreaties to curtail meat consumption. They embody the challenges entwined with shifting dietary patterns and implementing policy changes in countries heavily dependent on livestock industries.

As we savor the tenderness of a Filet Mignon, the richness of a Ribeye, or the primal satisfaction of a T-Bone, we must confront the reality that our indulgences contribute to the ferocity of the climate crisis. It is a paradox that demands our attention, for how can we reconcile our desires with the urgent need to safeguard our planet?

The path forward lies not in renouncing our passions but in embracing a more mindful approach. It beckons us to explore alternative sources of protein, to seek sustainability in our choices, and to support agricultural practices that mitigate environmental harm. The journey toward a more harmonious coexistence with our planet requires us to be conscious consumers, aware of the ripple effects our choices create.

So, as we revel in the melodies of country music, savor the warmth of a fine whiskey, and delight in the succulence of a perfectly grilled steak, let us not forget the responsibility we bear. Let us be inspired to find harmony between our passions and the well-being of our planet, for therein lies the truest satisfaction – a sense of fulfillment that transcends mere indulgence (Phillips, 2020).

In the tapestry of our lives, the threads of pleasure and responsibility can intertwine, forming a narrative that is both meaningful and sustainable. Let us embrace this challenge, for in doing so, we become not just consumers, but custodians of a world that beckons us to forge a brighter future. Beyond the mere indulgence of steak on demand, which, against all facts discussed earlier, is often perceived as a symbol of progress in various corners of the world, lies another aspect that warrants contemplation. The insatiable appetite for development and growth in contemporary societies has resulted in the proliferation of new cities, each adorned with aesthetically pleasing homes constructed primarily out of cement. However, in the pursuit of this seemingly endless progress, we must pause to ponder the hidden costs entailed in this ceaseless quest for urbanization and cement consumption.

Picture a bustling cityscape, with towering skyscrapers, sprawling highways, and majestic bridges, all made possible by the binding prowess of cement. This versatile substance, the backbone of construction, is responsible for the very foundation upon which our cities thrive.

To comprehend the sheer magnitude of cement's impact, let us delve into the realm of statistics. According to the United States Geological Survey (USGS), global cement production reached a staggering 4.1 billion metric tons in the year 2020 alone. China, a powerhouse of industrial might, reigns as the largest producer, responsible for a remarkable 58% of global cement production. Trailing closely behind are India, the United States, and several other nations that share in the cement-making endeavor (USGS, 2023).

Urbanization, the catalyst of change and progress, serves as the driving force behind the relentless demand for cement. As the United Nations reports, 55% of the world's population called urban areas home in 2018, a number projected to escalate to a staggering 68% by 2050. The construction of cities anew, the erection of residential edifices, and the need for robust infrastructure propel the insatiable thirst for cement.

The economic implications of this ubiquitous material reverberate across the globe, creating employment opportunities and fueling economic growth. In 2022 alone, the global cement industry reaped revenues amounting to approximately $ 400 billion. Cement, the lifeblood of urban development, leaves an indelible mark on the economic fabric of nations (Fortune, 2023).

But in the quest for progress, we must tread with caution. As cities burgeon, so does cement consumption. The insatiable hunger for infrastructure and housing begets an ever-growing demand for cement. The World Business Council for Sustainable Development (2023) sounds a warning, estimating that cement production may surge by 23% before 2030, an effort to satiate the voracious appetite of urbanization.

Yet, let us not lose sight of the darker side, the environmental impact that accompanies this march of progress. Cement production, like any other industrial process, is a source of negative environmental consequences. Climate change, air pollution, water consumption, habitat destruction – these are the burdens cement must bear.

Carbon dioxide (CO_2) emissions, the vanguard of climate change, weigh heavily on the cement industry's conscience. Calcination, the process through which limestone transmutes into lime, is the harbinger of CO_2 emissions. This chemical metamorphosis releases CO_2 as a byproduct, making cement production a substantial source of carbon emissions. Alas, the International Energy Agency reveals that the cement industry claims an unwelcome distinction, contributing approximately 8% of global CO_2 emissions (IEA, 2023; Tracy, 2023).

Air pollution, that insidious foe, also rears its noxious head in the realm of cement production. The crushing and grinding of raw materials, integral to cement manufacturing, engender dust and particulate matter emissions. These airborne particles, pollutants in their own right, tarnish the atmosphere, marring the purity of the air we breathe. Moreover, the high temperatures required within cement kilns become breeding grounds for nitrogen oxides, further fueling air pollution and endangering both human health and the delicate balance of our environment.

But it is not only the air that suffers under cement's reign; water, too, falls victim to its thirst. The production of cement guzzles substantial amounts of water, both as a coolant and as a vital ingredient in the process itself. Quarrying, the preparation of raw materials, and the cooling of equipment all lay claim to this precious resource. Regrettably, this voracious water consumption exacerbates the strain on already scarce water sources.

But fear not, dear readers, for in the face of this environmental conundrum, there is hope. Together, we can navigate the treacherous waters of cement production and embark on a journey towards sustainability. So, let us dive deeper into the depths of this issue, engaging in an interactive exploration of innovative strategies and groundbreaking advancements that offer glimmers of respite amidst the maelstrom.

Picture this: a world where cement plants capture their CO_2 emissions and either store them deep underground or repurpose them for other industrial processes. This is not a mere flight of fancy, but a promising avenue called Carbon Capture, Utilization, and Storage (CCUS). By seizing the pernicious gas and taming its destructive potential, we can turn it into a force for good, mitigating the industry's impact on climate change.

But that's not all, my friends! The quest for sustainable solutions leads us to the realm of alternative cementitious materials. These ingenious substitutes, such as fly ash, slag, or calcined clays, have stepped into the spotlight, offering a beacon of hope. By fashioning low-carbon alternatives with diminished carbon footprints, researchers and pioneers light the way towards a more sustainable future.

Energy efficiency and alternative fuels take center stage in our narrative, too. By bolstering energy efficiency in cement production and embracing biomass or waste-derived fuels, we unleash a dual assault on carbon emissions and our reliance on fossil fuels. It's a win-win situation, my friends – triumph for both the environment and our quest for sustainability.

Now, my dear readers, as we navigate the intricacies of this tale, let us not forget the perils of unsustainable growth in the context of urban development. The ticking time bomb threatens the very foundations of our cities, a relentless march of urbanization that engulfs our planet. But fear not, for sustainable urban planning, efficient resource management, and inclusive governance are the weapons we wield in this battle.

Let us envision a future where green building materials, such as alternative cementitious materials and low-carbon concrete mixes, mitigate the environmental impacts of cement usage. Integrated urban planning that emphasizes compact, mixed-use development and green infrastructure becomes our guiding star, curbing the demand for cement-intensive projects. The consequences of unsustainable growth may be far-reaching, but the power of sustainable urbanization knows no bounds.

Urban greening takes center stage, as we harness the power of trees, green spaces, and green roofs and walls to provide shade, evaporative cooling, and improved air quality. Cool roofs and pavements, with their reflective or light-colored materials, work their magic, reducing heat absorption and surface temperatures. Sustainable urban design principles come to our aid, optimizing material usage, considering the entire life cycle of a building, and integrating green spaces into the urban fabric.

My dear readers, we stand at a crossroads, facing the consequences of unsustainable growth and the challenges of cement production. But let us not despair, for the power to shape our future lies within our grasp. By embracing sustainable practices, fostering low-carbon alternatives, and enhancing industry-wide energy efficiency, we inch closer to a world that embraces sustainability in its very core.

So, my friends, I invite you to join me on this awe-inspiring journey – a journey that leads us towards a future where our cities thrive in harmony with nature and foster social well-being for all. The road ahead may be daunting, but with our collective resolve and unwavering determination, we can carve a path towards a sustainable tomorrow. Together, we can change the narrative and rewrite the tale of our planet's future. Let us embark on this adventure, for the sake of our planet and the generations to come!

And so, my friends, we arrive at the culmination of our odyssey – a voyage through the treacherous terrain of unsustainable growth and the formidable repercussions it unleashes upon our climate. Our expedition has taken us across the vast expanses of cement production and urbanization, as well as the intricate realms of plastics production, meat production, and fracking, unraveling the intricate tapestry of their environmental footprints. Yet, we must humbly recognize that our foray has merely grazed the surface of a sprawling and intricate web of interconnected challenges.

There exist untold tales, narratives yearning to be unearthed, and accounts that demand our unwavering attention. The abyssal depths of deep-sea mining, where the fragile equilibrium of ocean eco-systems teeters precariously on the brink of annihilation, implore us to delve further into the abyss. The Baltic Sea dead zones, asphyxiated by the surplus nutrients and perils of industrialized agriculture, murmur cautionary tales through the currents. Meanwhile, the relentless march of transportation, with its carbon emissions and ceaseless congestion, adds yet another enthralling chapter to the unfolding saga.

Alas, my friends, we cannot hope to encompass within these pages the entirety of unsustainable growth pursuits and their ferocious impact on our climate. We have merely sketched a few examples, seeking to shed light upon the dire urgency of our predicament. Yet, the world sprawls wide, and its tales are multitudinous. It now falls upon your discerning judgement, dear readers, to pursue further knowledge and immerse yourselves in the extensive literature that delves into these realms.

V. Artistic Echoes Of Climate Change: The Broken Balance

CHAPTER *Five*

Technology's Illusion: No Magic Wand For The Climate Crisis, Demanding A Revolution Of Policies And Mindsets

"The real problem is not whether machines think but whether men do."
– B.F. Skinner, behavioral psychologist

I. The Illusion Of Technology As A Magic Wand

On the 15th of September 2021, the German online magazine taz.de (2021), always keen on cutting-edge sarcasm, graced its readers with a headline that could make any cynic's heart skip a beat: "German engineered climate policy." How ingenious! It's as if the mere act of engineering, performed with that distinct German precision, can miraculously solve the climate crisis. Move over, global warming, the Germans are here to engineer the hell out of you!

And who better to lead this grand engineering symphony than the chairman of the Free Democratic Party (FDP), Christian Lindner himself? Yes, the FDP, that relatively small political party that fancies itself as the champion of the free market economy. Because, you know, who needs to worry about the environment when there are profits to be made and taxes to be slashed?

In the delicate dance of climate politics, Lindner emerges as the savior, armed with the holy trinity of solutions: the market, new technologies, and, of course, freedom. Ah, freedom! The sweet melody that dances off the lips of politicians when they find themselves in a tight spot. Can't regulate those pesky emissions? No worries! The invisible hand of the market will surely guide us to a greener future, just as it has done with all those other pressing global issues.

But wait, there's more! Lindner's understanding of technology seems to transcend the mere mortal realm. It is either a testament to his naivety, his sheer ignorance, or perhaps a cleverly disguised tactic to avoid any real debate on the climate crisis. For in his world, technology is the silver bullet, the magic wand that will whisk away our carbon sins and save us from ourselves. Hallelujah!

So here we are, my friends, caught in the crosshairs of political theater. The FDP, the party of big money, big capital, big oil, and big fossil, offering us a tantalizing promise wrapped in shiny new technologies and sprinkled with the sweet nectar of low taxes. It's a symphony of hope, a symphony that whispers, "Fear not, for technology will rescue us all!"

But deep down, amidst the grand promises and the virtuosic rhetoric, one cannot help but wonder if this is all just a well-choreographed dance, a cleverly constructed façade designed to delay, deflect, and distract from the urgent fight against the climate crisis. Or perhaps it's just an ode to the power of human ingenuity, fueled by the irresistible force of wishful thinking.

Before embarking on our journey of understanding, let us first immerse ourselves in the rich tapestry of contextual, historical, and contemporary perspectives.

In a world filled with ingenious minds and boundless curiosity, humans discovered a remarkable force that would forever change the course of their existence. They called it technology. The fusion of scientific knowledge and innovative tools allowed them to create practical solutions and improve their very way of life. From the humble discovery of fire to the extraordinary invention of the wheel, technology became the driving force behind societal transformation.

Through the annals of history, humanity faced countless challenges, but technology was always there, ready to lend a helping hand. In the 15th century, a revolutionary invention emerged from the depths of ingenuity: the printing press. With its arrival, the world was forever altered. Knowledge, once confined to the whispers of scholars and the privileged few, now cascaded across lands and oceans, sparking the fires of the Renaissance and the Scientific Revolution. The printing press had unveiled the secrets of the world, breathing life into new ideas and awakening the dormant minds of millions.

But that was only the beginning. A tidal wave of innovation surged forward during the 18th and 19th centuries, known as the Industrial Revolution. Steam power, mechanization, and the wondrous assembly line burst onto the scene, catapulting productivity to unprecedented heights. Humanity's collective potential had been unleashed, propelling economies forward and forever altering the course of history. With every clank and chug of machinery, the world roared with progress.

As time marched on, technology continued to reshape the human experience. In the realm of healthcare, miracles unfolded. Antibiotics, vaccines, and medical imaging emerged from the depths of scientific brilliance, pushing the boundaries of what was once deemed impossible. These technological marvels gifted humanity with longer lifespans, improved well-being, and the hope for a brighter future.

And then came the era of information and communication technologies. With the rapid advance of these remarkable tools, the world became a global village. The barriers of time and space crumbled, leaving a trail of connections in their wake. The internet, like a digital conductor, orchestrated a symphony of knowledge, spreading it to the farthest corners of the Earth. Boundless information flowed freely, and new industries sprouted like wildflowers in the spring.

In the realm of innovation, where dreams take shape and possibilities set ablaze, technology emerges logically as a formidable champion, ready to confront the impending climate crisis. Let us embark on an extraordinary journey where technology rises to the challenge, armed with the potent tools of renewable energy, carbon capture and storage, sustainable transportation, circular economies, and climate monitoring and prediction. Together, we shall witness the remarkable convergence of human ingenuity and environmental stewardship, forging a path towards a sustainable and resilient future.

Renewable energy, the noble champion of a greener future, strides forward with unwavering determination. Technology, the ever-faithful companion, tirelessly pioneers' advancements in solar, wind, and geothermal power. With each passing day, efficiency soars and affordability descends from lofty heights, bringing these transformative technologies within reach of all. The promise of cleaner, sustainable energy beckons, like a beacon of hope in a world veiled by the carbon-laden mists of the past. And as technology labors, driven by tireless minds and relentless innovation, energy storage solutions, like advanced batteries, gather strength, ready to embrace the challenge and pave the way for a reliable and widespread use of renewable energy.

But the battle against climate change demands more than just renewable energy alone. In the depths of ingenuity, technology unveils a new weapon: Carbon Capture and Storage (CCS). This formidable innovation takes aim at the very heart of the problem – carbon dioxide emissions. It slumbers now, awaiting the touch of technology's transformative hand, which will awaken its latent power to capture and store these greenhouse gas culprits, exiling them from the atmosphere. Through technology's advances, CCS may emerge more efficient, more cost-effective than ever before, heralding a new era where emissions are tamed, and the world breathes a collective sigh of relief.

Behold, as technology gallops forward, it unlocks the gateway to a realm of sustainable transportation. Electric vehicles, the noble steeds of a cleaner future, surge forth, driven by advancements that push the boundaries of what is possible. With each new breakthrough in battery technology and charging infrastructure, the limitations that once hindered their march are shattered. A world of zero-emission transportation beckons, where city streets thrum with the quiet hum of electric motors, and the air, once heavy with pollutants, dances with newfound clarity. And as technology charts the course to a greener horizon, public transportation systems, infused with efficiency and ingenuity, rise to the occasion, embracing the challenges of a cleaner, more sustainable future.

But technology's quest does not end there. In the realm of consumption and production, a revolution beckons – a circular revolution. With technological innovations as its guiding light, a transition unfolds, one that embraces sustainable materials, revolutionizes recycling processes, and slashes waste with the blade of ingenuity. Behold the wonders of biodegradable materials, born from the crucible of technology's brilliance, paving the way for a world where consumption no longer casts a shadow of destruction. And as technology dons the cloak of 3D printing and wields the power of smart packaging, the age of sustainable production dawns, where resources are cherished, waste is minimized, and the planet breathes a sigh of gratitude.

But to navigate the treacherous waters of climate change, knowledge is paramount. And in this realm, technology unveils a powerful ally: climate monitoring and prediction. With every whisper of wind, every rumble of thunder, technology stands watch, collecting data, monitoring climate patterns, and enhancing predictive models. Armed with this knowledge, policymakers, like skilled navigators, steer the ship of change with precision, making evidence-based decisions that hold the key to effective strategies for climate change mitigation and adaptation. Technology's hand guides them, illuminating the path forward amidst the fog of uncertainty, shaping a future where informed choices replace blind stumbles.

Amidst the burgeoning hope that technology might come to our aid, it is crucial to recognize that its transformative power does not manifest as a mere flick of a wand in a circus-like "simsalabim" fashion. Rather, the emergence, dominance, or establishment of technologies adheres to certain immutable laws. To gain a profound understanding of these principles, where better to turn than the champion league of capitalism's realm – Venture Capital? It is there, in the crucible of visionary investments, that the seeds of innovation take root, nurturing the growth of technologies that can shape the destiny of our planet and combat the challenges that lie ahead.

Beloved companions, allow me to share with you in this context the culmination of my two-decade-long odyssey, deeply entrenched in the vibrant realm of tech venture capitalism. Throughout this immersive journey as venture capitalist, I have served as a guiding light for young analysts, entrepreneurs, and enterprising visionaries, helping them navigate the vast expanse of technology and innovation. Central to this experience have been four cherished theories, akin to a steadfast compass, steadfastly steering us towards newfound horizons: the Gartner Hype Curve, the Disruptive Technology Theory, the Faster Better Cheaper Theory, and the Brave New World Theory. With their profound insights, they can illuminate our understanding of the path through the terrain of climate technologies, unveiling their potential to emerge as beacons of hope in our collective endeavor to confront and resolve the pressing climate crisis.

Let us embark on an exhilarating exploration of these theories and uncover how they might illuminate our path toward predicting mankind's ability to single-handedly conquer the climate crisis through technology.

First, we encounter the Gartner Hype Curve (2023) – an enchanting graphical representation of technology's life cycle. Picture a rollercoaster ride, starting with the Technology Trigger. This is where excitement and anticipation burgeon as emerging technologies capture our imagination. Think of Virtual Reality (VR), for instance. As VR emerged, there was an unprecedented frenzy surrounding its applications, propelling us to the Peak of Inflated Expectations. However, limitations and challenges soon emerged, plunging us into the Trough of Disillusionment. Fear not, for there is light at the end of this trough. As VR evolved and improved, it ascended the Slope of Enlightenment, enabling us to realize its true potential. We eventually reached the Plateau of Productivity, where VR became an indispensable tool in various industries.

Now, consider the climate crisis. Should technology alone possess the power to vanquish it entirely, we would witness the Gartner Hype Curve unfold before our very eyes. Initially, unbridled excitement and optimism would surround the potential of climate technology. Yet, as obstacles arise, disillusionment may cast its shadow, leading us to question the very foundations of our optimism. Fear not, for as the technology matures and its efficacy in addressing the climate crisis becomes evident, enlightenment will guide us toward widespread adoption and implementation. Ultimately, climate technology would ascend to the Plateau of Productivity, becoming an indispensable component of the climate crisis solution.

Ah, my discerning friends, I sense your question echoing in the very air. How swiftly will a climate technology ascend to the revered Plateau of Productivity, becoming an indispensable cornerstone of the climate crisis solution? The answer, my cherished companions, remains a mysterious tapestry woven with uncertainty. For when a climate technology ascends to that elusive Plateau, if at all it does, and where and how it shall reach such heights, no one can definitively foretell. The realm of emergence remains enigmatic, cloaked in the shadows of uncertainty, leaving us to rely on educated conjecture and fervent hope.

Yet herein lies the crux of the matter – when we anchor the survival of humanity upon the advent of new technologies, we enter treacherous waters. The very essence of our collective destiny becomes entangled with the capricious whims of innovation, a fate determined by myriad factors and unforeseen forces. Beware the charlatans who claim otherwise, for no one holds the key to absolute certainty.

Behold the litmus test of technology ventures, and therein lies the truth unveiled – the success rate, a stark reflection of the volatile landscape where triumph and tribulation intertwine. Let us tread carefully and embrace humility in the face of uncertainty, mindful that our future lies not solely in the hands of technology but in the profound choices we make as stewards of our planet.

Next, we encounter the Disruptive Technology Theory – devised by the brilliant Clayton Christensen in The Innovator's Dilemma (2016) and introduced by introduced by Joseph Schumpeter in his theory of creative destruction (1947). This theory reveals that new technologies often emerge in niche markets or lower-end segments, gradually disrupting established industries or technologies. The personal computer (PC) serves as a testament to this theory. In its nascent stages, the PC paled in comparison to mainframe computers used by large organizations. However, over time, PCs became more affordable, user-friendly, and increasingly powerful, disrupting the very foundations of the mainframe era and transforming industries and our daily lives.

Now, envision climate technology following a similar trajectory. Initially, climate technologies may emerge as specialized solutions, tackling specific aspects of the crisis. Yet, as these technologies evolve, become more accessible, and showcase their effectiveness, they would disrupt traditional energy systems, transportation, and other sectors contributing to greenhouse gas emissions. Industries would be reshaped, novel markets would sprout, and society would undergo transformative changes in how we confront the climate crisis.

When considering the potential trajectory of climate technologies, a host of critical questions emerge, beckoning us to explore the intricate landscape of innovation and its profound implications for the climate crisis.

One key query revolves around the adoption rate of these technologies – how swiftly can they gain widespread acceptance in the global market? Will their uptake follow a gradual progression, or can we anticipate a more rapid and widespread adoption?

Another crucial aspect is the level of investment and funding required to propel these technologies towards mainstream usage. Will private enterprises and public sectors unite in providing the necessary support to drive their implementation on a substantial scale?

In addition, the journey towards technological maturation becomes a pivotal concern. How long will it take for climate technologies to evolve into practical, efficient, and cost-effective solutions that can effectively address the pressing challenges of climate change?

Policy and regulation also play a pivotal role in shaping the fate of these technologies. What policy frameworks and regulations must be established to incentivize their adoption and encourage a global shift towards sustainable practices?

The integration of climate technologies into existing infrastructure and systems presents another set of challenges to navigate. What obstacles might arise in deploying and effectively implementing these transformative solutions?

Furthermore, global collaboration takes center stage as a critical factor in fostering the widespread adoption and scaling of these technologies. To what extent will international cooperation be essential in tackling the global climate crisis through unified efforts?

As we embark on this station of inquiry, uncertainties abound, underscoring the complexities and responsibilities entwined with the ascent of climate technologies and their potential to reshape the course of our planet's environmental future.

As we continue our journey, we come across the Faster Better Cheaper Theory (McCurdy, 2003) – a captivating notion that technological progress often yields innovations that are faster, superior in performance, and cheaper than their predecessors or competitors. Here, we encounter for example and for teaching purpose the evolution of digital cameras. Initially, these cameras were expensive, offered limited resolution, and lacked the features of their film counterparts. However, as technology advanced, digital cameras sprinted ahead, providing higher resolutions, improved image quality, and additional features – all at more affordable prices. This seismic shift led to the widespread adoption of digital cameras, leaving film cameras in the rearview.

The concept of "faster, better, cheaper" is essential in the context of climate technology because it envisions a transformational path where sustainable solutions become economically competitive with their conventional counterparts. Price parity plays a pivotal role in this equation, as it represents the tipping point where climate technologies become not only environmentally superior but also financially viable and economically attractive.

When climate technologies achieve price parity with traditional alternatives, it ignites a virtuous cycle of adoption and widespread implementation. As these technologies become economically competitive, they appeal not only to environmentally conscious consumers but also to cost-conscious businesses and industries. This growing demand leads to increased production and economies of scale, driving further reductions in costs and making the technologies even more affordable.

Price parity removes the financial barrier that has historically hindered the widespread adoption of climate technologies. As these solutions become accessible and economically viable, more businesses and individuals are incentivized to transition away from fossil fuels and embrace sustainable alternatives. This, in turn, accelerates the shift towards a low-carbon economy and strengthens the fight against climate change.

Moreover, price parity enhances the attractiveness of climate technologies for investors and policymakers. As these solutions become financially viable, investment flows naturally increase, driving further innovation, research, and development in the field. Policymakers are more likely to support the deployment of these technologies when they offer both environmental benefits and economic advantages, creating a conducive regulatory environment for their growth.

Overall, achieving price parity for climate technology is a critical milestone on the path towards a sustainable future. It propels the vision of "faster, better, cheaper" where climate solutions not only rival but surpass their conventional counterparts, providing a tangible pathway to combat the climate crisis while fostering economic prosperity.

So, my cherished companions, as we contemplate the potential of technology as a silver bullet to address the climate crisis, let us heed a crucial cautionary note – watch the cash flow, and watch the price parity. For without technology reaching that pivotal point, it shall only result in a financial drain. Yes, do you hear it? A lamentable burning of precious resources.

Lastly, we encounter the Brave New World Theory – an awe-inspiring concept inspired by Aldous Huxley's visionary novel Brave New World (1932). This theory shed slight on the potential ramifications of introducing groundbreaking inventions and novel product formats, driven by rapid technological progress. In this dynamic landscape, uncertainties abound, encompassing societal, ethical, business model, pricing, and other dimensions. To illustrate, let us examine the example of artificial intelligence (AI), an awe-inspiring innovation that sparks both enthusiasm and apprehension.

AI holds immense promise to revolutionize numerous industries, bestow greater productivity, and elevate our overall quality of life. Yet, alongside this optimism, concerns emerge regarding the displacement of jobs, ethical complexities, and the consolidation of power in the hands of a select few, casting a discernible shade of doubt.

Analogous to Columbus' arrival in the Americas, the Brave New World teems with potential and possibilities. As Bob Marley's immortal lyrics resonate, "Don't worry about a thing, cause every little thing gonna be alright," there is a desire to trust in a brighter future, despite the uncertainties that accompany innovation. In this ever-evolving landscape, courage, foresight, and thoughtful navigation become essential in ensuring that the transformative power of technology serves the greater good and leads us towards a promising horizon.

As we envision climate technology unfolding, we witness the potential for a brave new world. A world where clean energy sources, carbon capture technologies, and groundbreaking solutions seamlessly combat climate change. This future teems with restored environments, improved quality of life, and enhanced global cooperation. But we must tread carefully, for this brave new world may also harbor challenges and risks. Unintended consequences, unequal distribution of benefits, and the misuse of technologies may threaten our progress. Vigilance is paramount to ensure that the application of climate technology aligns harmoniously with our societal values and goals.

As we weave together the threads of these theories and examine their implications for the potential of technology to solve the climate crisis, a tapestry of insights emerges. Technological solutions will experience a thrilling journey – initial hype and anticipation, stumbling blocks and setbacks, and eventual maturation into effective tools. Climate technology will disrupt established industries, catalyze transformative changes, become faster, better, and cheaper over time, and wield both positive and negative impacts on society. However, let us not forget that technology alone may not be the sole panacea. Policy changes, behavioral shifts, and systemic transformations must complement technological advancements to forge a comprehensive approach to combat the climate crisis.

As we continue on our exhilarating journey into the brave new world of climate technology, it is crucial to explore additional rationales and theories that caution against hailing it as the panacea for solving the climate crisis. Beware, my fellow adventurers, for we must not succumb to the siren song of complacent politicians and lobbyists who peddle the notion that technology alone will be our saving grace.

It's easy to fall into the trap of relying on yet-to-be-developed technologies to solve the pressing issues at hand. One such example is the concept of carbon dioxide removal (CDR) technologies, which aim to extract carbon dioxide from the atmosphere and offset ongoing emissions. While some proponents argue that CDR technologies could play a significant role in achieving climate targets, it is essential to recognize the risks of relying too heavily on their potential.

Let's delve into a specific CDR technology that has garnered attention: bioenergy with carbon capture and storage (BECCS). This approach involves cultivating biomass crops, burning them for energy, capturing the emitted carbon dioxide, and storing it underground. Proponents suggest that BECCS could potentially achieve negative emissions by removing more carbon dioxide from the atmosphere than is released during the process.

However, we must exercise caution when embracing BECCS and similar CDR technologies as a "silver bullet" solution for climate change mitigation. There are several concerns that warrant consideration. Firstly, the scalability of BECCS at a global level is questionable. Implementing large-scale BECCS projects would require vast amounts of land, water, and resources, potentially leading to competition with food production and other land uses.

Secondly, the long-term environmental and social implications of widespread deployment of CDR technologies remain uncertain. There are concerns about potential eco-system disruptions, changes in land use patterns, and impacts on biodiversity. Moreover, the potential unintended consequences and risks associated with large-scale carbon capture and storage require careful evaluation.

A study by Williamson et al. (2020), published in Nature Communications, highlights the potential risks of relying on CDR technologies as the primary means of achieving climate targets. The authors argue that delaying emission reductions in favor of relying on future CDR technologies could lead to overshooting temperature targets, making it even more challenging to achieve long-term climate goals. This work provides a glimpse into the concerns associated with delaying action in the hopes of future technological fixes. It emphasizes the need for immediate and comprehensive action, combining emissions reductions, sustainable practices, and technological advancements to effectively address the climate crisis.

It is essential to recognize that relying solely on technology to address the climate crisis has its limitations. This is evident in various sectors where technological progress has not translated into substantial emission reductions due to underlying systemic challenges.

Take the transportation sector, for example. Despite advancements in electric vehicle (EV) technology, global emissions from transportation have continued to rise. The growing adoption of EVs did not prevent global CO_2 emissions from road transport from increasing by 1.5% in 2020, as reported by the International Energy Agency (2021) in their Global Energy Review. This illustrates that while EVs offer a promising solution, addressing the climate crisis requires more comprehensive strategies, such as improving public transportation, reducing overall vehicle usage, and implementing sustainable urban planning.

Let us further explore another intriguing vantage point presented in Nature Climate Change (Creutzig et al., 2018). The study delves into the realm of transformative technological innovations, with a specific focus on the transport sector's profound decarbonization potential. Their rigorous analysis highlights that placing exclusive reliance on technological advancements, such as electric vehicles (EVs), might not prove adequate in attaining ambitious climate objectives.

In essence, the researchers emphasize the paramount significance of synergizing technological progress with profound shifts in human behavior, thoughtful urban design, and progressive transport policies. By interweaving these multifaceted aspects, we can pave the way for substantial and meaningful reductions in greenhouse gas emissions, ultimately contributing to a more sustainable and ecologically harmonious future.

Through ongoing introspection and discernment, an abundance of insightful revelations shall grace the energy sector, casting a radiant light on the journey ahead. In this realm, the gradual shift from fossil fuels to renewable energy sources unveils the intrinsic constraints of technology in isolation. Despite commendable progress in renewable energy innovations, the persistence of soaring global CO_2 emissions from fossil fuels is undeniable.

The discerning observations from the International Energy Agency's Global Energy Review (2021) serve as a stark reminder of this reality. Despite a momentary decline attributable to the COVID-19 pandemic, global energy-related CO_2 emissions experienced an unexpected increase of 1.5% in 2020, perpetuating the pressing need for comprehensive and transformative solutions to overcome this challenge.

Allow the mind to envisage this profound perspective, as we embark on a journey of enlightenment, dear friends. Distinguished experts within the energy problem domain have devoted their expertise to scrutinize the velocity of the energy transition. Alas, their discerning analysis reveals a poignant truth – the present rate of renewable energy adoption fails to meet the imperative of constraining global warming to well below 2 degrees Celsius. In the wake of this realization, the erudite authors draw our attention to the pressing demand for transformative systemic alterations. Such a paradigm shift necessitates a harmonious interplay of dynamic policy reforms, judicious financial incentives, and profound shifts in human behavior. Embracing these holistic enhancements alongside technological progress in the energy sector becomes indispensable in steering us toward a sustainable future, resonating with ecological harmony and benevolence toward our shared planet.

Relying solely on future technologies to address the climate crisis is uncertain and may not align with the urgency of meeting climate goals. While technological advancements hold promise, their development, deployment, and scalability can be unpredictable and subject to various challenges. Encompassed within the realm of this thesis lie further data points, awaiting connection and illumination, serving as eloquent testaments to fortify its essence.

Again, consider the potential reliance on future breakthroughs in carbon capture and storage (CCS) technologies. CCS has been considered a promising solution for reducing greenhouse gas emissions from industries such as power generation and cement production. However, the widespread implementation of CCS at the required scale has faced numerous technical and economic challenges. The International Energy Agency (IEA) highlights in its report "Net Zero by 2050: A Roadmap for the Global Energy Sector" (2021) that the large-scale deployment of CCS technologies remains uncertain, and relying solely on their future development would be risky for achieving climate goals.

A pivotal aspect in this narrative centering on the reliance upon forthcoming strides in renewable energy technologies surfaces. While commendable progress has been achieved in harnessing sources like solar and wind, inherent challenges such as intermittency, grid integration, and energy storage demand unwavering attention and resolution. Placing sole faith in prospective advancements within these realms to attain widespread deployment and ensure steadfast reliability may inadvertently postpone the transition to renewable energy and impede the timely reduction of greenhouse gas emissions.

This perspective gains further credence through an enlightening study (Anderson et al., 2020), thoughtfully presented in Nature Energy. Their analysis scrutinized the feasibility of leaning heavily on speculative negative emissions technologies (NETs) to fulfill the ambitions of the Paris Agreement. In resolute conclusion, the researchers caution that relying solely on speculative NETs without immediate and substantive emission reductions in the present would herald perilous levels of global warming. Rather, they underscore the paramount urgency of undertaking near-term emissions reductions as the most dependable and prudent pathway towards achieving our climate objectives.

The Intergovernmental Panel on Climate Change (SR15, 2019) also highlights in its special report "Global Warming of 1.5°C" that the timely implementation of a wide range of mitigation options across all sectors is necessary to limit global warming. While future technologies can contribute to long-term decarbonization efforts, immediate actions are crucial to staying on track with climate goals.

Dear esteemed friends, as we approach the conclusion of this chapter, it would be remiss of me not to embark on a heartfelt discourse concerning an exceedingly paramount subject within the realm of climate technology – the phenomenon of green hydrogen. Permit me to share a wealth of cogitations inspired by a captivating piece of editorial prowess, authored brilliantly by Susanne Schäfer, and unveiled to the world through the esteemed German online publication "Focus" on the illustrious date of June 29, 2023 (2023). This masterful opus delved deeply into the intricacies of challenges, risks, and the enthralling odyssey of transforming from technological immaturity to full-fledged market adoption, all within the context of combating climate change.

Susanne's enlightening piece centered around the captivating topic of desalination to produce hydrogen and freshwater. She masterfully guided her readers through two distinct approaches adopted by Spain and Germany, shedding light on their unique struggles and potential solutions.

First, we embarked on a sun-drenched journey to Spain, a country determined to become a major player in the European hydrogen market. The Spanish government, like a visionary conductor, waved its baton and committed €1.5 billion by 2030, aiming to export two million tons of green hydrogen annually. Susanne highlighted the European Commission's support of Grupo Cobra, a Spanish company venturing into the construction of electrolyzers powered by renewable energy. However, a veil of mystery surrounded the source of freshwater required for hydrogen production.

As our minds danced with anticipation, Susanne led us to the captivating shores of Jávea, Spain, where a remarkable desalination plant operated by Acciona stood as a shining example of environmental friendliness. Though not directly involved in hydrogen production, this oasis of innovation served as a model for sustainable desalination. In collaboration with environmental scientists from the University of Alicante, Acciona built the plant with meticulous care, ensuring minimal impact on the precious marine eco-system. Groundwater wells near the coast acted as natural filters, eliminating the need for chemical pretreatment. The brine, diluted to perfection, found its safe haven in the sea, avoiding any harm to marine life.

But our adventure did not end there! Susanne whisked us away to Germany, where we discovered Boreal Light, a company on a noble mission to export desalination systems to remote and impoverished regions. Their innovative approach revolved around harnessing the power of solar panels and wind turbines to fuel the desalination process. This ingenious solution proved ideal for areas without reliable access to an electricity grid. Boreal Light even drilled shallow wells, extracting less saline water to reduce the energy requirements for desalination. They embraced UV light for disinfection, bidding farewell to chlorine, and the resulting brine found new purpose in irrigation or aquaculture.

As the article unfolded, an intriguing question arose: Could seawater itself be the magical solution for hydrogen production, eliminating the need for extensive desalination? Expert voices chimed in, explaining that current electrolysis technologies, especially those employing proton exchange membrane (PEM) electrolyzers, demanded highly purified freshwater to avoid membrane fouling. Hope, however, glimmered in the form of research projects like H_2 Mare in Germany, venturing into the uncharted waters of seawater electrolysis, albeit still in their nascent stages.

Throughout the mesmerizing tale, Susanne emphasized the paramount importance of finding environmentally sustainable solutions for desalination, a critical factor in freshwater and hydrogen production alike. The examples showcased by Spain and Germany unveiled diverse approaches, underscoring their commitment to minimizing the impact on marine eco-systems while simultaneously reducing energy consumption.

My cherished friends, allow me to expound on the reason behind my fondness for disseminating Susanne's remarkable episode. Amidst this enthralling odyssey, I beseech you to join me in a moment of introspection as we ponder the formidable obstacles that stand in the way of transforming hydrogen into an accessible and cost-effective energy commodity, serving the vital domains of heating, industry, and mobility. Just as ancient dragons vigilantly guard their priceless treasures, these hurdles manifest as formidable challenges, beckoning us to muster our collective resolve and conquer them with unwavering determination.

Freshwater availability emerges as the first barrier, as the production of hydrogen through electrolysis requires highly purified freshwater to prevent membrane fouling. Water scarcity in some regions intensifies the challenge of sourcing enough freshwater for large-scale hydrogen production.

As we inch forward, the cost of desalination looms before us, casting a shadow of doubt. The intricate process, often employed to produce freshwater for hydrogen production, can be an expensive endeavor. The cost of desalination technologies, including energy requirements and maintenance, bears weight on the overall affordability of hydrogen as an energy commodity.

Ah, energy requirements – the ever-present companion on this journey. The desalination process itself demands significant energy inputs, whether from renewable sources like solar panels and wind turbines or conventional electricity grids. The energy intensity of desalination influences the cost-effectiveness of hydrogen production, especially in regions where reliable and affordable electricity remains a distant dream.

As we venture further, a call for infrastructure development echoes in the air. Establishing a comprehensive hydrogen production, storage, and distribution infrastructure is pivotal in making hydrogen an accessible and affordable energy commodity. The construction of desalination plants and hydrogen production facilities requires substantial investments and meticulous planning, laying the foundation for a thriving hydrogen eco-system.

Technological advancements, like the flickering light at the end of a tunnel, hold the promise of overcoming these obstacles. Continuous research and development efforts become the guiding star, leading us toward more cost-effective and energy-efficient desalination processes. Exploring alternative methods of hydrogen production, with reduced reliance on freshwater, becomes the compass pointing to a brighter future.

But wait! We mustn't forget that hydrogen's journey doesn't end here. A tale within the tale unfolds, presenting key takeaways for the profitable production of hydrogen, destined to fuel our ambitions and energize our dreams.

Demand for hydrogen escalates as its potential as a clean energy carrier becomes ever more apparent. Like a phoenix rising, hydrogen emerges as a versatile solution across various sectors, from mobility to industry and housing.

The realm of cost reduction becomes our next battleground. The profitability of hydrogen hinges upon advancements in technologies such as electrolysis, steam methane reforming, and biomass gasification. These technological marvels pave the way for cost reduction, making hydrogen an economically competitive energy option.

Renewable hydrogen emerges as a beacon of hope, casting a vibrant glow upon the energy landscape. Producing hydrogen from renewable sources like wind, solar, and hydropower gains momentum, effectively reducing carbon emissions and enhancing the overall sustainability of the process.

As we tread this winding path, collaboration and infrastructure intertwine their destinies. The creation of a robust hydrogen eco-system necessitates collaboration between governments, industry stakeholders, and research institutions. Adequate infrastructure, encompassing hydrogen production, storage, and distribution facilities, becomes the backbone of profitability.

Policies flutter like flags in the wind, exerting their influence on the journey. Supportive policies, including research funding, carbon pricing mechanisms, and subsidies, serve as guiding stars, accelerating the growth of the hydrogen industry.

But hold on, dear adventurers! Our expedition is not yet complete. We must confront additional obstacles obstructing the realization of hydrogen's full potential as a commodity for heating, industry, and housing. These challenges, etched into the annals of our knowledge, demand our attention and concerted efforts.

High production costs cast a shadow upon hydrogen's profitability. Compared to conventional fossil fuels, the cost of producing hydrogen, particularly through clean and renewable methods, remains relatively high. Overcoming this financial barrier becomes paramount in establishing hydrogen as an economically competitive energy source.

The absence of infrastructure stands as a formidable hurdle. Developing a comprehensive hydrogen infrastructure, spanning production facilities, storage solutions, and distribution networks, requires substantial investments and time. Without a robust infrastructure in place, widespread adoption of hydrogen as a commodity is hindered.

Technological advancements weave their intricate tapestry of challenges. As we strive for efficiency and cost reduction, ongoing research and development efforts become essential. Technical challenges must be overcome, and hydrogen-related processes must be optimized to unlock the full potential of this miraculous energy carrier.

Safety concerns whisper cautionary words, reminding us of hydrogen's unique properties. Its high flammability and potential leak hazards demand our attention. Addressing safety concerns, developing robust safety standards and regulations, becomes paramount for wider acceptance and public confidence.

Regulatory frameworks, or the absence thereof, exert their influence on our course. Clear and consistent regulations governing hydrogen production, storage, transportation, and usage become the cornerstone of commercialization. Governments and regulatory bodies must establish supportive frameworks to facilitate the integration of hydrogen into existing energy systems.

Yet, in this tapestry of challenges, we can discern threads of hope and optimism, born from the concepts of Moore's Law, disruptive technology curves, and the Gartner Hype Curve. As we gaze into the crystal ball of possibility, we catch glimpses of what the future may hold for hydrogen as an energy commodity for heating, industry, and mobility. These glimpses, however speculative, guide our aspirations and illuminate the path ahead.

Technological advancements, driven by the force of Moore's Law, suggest a future of exponential growth. Hydrogen-related technologies, such as electrolysis, fuel cells, and hydrogen storage solutions, are poised to advance, both in performance and cost-effectiveness. As research and development continue their relentless pursuit, the emergence of more efficient and affordable hydrogen technologies becomes a tantalizing prospect.

Disruptive potential looms on the horizon. Like a whisper that grows into a roar, hydrogen possesses the power to disrupt the energy sector. In its early stages, hydrogen may appear as a niche or less efficient alternative to existing solutions. Yet, as it matures and economies of scale are achieved, hydrogen has the potential to disrupt traditional energy sources. It could become the cornerstone of our energy landscape, fueling our homes, industries, and transportation with newfound vigor.

The Gartner Hype Curve casts its spell, guiding our expectations. Hydrogen has experienced periods of hype, accompanied by high expectations and boundless enthusiasm. Yet, it is not immune to the disillusionment that often follows. As the technology matures and practical implementations come to fruition, hydrogen transitions from the peak of inflated expectations to a more stable phase of maturity. We come to understand its limitations and practical applications, paving the way for a sustainable future.

Infrastructure development remains a lighthouse, guiding our path. The future of hydrogen as an energy commodity hinges upon the development of a robust infrastructure. As the technology progresses and market demand grows, the focus on building a reliable and cost-effective hydrogen supply chain intensifies. Governments, industry stakeholders, and investors unite in their efforts to establish a comprehensive hydrogen infrastructure, unlocking the full potential of this wondrous energy carrier.

Integration with renewable energy sources becomes our guiding principle. As renewable energy generation flourishes, we face the challenge of balancing intermittent supply and demand. Hydrogen, acting as a versatile energy carrier, steps onto the stage. It allows excess renewable energy to be stored and utilized when needed, forging a symbiotic relationship with renewable energy sources. This harmonious integration enhances the sustainability and reliability of our energy system, paving the way for a brighter and greener future.

Esteemed friends, as we draw near to the conclusion of this extensive section delving into technology as our savior amidst the climate crisis, we find ourselves surrounded by a multitude of episodes, some intentionally revisited for emphasis. It is fitting to reflect upon our journey's inception, guided by the sagacious words of the German Politician Linder, who served as the catalyst and muse for this chapter. His mantra, echoing the belief that there is no need for action as technology and the free market shall pave the way to resolution, resonated deeply throughout our exploration.

Now, as I endeavor to distill the essence of this chapter's odyssey, I implore you to retain a discerning perspective on why such a viewpoint might be deemed naive, misleading, or simply unaware of the intricacies governing technology and business management. Indeed, as we unravel the complexities of our modern world, a more nuanced and holistic understanding of the interplay between technology, markets, and the pursuit of solutions emerges.

Technological solutions aimed at addressing the climate crisis face various challenges, including high costs, limited scalability, and uncertainties regarding their long-term impacts. While these challenges are not insurmountable, they require careful consideration and evaluation to ensure the effectiveness and sustainability of technological interventions.

Consider the challenge posed by high costs in deploying renewable energy technologies. While renewable energy sources such as solar and wind have become increasingly cost-competitive, the initial investment required for infrastructure development, such as solar panels or wind turbines, can still be significant. The International Renewable Energy Agency (IREA, 2020) discusses the importance of reducing the costs of renewable energy technologies in its publication "Renewable Power Generation Costs in 2020." Lowering the costs of renewable energy systems is crucial for their wider adoption and integration into existing energy systems.

Limited scalability is another challenge faced by technological solutions. Certain technologies that show promise at smaller scales may struggle to be implemented or scaled up to meet the demands of the entire global energy system. For instance, advanced nuclear power technologies, such as small modular reactors, have the potential to provide low-carbon energy. However, their scalability and deployment still require further research, development, and regulatory frameworks to ensure their safe and efficient operation. The International Atomic Energy Agency (IAEA, 2020) discusses the challenges and opportunities of advanced nuclear technologies in its report "Advanced Reactors: Technology Options for Generation IV and Small Reactors."

Uncertainties regarding the long-term impacts of technological solutions are also important to consider. The introduction of new technologies can have unintended consequences or environmental trade-offs that may not be immediately apparent. For example, the adoption of bioenergy with carbon capture and storage (BECCS) as a negative emissions technology raises concerns about the sustainability and potential competition for land resources. The Royal Society report "Greenhouse Gas Removal"(2018) highlights the uncertainties and potential risks associated with large-scale BECCS implementation.

Literature recognizes the challenges posed by high costs, limited scalability, and uncertainties surrounding the long-term impacts of technological solutions. The World Bank report "Invention and Global Diffusion of Technologies for Climate Change Adaption" (Touboul, 2020) emphasizes the importance of addressing these challenges to accelerate the adoption of low-carbon technologies. The report discusses the need for policy support, research and development investments, and collaborative efforts to overcome cost and scalability barriers.

Addressing the challenges of high costs, limited scalability, and uncertainties regarding long-term impacts requires a comprehensive approach that combines technological innovation with supportive policies, investment strategies, and rigorous assessment of environmental and social implications. By addressing these challenges, technological solutions can become more effective, accessible, and sustainable in mitigating the climate crisis.

In essence, the climate crisis requires immediate and comprehensive action to reduce greenhouse gas emissions. While technological innovations hold promise in addressing the issue, relying solely on future technologies is risky and uncertain. The scalability, costs, and long-term impacts of these technologies must be carefully considered. Moreover, technology alone is insufficient to meet climate goals. Holistic approaches that integrate changes in behavior, policies, and sustainable practices are crucial.

By acknowledging the complexities and challenges, we can develop effective strategies to combat climate change and create a sustainable future.

II. Demanding A Revolution Of Policies

Decades ago, in a far-off land whose name shall remain undisclosed for the sake of its oppressed citizens, an esteemed and cherished friend of mine found himself entangled in a political tapestry of tyranny and oppression. In this dictatorial regime, daring to challenge the ruling powers meant a perilous gamble with one's life – a chilling reality of capture, torture, and even brutal murder, not only for the dissenters but also for their hapless families. And yet, in the shadows of this darkness, there were revolutionary souls who, in the immortal words of my favorite hero, Bob Marley, dared to "get up and stand up for their rights."

In this turbulent backdrop, members of the opposition party approached my friend, beseeching him to honor his motherland by challenging the incumbent dictator in the upcoming sham election. After days of introspection and consultation with trusted confidantes like me and his family, he mustered the courage to reply with a resolute "No." The years went by, and I never probed him further on his decision. But it was during a recent encounter in Berlin, amidst a social gathering, that he reflected on that poignant episode, uttering, "You know, the cemeteries of this earth are full of heroes…" I nodded in understanding.

Yet, this sentiment underscores a profound paradox. Revolutionaries, who we so profoundly admire, have reshaped the course of history with their unwavering commitment, even at the cost of their own lives. But herein lies the challenge - while we revere these extraordinary souls, most of us are hesitant to pay the same heavy price they did. However, herein lies the intriguing paradox: without the presence of valiant heroes, the world would never have witnessed the birth of momentous revolutions. For it is these very revolutions that have forged the path to nationhood, giving rise to independent states like "The United States of America" and enshrining foundational documents such as the Magna Carta. The flames of revolution in France gave birth to the resplendent République Française, and without it, the principles of Liberté, Égalité, Fraternité might never have illuminated the world. In the annals of history, the indomitable spirit of Madiba paved the way for the liberation of a free South Africa and countless other transformative movements. We yearn for revolution but seek it at a discounted rate, one that doesn't entail such dire sacrifices.

So, dear reader, I extend a proposition to the rest of us - the ones who may not be Madiba, Steve Biko, Gandhi, Georg Elser, or the "Geschwister

Scholl." Fear not, for you, too, can be a hero and play your part in saving the very planet we call home. The good news is that this journey need not demand your life. Are you prepared to embrace the challenge and make your mark on the world? Let us embark on this collective endeavor and pave the way for a brighter, more sustainable future. The time has come to act as climate revolutionaries – let us begin.

Beloved Companions, as we have deeply pondered and diligently substantiated thus far, our planet finds itself perched at the edge of a precipice, teetering on the brink of a climatic catastrophe. The gravity of this existential threat demands a resolute and audacious response. In the face of such daunting challenges, the familiar path of incremental measures, those tentative strides tethered to the comforts of the status quo, can no longer suffice to shield us from the ominous specter on the horizon. The moment calls not for timidity but for the fervor of transformative change. Let us seize this opportunity to rise as one, to embrace boldness and ambition, and forge a path towards a sustainable and resilient future for generations to come.

The climate crisis is no ordinary challenge. It is an existential threat, casting a dark shadow over the future of humanity and our planet. Extreme weather events, rising sea levels, and disruptions to delicate eco-systems all bear witness to the harsh reality of what lies ahead if we fail to act decisively. The clock is ticking, and the consequences of inaction are far too grave to ignore.

The complexity of the climate system brings forth yet another danger - tipping points and feedback loops that can set off irreversible changes. Incremental adjustments, while well-intentioned, might not be enough to prevent these cascading effects that amplify global warming. Bold measures, those courageous strides taken with a firm commitment to avert catastrophe, are the only hope to steer clear of these critical thresholds.

It is now imperative to address the carbon budget, the finite and swiftly diminishing allocation we must adhere to in order to contain global warming within manageable thresholds. Incremental changes might eat away at this budget faster than we can imagine, leaving future generations scant opportunity to adapt and respond to a world fraught with challenges. It is imperative that we embrace ambitious measures to ensure a sustainable future, within the confines of the carbon budget we have left.

Yet, within this call for boldness, there lies promise. Innovation and technology can be harnessed to pave the way forward. The key lies in unlocking their potential to accelerate the transition to renewable energy, carbon capture, and other sustainable technologies. It is not just about embracing change; it is about creating an environment that nurtures research and development, driving the progress we so desperately need.

Bold climate measures are not just about saving the planet; they offer co-benefits that resonate on multiple fronts. For instance, promoting public transportation and active mobility doesn't just cut emissions; it also improves air quality, enhancing public health and well-being. Investing in green infrastructure, besides being an environmental boon, can create job opportunities and invigorate economies.

Leadership plays a crucial role in this struggle. When leading countries and regions take decisive action, their actions reverberate across the globe, catalyzing a sense of shared responsibility and encouraging a united front in tackling this global challenge. It is a call for cooperation, for a collective endeavor that transcends borders and ideologies.

Resilience and adaptation are the pillars that will enable us to weather the storm that lies ahead. Bold measures can fortify societies and eco-systems against the impacts of climate change, reducing vulnerabilities and minimizing the costs of future disasters. It is a pragmatic approach to safeguarding our future and the future of generations yet to come.

In this battle against time, in this war to secure a sustainable and resilient future, we must cast aside the comfort of the familiar. Incremental changes, while commendable, may not save us from the precipice. We must muster the courage to embrace bold and ambitious measures. Only then can we navigate the treacherous waters of the climate crisis and forge a path towards a brighter and more sustainable future.

My dear friends, the poignant inquiry that echoes within us now revolves around the fortitude of policy frameworks - the bedrock of our climate revolution. How do these resolute structures empower and embolden individuals and societies to partake as a true revolutionary in the grand endeavor that lies before us?

Ladies and gentlemen, let us now delve first into the intricate tapestry of a comprehensive policy framework – a dynamic blueprint that resonates not only with the intellect but also with the very essence of our shared humanity. Just as an artist dissects a masterpiece to reveal its nuances, brushstrokes, and hues, we shall study strategies, unraveling the complex strands that weave together to craft a powerful symphony for addressing the profound challenge of our age – the climate crisis.

Imagine, if you will, standing before a canvas bearing the aspirations of generations, where each stroke, each hue, contributes to an artistic vision larger than life. Such is the essence of a comprehensive policy framework. It is the crucible wherein aspirations are melded with pragmatism, dreams are translated into action, and the tides of change are harnessed to steer our societies toward sustainable and resilient shores.

At the heart of this framework lies a collection of interwoven threads, each endowed with singular significance yet harmoniously entwined, converging toward a singular goal – the pursuit of transformative change. These threads, like melodies in an orchestra, beckon us to unearth their individual voices while rejoicing in the symphony they collectively create.

Picture now, if you will, the first thread - clear and ambitious goals. Like celestial stars guiding explorers on an uncharted sea, these goals illuminate our path toward a common vision. Whether the reduction of emissions to a whisper, the embrace of renewable energy's embrace, or the guardianship of irreplaceable eco-systems, these aspirations crystallize the essence of our endeavor.

The second thread, akin to the warp and weft of a legal tapestry, introduces us to the realm of regulatory mechanisms. These are the watchful guardians, the keepers of accountability and adherence. They mold the fabric of compliance, binding individuals and entities to the symphony's rhythmic harmony, fostering a balance between sustainable practices and incentivized alignment.

Economic instruments form the next thread, a melodic dance between incentives and disincentives. The allure of subsidies kindles the fire of green innovation, while the solemn weight of carbon pricing reverberates as a reminder that actions have consequences. This delicate dance, a choreography of influence, guides behaviors toward a more sustainable choreography.

Innovation and research, the virtuoso thread, beckon us to an eco-system of knowledge and pioneering spirit. Investment in green technologies, a crescendo of collaboration between academia and industry, culminates in a symphony of solutions. These are the notes that resonate as innovation and vision collide, creating a harmonious chorus of progress.

Stakeholder engagement, akin to a convivial gathering of diverse voices, manifests as the next thread. In this grand assembly, affected communities, industries, and interest groups intermingle in dialogue, shaping policies that resonate with real-world intricacies. The chords of communication resonate, nurturing ownership and commitment, ensuring our symphony sings in unison.

Education, a thread of enlightenment, invites us to enrich our understanding of urgency and consequence. Initiatives, programs, and campaigns empower the populace, igniting a spark of awareness that illuminates our collective path, emboldening each step towards a sustainable future.

Adaptability and resilience, as the flexibility thread, mirror the ebb and flow of a dynamic world. These threads are pliable yet unyielding, bending in the face of unforeseen challenges, yet resolute in their intent. Regular reviews, the conductor's baton, ensure our symphony maintains relevance and efficacy.

In this symphony of thought, each thread is interwoven, a cog in the grand machine of a comprehensive policy framework. Through discerning eyes and a holistic embrace, we dissect the core, appreciating the sum of its parts, the resonance of its chords. For it is through this comprehensive evaluation that our strategies refine, gaps are addressed, and the foundation of resilience is fortified. In our quest to create a resilient and effective framework, we write our chapter in the narrative of humanity, a symphonic testament to the power of collective action, propelling us towards a future that harmonizes with the rhythms of nature and the dreams of generations yet unborn.

Scientists and multilateral organizations have for illustration purposes laid the groundwork for a comprehensive climate policy framework that spans multiple sectors (EGR20, 2020), employing a cross-cutting approach guided by precise and measurable benchmarks. This all-encompassing framework delves into various sectors, sub-sectors, and key indicators, which are meticulously tracked and compared against the performance of the most successful models. To illustrate this, let us consider the example within the energy generation sector.

Within the realm of energy generation, a harmonious melody emerges. The sub-sector in focus entails a dedicated effort as proposed by scientists and problem domain experts to elevate the presence of renewable electricity, complemented by the implementation of support policies designed to expedite the deployment of renewable energy sources. This subsector's evolution is further illuminated by historical growth patterns that showcase the progressive integration of renewable electricity into the broader landscape of total electricity generation. By contextualizing these figures on a global scale, we gain a panoramic view of the strides that have been achieved, lending credence to the journey of advancement.

Interestingly, various nations have embraced this approach, embedding it within their distinct frameworks and enacting specific projects tailored to align with their unique trajectories. These projects serve as poignant exemplars of how this comprehensive methodology can be seamlessly integrated into real-world applications. As we immerse ourselves in these succinct yet illuminating use cases, it becomes evident that the potency of a comprehensive climate policy framework is not confined to theory; it translates seamlessly into tangible actions that reverberate across the global stage.

In essence, this approach, masterminded by scientific minds, serves as a lodestar guiding policymakers and stakeholders alike. It showcases the harmonious synchronization of sectors and sub-sectors, punctuated by meticulous tracking and measured progress against benchmarks. As we embark on this journey through a tapestry woven with threads of interconnected pursuits, it becomes unmistakably clear that the comprehensive framework is a creation far beyond the realm of theory. It stands as an embodiment of transformation, a maestro orchestrating a symphony of advancement in our united quest for a future imbued with sustainability and resilience. Its essence is not confined to abstraction; it is a catalyst of change, harmonizing diverse elements into a crescendo of progress.

Ladies and gentlemen, let us go now imagine for a moment, a mosaic of experiences, a rich tapestry woven from the threads of successful models and case studies, each offering a glimmer of wisdom and hope in the realm of climate policies.

This endeavor is not a mere replication of past achievements, but a quest to unearth invaluable insights, a quest that seeks to illuminate the path towards a comprehensive climate policy framework. As we delve into the pages of innovation and cooperation, we are met with stories that resonate with guidance and aspiration, stories that remind us that the symphony of collective action is not an unattainable crescendo, but rather a composition that we all can join.

Picture, if you will, the pioneering journey undertaken by Sweden, a nation that has etched its name in the annals of progressive energy transition. Enshrined within the pages of the Energy Agreement of 2016, Sweden's resolute march towards 100% renewable energy by 2040 stands as a testament to the power of a holistic policy framework. Here, we witness the harmonious convergence of clear objectives, economic tools, and the resonance of stakeholder engagement. The Swedish saga underscores the transformative potential that emerges when policies are guided by a lucid vision, and economic incentives become the bridge that beckons us toward greener horizons (IRENA, 2020).

Now, let us set sail across the vast expanse of the Pacific Ocean, where the nation of Fiji invites us to glimpse its masterpiece – the Climate Change Act (2021). This legislative magnum opus not only embeds climate adaptation and mitigation into the very fabric of its national identity, but it also extols the virtues of consultation and community engagement. In this case study, we discern the symphony of stakeholder resonance and capacity-building, a narrative that intertwines the mechanics of law with the resilience of communities on the front line of climate challenges.

As our journey continues, we alight upon the captivating landscapes of Costa Rica, a nation that whispers a tale of transformation toward carbon neutrality. The National Decarbonization Plan (2019), a guiding star, charts a course toward net-zero emissions by 2050, propelled by bold investments in renewable energy, electrified transportation, and the rekindling of lush forests. In this case study, the vitality of innovation and research unfurls before us, a testament to the remarkable synergy that emerges when policies nurture a culture of exploration and collaboration across academia, industry, and government.

Yet, the crescendo of our exploration finds its zenith in the European Union's resounding Green Deal. An ambitious opus that interlaces policies across sectors and borders, the Green Deal envisages a harmonious, climate-neutral continent by 2050. The EU's symphony showcases the potent melody of legal mechanisms, economic instruments, and the collective harmony of stakeholder engagement. This grand arrangement encapsulates the essence of a comprehensive policy framework, spanning a vast and intricate landscape that mirrors the very complexity of our climate challenges (Fetting, 2020).

With a sense of anticipation, we now unveil remarkable instances that illuminate how the noble endeavor of combatting the climate crisis can seamlessly interlace with the fabric of our cities and lives. These instances stand as beacons, illuminating the path toward a more sustainable world, their simplicity and efficacy echoing the profound impact that conscientious action can bring to fruition.

In the grand tapestry of urban evolution, Barcelona emerges as a vibrant chord resonating with the symphony of sustainability. Its Green Infrastructure and Biodiversity Plan (BGIBP, 2020) orchestrates a harmonious union, weaving together diverse elements of the urban landscape into a masterpiece of connectivity. Within this melodic canvas, a new composition emerges – the Barcelona Tree Master Plan for 2017-37. This intricate score, meticulously composed, aims to extend nature's embrace through an array of actions, nurturing the resilience of trees against the capricious forces of climate.

This botanical opus is a testament to foresight, as it harmonizes the natural rhythms of tree growth with the cadence of water scarcity and heat stress. The symphony crescendos with the diversification of tree species, creating a symphonic mosaic that enhances ecological balance. The ballet of water management takes center stage, orchestrated through the intricate choreography of runoff water and the precision of automated irrigation. Barcelona's streets bear witness to this verdant concerto, adorned with trees that offer solace amidst urban bustles.

The canvas of the Tree Master Plan is generously adorned with a budget of EUR 9.6 million annually, with the lion's share of EUR 8.3 million earmarked for the cultivation of this arboraceous composition. The remaining EUR 1.3 million are tenderly invested in the nurturing of soil and water management – a cornerstone of resilience against urban density and the reverberations of the heat island effect.

As Barcelona's architectural marvels ascend toward the sky, their shadows cast a complex choreography of climate challenges. Amidst this urban ballet, Barcelona takes a bold step forward, embracing the rhythm of sustainable city hood. An ode to biodiversity, the city's embrace of trees transcends mere aesthetics, becoming a vibrant passage in nature's song.

Across the Alpine landscape, Basel, Switzerland, unfurls its own melody – a harmonious blend of architecture and ecology. Here, rooftops become canvases for a symphony of green roofs, a poignant duet between mitigation and adaptation. Basel envisions a future where rooftops become the guardians of energy savings, orchestrating a ballet of temperature moderation while alleviating flood risks.

The composition of green roofs is guided by an ensemble of financial incentives and regulations, each note a step toward transforming architecture into a living, breathing entity. The Energy Saving Fund adds resonance to the rooftop symphony, infusing it with life through the gentle levy of energy bills. Basel's rooftops, once barren, now harmonize with biodiversity, each roof an instrument of adaptation (Talbot, 2021).

From Basel, the melody shifts to Copenhagen, a city sculpted by resilience. The Cloudburst Management Plan (2012) becomes the score, an orchestration against the crescendo of intensified rainfall and climate change. This urban cantata aims to counteract the symphony of damage with an intricate ballet of adaptation.

In Copenhagen's dance with the rain, a medley of solutions emerges – storm water roads and pipes, retention roads and areas – all conduits for the orchestration of water's journey. The dance of adaptation intertwines with traditional sewerage, as the city weighs the resonance of cost against the harmonies of sustainable solutions. As Copenhagen steps into the future, it orchestrates a symphony of resilience, composing a future where urban spaces are transformed into conduits of climate adaptation.

Across these landscapes, Barcelona, Basel, and Copenhagen resonate as symphonies of transformation, their unique harmonies weaving a melody of resilience and adaptation. The chords of green roofs, resilient trees, and adaptive infrastructure blend seamlessly, creating a resonant symphony – a powerful reminder of the boundless potential within the realm of human ingenuity and commitment.

In closing, these narratives of innovation, dedication, and harmonious collaboration stand as luminous beacons guiding us through the labyrinthine corridors of policy design. They remind us, amidst the varied contexts that paint the global canvas, that the foundational principles underpinning a comprehensive climate policy framework remain unwavering. We unravel wisdom from this intricate tapestry of global experiences, and in doing so, we find ourselves inspired to tailor our strategies to meet the unique aspirations and challenges of our own societies.

Just as a symphony marries diverse instruments to create a harmonious composition, so too shall our endeavors harmonize into a resounding crescendo of progress in the face of the ever-pressing climate crisis. As we reflect on the stories we've uncovered today, let us remember that the journey of transformation begins with each step, each note, each policy, resonating in harmony with the symphony of a sustainable future.

III. Shift In Mindsets: Individual And Collective Responsibility

In the grand theater of humanity's trials, the climate crisis emerges as an adversary of unparalleled proportions. Its resolution, however, remains impervious to the solitary embrace of technological panaceas. Rather, it beckons for an elemental recalibration in the contours of our cognition, behavior, and communion with the world that envelops us. Thus, the crux of this treatise resides in the assertion that the exigencies of the climate crisis proffer a dual metamorphosis: a profound transmutation in individual and collective mindsets.

On the personal plane, this thesis accentuates the imperative of agency and accountability in fashioning a legacy conducive to a sustainable tomorrow. Each constituent of humanity, it avers, occupies a pivotal role, transcending a mere acknowledgment of their involvement. This role unfurls as an active embrace of lifestyles steeped in sustainability – a tapestry woven with meticulous choices dictating our resource consumption, modes of travel, and the cadence of our quotidian existence. This recalibration obliges an exodus from the prevalent culture of excess and profligacy, towards a more pensive reverence for the finite endowments of our planetary abode.

Beyond the individual, the thesis casts a luminous spotlight upon the theater of collective responsibility. The contours of the climate crisis demand a collection of concerted endeavors, a symphony that swells beyond the solitary strums of individual endeavors. To orchestrate such a harmonious ensemble, the cultivation of an ethos steeped in environmental mindfulness and guardianship becomes imperative across diverse strata of society's fabric. Engraving a profound comprehension of the interwoven tapestry of human existence and the natural world takes center stage in this endeavor. This cultivation unfurls through the power of pedagogical pursuits and campaigns of consciousness. By disseminating cogent and gripping elucidations of the climate crisis, these endeavors stoke the flames of urgency, infusing it into the collective consciousness as a palpable, proximate crisis that reverberates within the chambers of daily existence.

In this pursuit, the imperative to kindle environmental consciousness cascades down to the cellular structure of communities, organizations, and societies. Their engagement assumes the mantle of custodianship, weaving sustainable practices into the very sinews of operational and decision-making paradigms. Embracing an all-encompassing vista, these entities traverse the terrain of adopting eco-friendly technologies, advancing the cause of circular economies, and paring the excesses of wastefulness. This endeavor, propelled by a holistic ethos, exalts the longevity of ecological well-being over the ephemeral siren calls of transient gains. As they chart this course, they engender an exemplar, a beacon for emulation.

Central to this narrative of environmental enlightenment is the symphony of interconnectedness – an ethos that reverberates with the understanding that our planet's fate finds itself entwined with the warp and weft of every individual's action, every community's aspirations. This recognition casts an incandescent glow upon the mantle of shared destiny, a potent elixir that galvanizes collective action. Within this crucible of shared purpose, the individual finds themselves not as a solitary note, but as part of a sonorous symphony, an orchestra of endeavor serenading a common destination. Such realization, akin to an inner crescendo, propels individuals toward behaviors that breathe life into transformative change.

In the moments that lie ahead, let us embark upon an odyssey of contemplation, delving into the mystical alchemy that orchestrates the metamorphosis of individual paradigms into vessels of enduring awareness and communal accountability. This symphony, kindled by the fervent embers of environmental consciousness and custodianship, beckons us to equip each individual with the illuminating torch of knowledge, forging within their grasp the tools to carve resolute choices from the crucible of possibility. And as we weave this intricate tapestry, interwoven with threads of profound interconnectedness and shared destiny, we bestow upon society the canvas upon which to unfurl its sails. With these billowing banners, we gather the tempestuous winds of motivation and collective volition, indispensable for charting a steadfast course through the turbulent seas that beset us. Thus fortified, we hoist a mantle of efficacy as we navigate the tumultuous currents, steering towards the heart of the climate crisis with unwavering purpose.

Here we stand at the precipice of a narrative, poised to embark on a journey that traverses the intricacies of sustainable living, a voyage that carries us through the corridors of conscious choices and ecological stewardship. In the chapters ahead, we'll illuminate ten distinct avenues, each a facet of the profound thesis that embracing sustainable lifestyles and choices is pivotal in the relentless quest to confront the climate crisis head-on.

What lies before us are not mere abstract concepts, but concrete examples – ten fragments of a larger puzzle, each piece meticulously shaped to articulate the intricate dance between individual actions and the greater environmental narrative. In the spirit of exploration, let us dive deep into these exemplars, immersing ourselves in stories that illustrate the transformative power of human agency.

First, let us traverse the terrain of Transportation Preferences, where individuals, armed with a resolve to reduce carbon footprints, opt for eco-friendly modes of mobility. Bicycles, footsteps, and the rhythm of public transit replace the cacophony of personal vehicles powered by fossil fuels, painting a canvas of cleaner air and minimized congestion.

Consider the story of Alex, a dedicated city dweller who decides to embark on a transformative journey by altering his transportation preferences. Living in a bustling metropolis known for its notorious traffic snarls and smog-choked air, Alex becomes acutely aware of the environmental toll that vehicular congestion and emissions exact upon his beloved urban landscape.

With a resolute determination to make a meaningful impact, Alex embraces a lifestyle choice that transcends convenience and conventional norms. He opts to forego the familiarity of his personal car, and instead, embarks on a daily commute that melds seamlessly with the rhythm of his surroundings. Donning his helmet and mounting his trusty bicycle, Alex embarks on a two-wheeled odyssey to work.

As Alex pedals through the city's labyrinthine streets, a metamorphosis unfolds. He becomes an emissary of change, effortlessly weaving through the very gridlock that once imprisoned him in a metal cocoon. Passersby witness his earnest dedication, and some are moved to reassess their own commuting habits. The cumulative effect is a ripple of inspiration that slowly permeates the urban fabric.

Simultaneously, Alex incorporates public transportation into his repertoire. The hum of the subway, the rhythm of the bus, and the shuffle of footsteps become his urban symphony. With each choice to forego his car's ignition, Alex is casting his vote for a greener, more sustainable future. His decision translates into tangible benefits for the city's eco-system – a reduction in noxious fumes, a reprieve from the cacophony of horns, and a mitigation of the carbon emissions that once loomed like a specter.

Yet, Alex's impact extends beyond immediate environmental benefits. His actions reverberate through the community, sparking conversations and nudging others to reconsider their own commuting habits. Colleagues, neighbors, and even local businesses take note, spurred by Alex's example to explore alternative modes of transportation. The city's streets gradually undergo a subtle transformation, as an increasing number of bicycles are seen perched on racks, and commuters gather at bus stops, engaging in camaraderie that is as much about shared purpose as it is about convenience.

As Alex's journey continues, he finds himself not only contributing to the alleviation of traffic congestion and reduction of carbon emissions but also becoming a catalyst for change. His choice to pedal and ride the rails ushers in a nuanced transformation – a transformation that transcends mere movement and reshapes the very contours of his urban habitat. With each turn of the pedal and each station stop, Alex becomes an unsung hero of his city, a testament to the power of a single individual to ignite a chain reaction of positive change.

As we venture further, the landscape of Energy Consumption unfurls, where the switch to renewable sources like solar and wind power takes center stage. Homes and businesses alike embrace these dynamic energies, reducing reliance on non-renewable fossil fuels and setting a precedent for a more sustainable energy mix.

Meet Sarah, a forward-thinking homeowner who harbors a deep reverence for both her abode and the planet it rests upon. Intrigued by the prospect of transforming her dwelling into a beacon of sustainability, she embarks on a quest that redefines not only her energy consumption but also her relationship with the environment.

In a sun-soaked ritual of transformation, Sarah makes the pivotal decision to adorn her roof with a symphony of solar panels. As these silent sentinels harness the boundless energy of the sun's rays, they initiate a subtle metamorphosis. Her once-passive rooftop awakens into a dynamic powerhouse, converting photons into electricity with a graceful efficiency that belies its newfound significance.

Sarah's home becomes more than just a sanctuary; it becomes a microcosm of change. Each kilowatt-hour of energy produced from her panels is a whisper of commitment to a cleaner, greener world. But Sarah's journey doesn't end here – it's just the prologue to a narrative of profound impact.

The canvas of her rooftop energy production becomes an artistic tapestry of surplus. On days when the sun's generosity outpaces her consumption, Sarah's surplus energy flows back into the grid, a selfless offering that reverberates far beyond her immediate surroundings. In this act of reciprocity, her home becomes a miniature power station, contributing to the broader energy needs of her community.

As Sarah's panels hum with purpose, they cast a luminous ripple effect. Her embrace of solar energy reverberates through her neighborhood, stimulating curiosity and inspiring others to consider their own energy choices. Conversations ignite at neighborhood gatherings, as Sarah's journey serves as an ambassador of possibility, heralding a future where rooftops bear witness to a harmonious collaboration with nature.

But Sarah's influence extends beyond her immediate community. Her modest contribution to the grid joins a collective crescendo that challenges the dominance of fossil fuels. As more homeowners follow her example, the energy landscape begins to shift. A growing chorus of solar panels orchestrates a harmonious transition towards a more sustainable energy mix, where the sun's benevolence takes center stage, reducing the stranglehold of carbon emissions.

In the grand tapestry of energy consumption, Sarah's decision to harness solar power isn't just a pragmatic shift; it's an embodiment of stewardship. Her rooftop solar panels transform her dwelling into a nexus of positive change, bridging the chasm between individual choices and global impact. Through each sunrise and sunset, her home becomes a silent protagonist, an unwavering reminder that progress is not just a grand gesture – it's a symphony of small, intentional choices that collectively compose a brighter, more sustainable future.

Dietary Choices beckon us next, inviting exploration into the profound impact of embracing plant-based diets or curbing meat consumption. The curtain rises on the colossal role that livestock production plays in greenhouse gas emissions, and how conscious culinary decisions can wield the power to sculpt a more harmonious relationship with the planet.

Allow me to introduce Emily, a conscientious individual who embarks on a culinary journey that transcends mere sustenance, infusing her dining choices with a profound sense of purpose and ecological mindfulness.

Emily's transition to a vegetarian diet unfolds like a gradual transformation, akin to the unfurling of delicate petals. With each meal, she forges a new connection between her plate and the planet. Her decision to abstain from meat isn't just a dietary shift; it's a conscientious pivot towards a more sustainable future.

As Emily navigates the aisles of the grocery store, her cart brims with an array of vibrant vegetables, legumes, and whole grains. Her culinary canvas broadens as she discovers a tapestry of flavors, textures, and colors that enrich her palate and nourish her body. Yet, her choice isn't solely about personal well-being; it's about embracing a lifestyle that resonates with the symphony of the natural world.

Emily's journey leads her to the heart of a complex web of ecological interdependence. She learns that the production of meat exacts a formidable toll on the environment. The land required for livestock grazing, the water consumed in animal agriculture, and the energy expended in feed production collectively compose a narrative of environmental strain. With each bite of a vegetarian meal, Emily unburdens the earth from the weight of resource-intensive meat production.

As Emily's vegetarian odyssey unfolds, her carbon footprint undergoes a profound transformation. The emissions associated with livestock farming, including methane from cattle and the energy required for feed and processing, steadily diminish. Her choice radiates like a pebble dropped into a pond, creating ripples that reverberate far beyond her individual sphere.

In her newfound culinary path, Emily is not alone. Her dining choices become a catalyst for conversation and exploration among friends, family, and colleagues. As they gather around shared meals, they are inspired to reconsider their own dietary habits, awakening to the potential of sustainable dining choices. In this communal exchange, a tapestry of change takes shape, woven from a collective commitment to reduce the environmental impact of food consumption.

Emily's embrace of a vegetarian diet is a testament to the symbiotic relationship between personal choices and global impact. With each meal she savors, she fosters a legacy of stewardship – a narrative of mindful dining that transcends the immediate satisfaction of taste to champion a more harmonious coexistence with the planet. In the silent symphony of her dietary choices, Emily adds a melodic note to the chorus of individuals striving to mitigate greenhouse gas emissions and nourish a sustainable future.

Waste Reduction emerges as a protagonist in our narrative, urging us to witness the transformative potential of shunning single-use plastics, composting organic remnants, and embracing responsible recycling. Landfills stand as testimonies to unchecked consumption, but our journey reveals the promise of redefining waste into a resource.

Step into the lives of the Johnson family, a dynamic unit that transforms their home into a laboratory of waste reduction, embodying the ethos that small changes can yield profound ecological impact.

In their bustling kitchen, the Johnsons orchestrate a daily symphony of sustainability. With every slice of fruit and chop of vegetables, they carefully segregate kitchen scraps, funneling them into a compost bin. What once would have been discarded as mere refuse now assumes a role in a grand ecological narrative. As the compost pile thrives, it metamorphoses into rich, nutrient-dense soil that nurtures their garden, a microcosm of circularity that mirrors the cycles of nature.

This composting ritual embodies a commitment that transcends the immediacy of convenience. By diverting organic waste from landfills, the Johnsons contribute to a reduction in methane emissions – potent greenhouse gases (GHGs) that arise during the decomposition of organic matter. In this seemingly mundane act, they foster a micro-climate of change that resonates far beyond the boundaries of their backyard.

Beyond the kitchen, the Johnsons navigate the landscape of consumption with discernment. Their household brims with an assortment of reusable containers, which have become a hallmark of their conscious choices. Rather than succumbing to the allure of single-use plastics, they deftly employ these containers to store leftovers, pack lunches, and procure groceries in a dance of mindful consumption.

This shift is not merely an aesthetic alteration; it's a rebellion against a culture of disposability. By opting for reusable containers, the Johnsons curb the influx of plastic waste into their lives. With each carryout container or plastic bag they eschew, they contribute to the reduction of plastic pollution, mitigating the environmental degradation that plagues oceans, landfills, and eco-systems.

The Johnsons' pursuit of waste reduction cascades into the fabric of their community. Friends and neighbors, intrigued by their dedication, join in the crusade. Conversations emerge over garden fences and at local gatherings, generating a collective momentum that ripples through the neighborhood. What began as a family initiative evolves into a communal ethos, a testament to the power of example and shared responsibility.

As the Johnsons navigate their daily lives, their commitment to waste reduction reverberates on multiple levels. They diminish the burden on landfills, mitigating the release of methane, a potent driver of climate change. Their embrace of reusable containers fosters a culture of sustainable choices, exerting gentle pressure on industries to rethink their packaging practices. In their mindful quest for waste reduction, the Johnsons echo the sentiment that every discarded peel, every composted leaf, and every reusable container is a brushstroke in the masterpiece of responsible stewardship – a masterpiece that inspires others to join the canvas and create a more sustainable world.

As we venture into the realm of Water Conservation, a symphony of drops resonates, revealing the possibilities born of fixing leaks, employing water-efficient appliances, and cultivating landscapes that mirror nature's elegance. Each droplet becomes a testament to our collective ability to shepherd one of our most precious resources.

Enter the Thompson residence, a haven where water conservation isn't just a notion, but a way of life intricately woven into the fabric of daily existence. In their pursuit of stewardship, the Thompsons transform their home into an oasis of resource-conscious practices, each droplet a testament to their commitment.

As the Thompsons step into their bathroom, a symphony of water-efficient fixtures greets them. Faucets, showerheads, and toilets have all been thoughtfully curated to minimize water consumption without compromising the quality of their experience. With each turn of the tap, they witness the fusion of innovation and mindfulness, where drops become a currency of conscience.

Beyond the walls of their home, the Thompsons turn their gaze skyward. A collection system adorns their rooftop, poised to capture the life-giving rain that graces their land. This harvested rainwater is channeled into cisterns, a reservoir of liquid gold that holds the promise of sustenance. With precision, the Thompsons direct this water towards their garden, where it nourishes the earth with a sense of purpose that mirrors nature's cycle.

In their backyard, the Thompsons cultivate a landscape that dances in harmony with the rhythm of their region. Drought-resistant plants adorn the terrain, their vibrant hues an ode to nature's resilience. These plants, having evolved to thrive in arid conditions, require minimal irrigation, ushering forth a garden that is both a sanctuary of beauty and a refuge of resourcefulness.

The Thompsons' commitment to water conservation is more than a domestic endeavor; it is an emblem of solidarity with a changing world. They reside in a region where changing climate patterns have ushered in an era of water scarcity, transforming once-reliable sources into dwindling reserves. Amid this backdrop, their practices become a testament to adaptability and foresight.

As the Thompsons unveil their journey of water conservation, their ripple of influence extends outward. Neighbors, friends, and passersby bear witness to their efforts, sparking conversations that stretch beyond picket fences. The community takes note, engaging in a dialogue that spreads awareness and kindles collective action. Soon, a tapestry of water conscious households emerges, a mosaic of transformation that reflects the power of collective responsibility.

The Thompsons' household is a living tableau, where water-efficient fixtures, rainwater collection, and drought-resistant landscaping converge to tell a story of mindfulness and harmony. Each drop conserved, each plant nurtured, is a brushstroke in a portrait of resilience. Their journey speaks to a timeless truth: that in the face of changing climate patterns, every gesture towards water conservation is a lifeline, a promise, and a legacy of stewardship for future generations.

Consumer Habits emerge as actors on our stage, shedding light on the virtuous cycle of conscious choices. Supporting brands that champion ethical and sustainable practices, individuals emerge as architects of market trends, steering industries towards greener pastures.

Picture Elena, a dynamic young professional whose passion for style converges harmoniously with her commitment to the planet. In her quest for authenticity, she transforms her role as a consumer into a catalyst for change, igniting a shift in the landscape of fashion.

Elena's wardrobe isn't just an assemblage of garments; it's a testament to her values. With an artisanal eye, she seeks out clothing brands that resonate with her ethos of sustainability and ethical production. Each purchase is imbued with intention, a conscious choice to support those who prioritize the well-being of both people and the planet.

As Elena peruses the aisles of boutiques and scrolls through online catalogs, she unveils stories behind the seams. She seeks brands that champion eco-friendly materials, minimizing the footprint left by production processes. Her selections reflect her discernment, choosing fabrics that tell tales of organic growth and responsible cultivation.

Yet, Elena's choices transcend the realm of aesthetics. Each garment she selects is a vote cast for a world where the allure of fashion is intertwined with a commitment to ethical labor practices. She navigates the terrain of her choices with the foresight that by patronizing brands that uphold fair wages and safe working conditions, she contributes to a tapestry of change that spans continents.

Elena's impact is not confined to her own closet; it reverberates across the industry. As demand for sustainable and ethically produced clothing surges, brands take notice, recalibrating their practices in response to the collective crescendo of conscious consumers like Elena. Supply chains are scrutinized, production processes evolve, and the compass of the fashion industry aligns itself more closely with the principles of responsibility and accountability.

Elena's choices cascade into a symphony of transformation. Conversations among her peers evolve, as they inquire about her selections and embark on their own journeys of mindful consumption. Her social media platforms become a virtual runway, showcasing not just chic ensembles, but a narrative of empowerment and change. The ripple of influence extends far beyond Elena's immediate circle, as her choices ripple through networks and communities.

In the grand narrative of consumer habits, Elena stands as a beacon of possibility. Through her discerning choices, she transforms fashion from a mere transaction to a conversation that spans the globe. Each garment she adorns becomes a message, a statement that style and sustainability are not mutually exclusive. Elena's impact radiates – a constellation of choices that not only elevates her own wardrobe but alters the trajectory of an entire industry. In each thread, each stitch, each conscious purchase, Elena weaves a narrative that resonates with empowerment, transformation, and the potent potential of consumer habits as agents of positive change.

The sartorial world of Clothing and Fashion unveils its secrets next, juxtaposing the allure of secondhand or sustainably produced attire with the grim environmental toll of fast fashion. Threads of change are woven as we contemplate the transformative sway of each purchase.

Visualize the campus of an illustrious university as it becomes the stage for a transformative narrative, one scripted by a visionary college student named Maya. In her pursuit of sustainable style, she orchestrates an event that not only challenges conventions but redefines the very fabric of clothing and fashion.

Maya's endeavor is a clothing swap, a vibrant tapestry of exchange where garments take on new life in the hands of their new owners. The campus buzzes with anticipation as students congregate, each carrying a bag of well-loved clothing items that have outlived their initial chapters. In this gathering, fashion becomes a vehicle of connection, a medium through which stories are shared, friendships are forged, and an ethos of conscious consumption is woven into the communal psyche.

As the event unfurls, a symphony of fashion narratives emerges. Garments, once cherished but now awaiting rejuvenation, find themselves in the arms of new custodians. Jeans, dresses, shirts – each item carries a legacy, an imprint of the lives they've adorned. As students exchange articles of clothing, they create a mosaic of interconnectedness, each piece an emblem of shared responsibility.

The impact of Maya's initiative extends beyond the mere exchange of fabrics. It is a deliberate effort to curb the rampant textile waste that plagues the fashion industry. The clothing swap breathes life into Maya's vision of a world where garments are cherished, where their value surpasses the fleeting trends that dictate conventional fashion cycles. In this endeavor, she champions a philosophy where the allure of clothing is intertwined with an appreciation for its inherent worth.

As the campus echoes with laughter and camaraderie, the clothing swap transcends its pragmatic purpose. It becomes a platform for dialogue, an avenue for students to explore the connections between style and sustainability. Conversations ripple through the event – discussions about the impact of fast fashion, the resources embedded in each piece of clothing, and the potential for transformative change through mindful choices.

The echoes of Maya's initiative resound well beyond the event's confines. The clothing swap catalyzes a cultural shift on campus, nurturing a generation of students who approach fashion with an empowered perspective. Peers are inspired to examine their own closets, to scrutinize their consumer habits, and to explore alternatives that mirror Maya's ethos of mindful consumption.

Maya's clothing swap stands as a testament to the transformative power of innovation and intention. Through her initiative, clothing and fashion evolve from expressions of personal style to vehicles of collective change.

In each exchanged garment, Maya weaves a narrative of resilience, where the act of donning a piece of clothing becomes an act of stewardship, a celebration of connection, and an affirmation of the potential for individuals to shape the trajectory of an entire industry.

Home Design beckons us to explore the frontiers of architecture, where eco-friendly paradigms redefine construction. Concepts like energy efficiency, natural illumination, and insulation shape structures that marry human comfort with the embrace of the Earth.

Step into the visionary world of David, an architect whose blueprints paint a portrait of innovation and sustainability. In his pursuit of redefining home design, he crafts a masterpiece that not only shelters its inhabitants but nurtures a harmonious coexistence with the environment.

David's canvas is a plot of land, a blank slate where his vision unfurls. With meticulous care, he envisions a net-zero energy home that stands as a testament to human ingenuity and ecological mindfulness. The design is an intricate ballet of elements – passive solar principles, cutting-edge insulation techniques, and a symphony of energy-efficient appliances.

As the home takes shape, its orientation becomes a choreography of connection with the sun's path. Large windows facing south welcome the embrace of sunlight, warming the interiors during winter while minimizing heat gain in the summer. The architecture becomes a bridge between the built and natural worlds, a vessel that channels nature's energies to minimize the need for external heating or cooling.

In the embrace of this innovative home, advanced insulation becomes an unobtrusive guardian, a silent sentinel that ensures a stable internal climate. The insulation transcends mere materiality; it is a promise that the home's warmth isn't the result of energy consumption, but an orchestration of nature's grace. As seasons shift, the insulation's guardianship remains unwavering, a testament to David's meticulous attention to detail.

Within the home's sanctuary, energy-efficient appliances engage in a symphony of conservation. The dishwasher, the refrigerator, the lights – each element bears the mark of conscious choice, whispering a commitment to a future where functionality is synonymous with sustainability. The home itself becomes a haven of education, a living laboratory where inhabitants witness the marriage of convenience and environmental stewardship.

David's net-zero energy home isn't just a structure; it's a declaration of principles that resonates beyond its walls. Its very existence becomes an invitation to rethink the relationship between shelter and the earth. Its architectural vocabulary weaves a narrative that sets a precedent, challenging conventions and sparking a paradigm shift in construction practices.

The impact of David's creation cascades beyond its four walls. As the community bears witness to his innovation, a ripple of curiosity spreads. Conversations among builders, designers, and homeowners evolve, punctuated by discussions of passive solar design, insulation techniques, and energy-efficient technologies. The net-zero energy home becomes a catalyst, birthing a movement that champions sustainability as the cornerstone of construction.

In the grand tapestry of home design, David's net-zero energy creation is a resounding symphony of purpose. It marries the elegance of architecture with the ethics of sustainability, proving that spaces need not merely shelter – they can inspire, educate, and embody a future where innovation and ecological mindfulness coalesce. With each sunbeam that caresses the home's windows, with each whisper of wind harnessed for power, David's architectural opus narrates a story of possibility, where a single home becomes a cornerstone of transformative change.

Education and Advocacy beckon us to raise our voices, wielding awareness as a formidable weapon. Through climate education and advocacy, individuals embark on a journey of enlightenment, kindling sparks of change within their communities and beyond.

Allow me to introduce Emma, a passionate advocate who transforms her role as an individual into a force of change through education and advocacy. In her pursuit of a more sustainable world, she becomes a beacon of enlightenment within her community, igniting a fire of awareness and action.

Emma's journey begins with an earnest desire to bridge the gap between knowledge and action. Armed with a wealth of research and information, she organizes workshops and seminars on climate change and sustainability. Her gatherings are more than mere lectures; they are interactive forums where ideas are exchanged, questions are raised, and the complexities of environmental challenges are dissected with precision.

At these events, individuals from various walks of life congregate, each drawn by the allure of understanding and the promise of impact. Emma's passion is contagious, her enthusiasm infectious, as she presents a roadmap to a more sustainable future. Attendees leave with newfound insights, armed with the knowledge to make informed choices that reverberate beyond their immediate sphere.

Emma's advocacy extends beyond the confines of lecture halls. She becomes a vocal proponent of policy change, leveraging her voice to engage with local government officials and decision-makers. Her advocacy resonates through op-eds, letters to the editor, and public speeches, each sentence a plea for a future defined by ecological stewardship.

In one instance, Emma rallies her community to advocate for the adoption of renewable energy initiatives within the city. Her impassioned call to action is echoed by her fellow citizens, and soon, the movement gains momentum. Meetings with local representatives become more than mere discussions; they are dialogues of potential, where the power of collective advocacy becomes palpable.

Emma's efforts culminate in the adoption of policies that prioritize sustainability. Rooftops adorned with solar panels become a common sight, and wind turbines grace the skyline. The city's commitment to renewable energy becomes an emblem of change, a testament to the potency of education and advocacy.

Yet, Emma's impact is more profound than policy alone. Her advocacy has woven a tapestry of collective action, a web of informed individuals who have been inspired to become ambassadors of change. Conversations ripple through the community, households adopt ecofriendly practices, and businesses pivot towards sustainability. The landscape is transformed, not just through policy shifts, but through the shared consciousness of a community awakened.

Emma's journey is a testament to the power of education and advocacy. Through her tireless efforts, she has breathed life into the abstract notions of climate change and sustainability, making them tangible and urgent. Her story illuminates the path from knowledge to action, showcasing how a single individual can kindle a flame that transforms not just attitudes, but entire communities. In each workshop conducted, each policy advocated, Emma etches a narrative of possibility – a story that underscores the potential for individuals to rewrite the future through the indomitable forces of education and advocacy.

Lastly, our narrative culminates in the embrace of Community Engagement, where individuals join hands in local initiatives. Community gardens bloom, clean-up campaigns sparkle, and conservation projects carve pathways towards shared responsibility, where the sum of individual efforts paints a tapestry of collective impact.

Let's delve into the story of Alex, a community catalyst who embodies the spirit of shared responsibility through his unwavering commitment to local environmental initiatives. Alex's journey is a testament to the transformative power of community engagement, where the threads of individual effort weave a tapestry of collective impact.

In the heart of his neighborhood, a once-neglected plot of land transforms into a vibrant community garden, a sanctuary of green amidst the urban sprawl. Alex becomes its steward, dedicating hours of toil to tend the earth, coaxing life from the soil and nurturing a cornucopia of fruits and vegetables. Each seed sown, each harvest reaped, is a testament to his belief in the potential of shared spaces to foster interconnectedness and rekindle our relationship with nature.

Alex's garden is more than a testament to his green thumb; it becomes a communal oasis, a locus of connection and education. Families gather, children discover the magic of growth, and neighbors exchange gardening tips. The garden breathes life into a sense of shared purpose, where a diverse array of individuals converges to cultivate not just crops, but a culture of unity and responsibility.

Beyond the garden's borders, Alex extends his influence to local clean-up campaigns. Armed with gloves and garbage bags, he rallies his neighbors to scour parks, streets, and waterways, transforming discarded refuse into tokens of rejuvenation. With each piece of litter collected, a dialogue unfolds, not just about waste, but about the obligation to care for the spaces we all call home.

In another corner of the community, Alex joins hands with conservationists in restoring a local nature reserve. Together, they mend trails, plant native flora, and protect the habitats of local wildlife. Alex's dedication illustrates the synergy of collective effort, where each individual's contribution reverberates through the intricate web of nature, reaffirming our shared responsibility as custodians of the Earth.

Alex's initiatives, whether through community gardens, clean-up campaigns, or conservation projects, foster a sense of unity that transcends the immediate tasks at hand. Conversations are ignited, relationships are forged, and a collective consciousness awakens. Through his endeavors, Alex galvanizes his community to embrace an ethos of stewardship, proving that participation in local environmental initiatives is not just about tangible results, but about forging connections that bind us to the Earth and to one another.

In the grand narrative of community engagement, Alex stands as an embodiment of the power of collective responsibility. His actions echo through the community, leaving an indelible mark on the landscape and the hearts of those he touches. With each seed planted, each piece of litter picked up, and each conservation effort undertaken, Alex weaves a narrative of shared purpose, illustrating the extraordinary impact that can emerge when individuals come together to nurture the world they inhabit.

With these ten examples as our guideposts, we descend into the heart of sustainable choices, each story a testament to the inextricable link between individual actions and the broader eco-system. As we delve into the depths of these narratives, we discover the harmonious cadence between personal empowerment and global transformation – a symphony of agency that underscores the fundamental truth: that in the face of the climate crisis, the choices we make today resonate as ripples of change that stretch far beyond the horizon.

Embedded within this symphony is the motif of environmental mindfulness, a melody that calls us to tune into the world around us with heightened sensitivity. It beckons us to tread upon the Earth with a conscious stride, to weigh our actions against their ecological consequences. The tales of transportation preferences, energy consumption, and waste reduction illustrate the art of being attuned to the environment, as individuals, armed with knowledge, make deliberate choices that safeguard the delicate balance of our planet.

These stories intertwine seamlessly with the cadence of pedagogical pursuits and campaigns of consciousness. Education and advocacy emerge as powerful instruments, capable of orchestrating shifts in collective awareness. Through workshops, seminars, and community engagement, individuals sow seeds of understanding that blossom into gardens of change. Like an ensemble of teachers conducting a symphony of enlightenment, these examples resonate with the vital role of imparting knowledge and nurturing an awakened populace that can catalyze widespread transformation.

The thread of interconnectedness weaves through our exploration, binding the choices of one to the destiny of all. In the narratives of dietary choices, consumer habits, and community engagement, we perceive the delicate threads that bind us to one another and to the Earth.

These stories reflect a profound recognition that we share not just a planet, but a fate intertwined with every living being. The concept of interconnectedness radiates, reminding us that our actions, however seemingly small, reverberate across the grand tapestry of life.

At the heart of these narratives beats the rhythm of shared purpose. Whether in the context of water conservation, home design, or clothing and fashion, we witness individuals uniting around a common vision. This shared purpose unfurls as a potent motivator, a call to action that transcends individual concerns and cascades into collective endeavors. These stories echo the refrain that when we join hands with a shared resolve, we become architects of change, shaping a world where sustainability is not just a choice, but a shared destiny.

In the symphony of environmental mindfulness, pedagogical pursuits, interconnectedness, and shared purpose, we find our anthem of defiance against the climate crisis. These concepts emerge as pillars of hope, guiding us through the labyrinth of challenges that lie ahead. As we navigate this intricate interplay, let us remember that each choice, each step, each spark of awareness has the power to radiate as a transformative force, crossing the boundaries of time and space to paint a portrait of a world reimagined, revitalized, and reinvigorated.

IV. Artistic Echoes Of Climate Change: Echoes Of Extinction

CHAPTER

Six

Greenwashing Beware: Unveiling Scams That Swindle With Biomass, Solar, Electric Vehicles, and Strategic Metals

"If we destroy nature, nature will destroy us."
– Swahili proverb

I. Greenwashing Galore: Multinational Corporations' Environmental Commitments or Fairy Tales?

Ah, greenwashing. The art of painting a rosy shade of green over a business's true colors. It's a clever marketing tactic, designed to dupe consumers into believing that a product or service is environmentally friendly, when in reality, it's as green as a grasshopper on an asphalt road.

You see, greenwashing has been around for quite some time. The term was coined in the 1980s by a sharp-eyed environmental activist named Jay Westerveld. He had a keen eye for spotting hotels that touted their eco-consciousness by asking guests to reuse towels, while behind the scenes, they continued to guzzle resources like there was no tomorrow.

The deceptive nature of greenwashing knows no bounds. It can take the form of lofty claims like "100% natural" when the product is actually riddled with harmful chemicals. Or companies may plaster their advertisements with pictures of serene forests and frolicking animals, all while wreaking havoc on the environment through other unsavory practices.

Why do they do it, you ask? Well, it's all about the bottom line, my friends. Greenwashing allows companies to cash in on the growing concern for the environment without lifting a finger to make meaningful changes. It's like putting a band-aid on a broken dam and hoping nobody notices the flood.

But here's the rub. Greenwashing is more than just a harmless fib. It's a serious problem because it tricks consumers into thinking they're making eco-friendly choices when they're really just buying into an illusion. It gives them a false sense of security, like a warm blanket on a chilly night, preventing them from taking real action to reduce their environmental impact.

To protect ourselves from the siren song of greenwashing, we must remain ever vigilant. Skepticism should be our faithful companion when faced with exaggerated or vague environmental claims. Look for independent certifications or third-party verifications that cut through the smokescreen of deception. And for goodness' sake, do your own research! Don't just take a company's word for it. Dig deep, my friends, and uncover the true impact of the products and services you consume. Support those businesses that walk the talk of sustainability, rather than those who merely talk the talk.

But here's the kicker, folks. As much as we may decry greenwashing and demand action on climate change, there's an inconvenient truth we must confront. A recent article by McVeigh (2023) laid bare the hypocrisy that lurks within us all. It revealed that while many Europeans claim to be alarmed by the climate crisis, their support for climate action starts to waver when it threatens to disrupt their comfortable lifestyles.

According to a YouGov survey in seven European countries, respondents were all for measures that required minimal sacrifice. Tree-planting programs? Sure, sign them up! Growing their own plants? Why not! But when it came to more substantial changes like giving up meat and dairy products or ditching fossil-fuel cars, support dwindled faster than an ice cube in a sauna (Henley, 2023).

It seems we're all in favor of saving the planet, as long as it doesn't inconvenience us too much. We want to combat climate change, but only if it doesn't mean sacrificing our creature comforts. Ah, the sweet aroma of hypocrisy!

So, dear audience, let us reflect on this grand theater of greenwashing. As we navigate the treacherous waters of eco-conscious consumerism, let us remain vigilant against the lies and manipulations of those who seek to exploit our desire for a greener world. And let us not forget to look within ourselves and confront the inconvenient truths that lie beneath our own actions and choices.

The greenwashing game extends its reach to the energy sector as well. Brace yourselves, dear audience, for a tale of deceptive practices and false promises in the realm of renewable energy.

Many of the "green" energy tariffs we encounter are nothing more than cunning façades, hiding behind a veil of fake certifications and legal loopholes. These energy providers boldly claim to offer 100% renewable energy, enticing us with visions of a cleaner, greener future. But alas, the truth is far from the sparkling image they project.

You see, these suppliers engage in a clever sleight of hand. They purchase their energy from the wholesale market, matching it to customer energy use, but there's a catch. Instead of ensuring it comes from a genuine renewable generator, they dip their fingers into the murky pool of unsustainable fossil fuels and nuclear plants. The audacity!

And here's the real kicker, my friends: There's no way for us, the unsuspecting customers, to know where our energy truly comes from. It's like trying to navigate a maze blindfolded, only to find out that the path is littered with dirty secrets.

To add another layer to this grand charade, energy suppliers resort to a clever trick called Renewable Energy Guarantee of Origin (REGO) certificates. These little pieces of paper are issued by renewable generators, but they are separate from the energy itself. So, what do the sly suppliers do? They buy these certificates for as little as 20p, without actually investing in the green power they claim to provide (IEA, 2014). It's a classic case of buying the label without the contents, like purchasing an empty box labeled "happiness."

As the spotlight has shone on this deceitful tactic, these cunning energy suppliers have devised a new plan. They turn their gaze towards the European Guarantees of Origin (GoO) certificates, seeking solace in the arms of continental trickery. Each EU member state must have a GoO scheme to prove their utilization of green power. But lo and behold, UK companies can bypass the REGO certification entirely and directly purchase certificates from member states. It's a labyrinth of deception, where one false trail leads to another (Wimmers et al., 2020).

But let us not confine our journey to the energy sector alone. Oh no, my friends, greenwashing knows no bounds. In the realm of automobiles, car manufacturers have mastered the art of misleading consumers with their claims of eco-friendliness. They parade their electric and hybrid vehicles as the epitome of green transportation, yet their reliance on fossil fuels remains steadfast.

These cunning manufacturers have been caught in the act, exaggerating fuel efficiency or manipulating emission testing to create an illusion of lower emissions. It's a theatrical performance, where smoke and mirrors take center stage, while the true impact on our environment remains obscured.

And what about those companies that claim to offset their carbon emissions through noble endeavors like reforestation or renewable energy projects? Beware, for their promises may be as thin as the morning mist. They may invest in projects that offer little or no environmental benefit, like a magician pulling a rabbit out of an empty hat.

For instance, a company may boast of offsetting its emissions by investing in a hydroelectric project. But if that project was already in the works, with funds secured and plans in motion, the company's investment adds no real value to our battle against climate change. It's like trying to douse a fire with an empty bucket.

So, dear audience, beware the alluring call of greenwashing in the energy and automotive industries. Let us pierce through the veil of deception, demand transparency, and support companies that genuinely walk the path of sustainability. Remember, it is our collective voice and conscious consumer choices that can shine a light on the true heroes of environmental responsibility.

Ah, the world of greenwashing expands its deceptive tendrils into various industries, leaving a trail of misleading claims and shattered illusions. Let us explore some notable examples cross different sectors, where the green façade conceals a darker reality.

In the automotive industry, we have witnessed some truly remarkable acts of deception. Take Volkswagen, for instance, who graced us with their "Clean Diesel" scandal. Oh, what a grand performance it was! They installed illegal software in their diesel cars to cheat emissions tests, all while promoting their technology as environmentally friendly (Hotten, 2015, ADAC, 2023; Traufetter, 2023; Zeit, 2015; Kell, 2022; Aurand et al., 2018). Bravo, Volkswagen, for your masterful manipulation of both machines and minds!

But let us not forget Ford, who graced our screens with their "Green" ads. They painted a picturesque image of hybrid and electric vehicles, conveniently omitting the carbon footprint that comes hand in hand with their battery production and disposal (Johnson, 2004; Briscoe, 2021/2023; Mogensen, 2018). Oh, the beauty of selective storytelling, where only half the truth is worthy of our attention!

And who can overlook Toyota, the master of hybrid deception? They dazzle us with their "Hybrid Synergy Drive" claims, presenting their vehicles as eco-friendly and efficient. But critics, ever the skeptics, raise concerns about the energy-intensive manufacturing process and the difficulties of battery recycling. It seems even hybrids have a touch of the smokescreen, my friends (Bleakley, 2023; The Guardian, 2023).

Now, let us turn our attention to the glamorous world of fashion, where greenwashing parades itself in a dazzling display of half-truths. H&M, the siren of sustainability, introduces us to their "Conscious Collection." They whisper sweet nothings of sustainability and environmental friendliness, while critics point out their continued reliance on non-renewable resources and the collection's mere token presence in their vast repertoire (Balch, 2013; Butler, 2022). Oh, applause to H&M for their masterful masquerade – truly, they are the epitome of virtue in the fashion charade.

But fear not, for Patagonia, the self-proclaimed guardian of ethics, joins the stage with their "Responsible" claims. Yet, beneath the surface, questions arise about their use of virgin materials and the opacity of their supply chain. Ah, the allure of the great outdoors, tarnished by the shadows of suspicion!

Nike, a household name in the realm of sports and fashion, graces us with their "Considered" line. They pledge allegiance to environmental friendliness and sustainability, while critics raise concerns about their continued reliance on non-renewable resources. Oh, Nike, your intentions may be good, but your execution leaves room for doubt (Kwasniewski, 2013; Balch, 2012).

But let us not forget the financial industry, where greenwashing is an art form. Wells Fargo, in a performance worthy of an Oscar, claims to invest in environmentally friendly projects while simultaneously pouring money into fossil fuels. It's like a magician performing an elaborate trick, distracting us with one hand while the other engages in a dance with the devil (Skinner, 2021; Hahn, 2021; Leavitt, 2021).

JPMorgan Chase, not to be outdone, pledges to stop financing certain fossil fuel projects, all while their overall record on climate change remains questionable. It's a masterstroke of selective investment, where they choose which fossil fuel projects to support, leaving us to question the true depth of their commitment (Skinner, 2021; Hahn, 2021; Leavitt, 2021).

And then we have BlackRock, the behemoth of the financial world, accused of playing the greenwashing game. They parade their "sustainable" funds, while critics argue they continue to invest heavily in fossil fuels (Willmroth, 2021). Oh, BlackRock, the guardian of our finances, or perhaps just another wolf in sheep's clothing?

But let us not forget the chemical industry, where greenwashing is as common as the scent of a chemical cocktail. BP, with their "Beyond Petroleum" campaign, promised a world beyond oil, only to find themselves knee-deep in one of the worst environmental disasters in history. It's like a tragic comedy, where irony and devastation dance hand in hand (Birnbaum et al., 2010; Fischer, 2023).

And who can overlook Monsanto, the champion of the "Green Revolution." They promised increased food production and reduced hunger, all while relying heavily on pesticides and fertilizers, causing environmental and health problems in their wake. It's a tale of good intentions gone awry, with consequences that stretch far beyond our fields (Pearce, 2009; Dr. Leu et al., 2023).

Now, as we traverse the terrain of greenwashing, let us not forget the wood and furniture industry, where green claims wilt in the face of scrutiny. Asia Pulp & Paper, once adorned with the title of "sustainability," found themselves accused of deforestation and illegal logging, tarnishing their green reputation. It's a story of forests sacrificed in the name of profits (Duff, 2011).

And who can overlook IKEA, the grand purveyor of affordable furniture, accused of purchasing wood from illegal and unsustainable sources in Russia. The environmental watchdogs raised their voices, demanding accountability, and stricter sourcing policies. Oh, IKEA, your furniture may be flat-packed, but your green promises have proven far more elusive (oekoreich, 2022; Schmidt et al., 2022).

The art of greenwashing knows no bounds! Let us venture into the realm of hospitality, travel, and hotels, where deception and green fantasies intertwine.

Marriott International, the maestro of vagueness, launched their "Environmentally Conscious Hotel" program. A round of applause, please! But wait, critics pointed out the lack of measurable goals or metrics, leaving us to wonder what "conscious" really means in Marriott's world (Caradonio, 2022; Marcus, 2021; Clawson, 2020). And oh, how delightful to discover their membership in the American Legislative Exchange Council (ALEC), a group known for denying climate change. A truly harmonious symphony of contradictory actions!

Next, we have Norwegian Cruise Line (NCL), sailing through the sea of greenwashing accusations. They promised us the latest and greatest in environmentally friendly technology, only to find themselves sued by environmental organizations. The allegations flew, claiming false and misleading claims about their environmental practices (Klawans, 2023). Oh, NCL, the irony of polluting the oceans while boasting of your green superiority!

But fear not, for Hilton Worldwide comes to the stage with their "Travel with Purpose" program. The curtain rises, and the claims of reducing environmental impact and promoting sustainability resound. But alas, the critics raise their voices once more, decrying the lack of concrete goals and metrics (Tark et al., 2021). And what's this? Hilton's membership in the International Air Transport Association (IATA), a group that opposes climate policies? Oh, the contradictions are as vast as the hotel chain itself!

Now, let us embark on a journey through the skies, where the airline industry spreads its wings of greenwashing. KLM, the Dutch airline, presented us with their "Flying Green" program, promising to offset CO_2 emissions and make flights "carbon neutral." Ah, what a noble cause! But alas, environmental groups like Greenpeace saw through the façade. They argued that offsetting emissions did not address the true problem: excessive CO_2 emissions from aviation. And let's not forget the selective focus on biofuels, conveniently ignoring other harmful emissions (ClientEarth, 2023; Sterling, 2023; Symons, 2023). Oh, KLM, your green flight may have hit some turbulence!

Now, let us quench our thirst with some examples from the beverage and food industry, where greenwashing flows like a polluted river.

Fiji Water, the exotic brand of bottled water, captured our attention with claims of environmental friendliness and carbon negativity. How refreshing! But critics swiftly pointed out the irony of transporting water from Fiji to various corners of the globe, leaving behind a substantial carbon footprint. And let's not forget the whispers of poor labor practices in Fiji, a sour note in the symphony of sustainability.

McDonald's, the fast-food giant, joined the greenwashing party with their "green" restaurants, promising a more environmentally friendly dining experience. Ah, the allure of greener pastures! But alas, the truth emerged, revealing that these new restaurants were not significantly more sustainable than their predecessors. And critics pointed out the company's role in deforestation and climate change, leaving us to ponder the true shade of their golden arches (McGuire, 2021; Perkins, 2021).

Coca-Cola, the titan of the beverage industry, swirled into the greenwashing scene with their "World Without Waste" campaign. Oh, what a splendid vision! But critics raised their voices, reminding us that plastic bottles often go unrecycled, littering landfills and oceans. And let's not forget their water usage and the exacerbation of water scarcity in various parts of the world. A tale of empty promises in a sea of sugar-laden beverages.

But wait, there's more! Let us delve into the mining industry, where greenwashing tactics dig deep into the earth.

Chevron Mining, accused by Earthworks, attempted to sell us the idea of a "green" molybdenum mine. But Earthworks swiftly exposed the mine's history of environmental violations and its impact on polluting the nearby Red River. Oh, the audacity of greenwashing in the depths of the mining world!

Rio Tinto, a mining giant, sought to preserve cultural heritage until they destroyed a 46,000-year-old Aboriginal sacred site. What a striking contradiction! Promoting preservation while erasing history. A harsh lesson in the art of greenwashing.

And lastly, Freeport-McMoRan, the Indonesian government's target, promised a new smelter to reduce pollution, only to be accused of greenwashing. The government claimed they used their environmental commitments to gain support while failing to fulfill their promises. A dance of empty words in the world of mining.

So, my astute audience, let us remember to peel back the layers of greenwashing across industries, for it is only through our skepticism and demand for truth that we can uncover the genuine shades of green and hold companies accountable for their deceitful performances.

Ah, the world of technology and digital infrastructure, where greenwashing thrives amidst the bits and bytes. Let us embark on a journey through the land of false claims and digital illusions.

Apple, the tech giant, graced us with their green energy claims. In 2012, they proudly declared that their data centers were powered by 100% renewable energy. What a shining beacon of sustainability! Or so we thought. Alas, it was later revealed that only a mere 16% of their data centers actually ran on renewable energy. Oh, the magic of greenwashing! Apple eventually made efforts to rectify their energy usage, but the initial exaggerations left a stain on their reputation.

Next, we have Amazon, the e-commerce titan, with their grand announcement of the "Shipment Zero" initiative. It promised to make all Amazon shipments net-zero carbon. How noble, how inspiring! But hold on, dear audience, for there were no details provided on how this goal would be achieved. And what's this? Many of their products were still packaged in non-recyclable materials, despite claims of using recycled materials. Oh, the irony of greenwashing, where words speak louder than actions!

And let us not forget Google, the master of carbon neutrality claims. In 2019, they proudly proclaimed that all their data centers and offices had achieved carbon neutrality. How commendable! But wait, a controversy emerged as critics pointed out their heavy reliance on carbon offsets. These offsets merely shifted the carbon emissions elsewhere, without actually reducing them. Oh, the illusions of greenwashing, where smoke and mirrors replace genuine environmental action!

So, my enlightened audience, let us not be swayed by the glowing screens and captivating marketing of the digital world. Behind the sleek designs and innovative technologies, greenwashing may lurk, trying to deceive us. It is up to us to demand transparency, to hold these digital giants accountable, and to seek genuine efforts in reducing their environmental impact. Only then can we separate the pixels of greenwashing from the true shades of sustainability.

II. Biomass: Burning Our Way To A Greener Future?

In the realm of renewable energy, one source has captured both attention and controversy: biomass power or bioenergy. The promise of using organic materials such as wood pellets as a carbon-neutral and renewable fuel has attracted investors, policymakers, and environmentalists alike. Yet, beneath the surface, a heated debate rages on about whether biomass power truly lives up to its claims of being climate-friendly and aligned with Environmental, Social, and Governance (ESG) principles. In this narrative, we delve into the intricacies of this contentious topic, exploring the misleading nature of biomass power as a climate-friendly investment and proposing alternative approaches to ensure a truly sustainable and low-carbon energy future.

Enter the stage, the Potsdam Institute for Climate Impact Research (PIK), whose recent study has thrown a spotlight on the environmental impact of bioenergy. While their study carries weight and significance, it is imperative that we subject their findings and arguments to thorough scrutiny. Only through critical analysis can we unravel the complexity and true nature of biomass power's climate credentials.

The study, first and foremost, highlights the crucial role of land-use regulations in determining the climate compatibility of bioenergy. Unregulated land-use practices, such as large-scale deforestation to make way for biomass cultivation, can paradoxically result in higher carbon dioxide (CO_2) emissions than burning fossil fuels themselves. The implications are clear: sustainable land-use policies must be diligently implemented to mitigate the negative environmental consequences associated with bioenergy production.

Beyond direct environmental impacts, the study delves into the web of indirect effects spawned by bioenergy production. These ripple effects can be far-reaching, encompassing the displacement of food production and the encroachment of agricultural activities into natural areas. In a globally interconnected market, where the dynamics of food and bioenergy are deeply intertwined, relying solely on national regulations may prove insufficient to effectively control these indirect effects. A comprehensive approach, one that takes into account various factors, policy frameworks, and sustainability considerations, is paramount to successfully address these challenges and minimize adverse impacts.

An essential element emphasized by the researchers is the necessity of a pricing mechanism for emissions resulting from land-use changes. Without such carbon pricing, bioenergy risks becoming as environmentally damaging as its fossil fuel counterparts. By internalizing the environmental costs associated with different energy sources, a pricing system for emissions from land-use changes can promote sustainability and guide investment decisions toward genuinely climate-friendly alternatives.

To mitigate the high emissions resulting from land-use changes linked to energy crop production, the study posits the need for a global and comprehensive management system. This holistic approach entails considering diverse elements, including both direct and indirect effects, policy frameworks, and the overall sustainability of bioenergy production. By adopting a systematic approach, we can begin to disentangle the intricate complexities of biomass power and pave the way for a more sustainable energy future.

However, it is crucial to acknowledge that the study's findings represent a single perspective. As with any research, further analysis and investigation are required to comprehensively assess the validity and implications of these claims. While the study raises valid concerns about the potential environmental risks of bioenergy, it does not inherently brand bioenergy as a fraudulent concept. Instead, it underscores the need for careful planning, sustainable practices, and effective policies to ensure that bioenergy production genuinely contributes to the transition toward a cleaner and more sustainable energy system.

To chart a path forward for bioenergy that aligns with climate friendly and ESG principles, alternative measures can be considered:

Firstly, implementing stringent land-use regulations to prevent deforestation and protect natural eco-systems is of paramount importance. Clear guidelines should be established to restrict the expansion of bioenergy crops into environmentally sensitive areas or agricultural lands crucial for food production. This approach will help preserve biodiversity, protect water resources, and safeguard local communities reliant on these eco-systems.

Secondly, incorporating robust sustainability criteria for biomass sourcing is imperative. Biomass should only originate from well-managed and certified sources, such as sustainably managed forests, agricultural residues, or energy crops cultivated on degraded or marginal lands. Implementing rigorous certification systems can verify compliance with sustainable sourcing standards and ensure that bioenergy production is genuinely environmentally responsible.

Thirdly, conducting comprehensive lifecycle assessments of bioenergy production is essential. These assessments should go beyond direct emissions from combustion and include the full spectrum of indirect emissions associated with land-use change, transportation, and processing. Developing transparent and standardized methodologies for conducting lifecycle assessments will facilitate accurate accounting and informed decision-making.

Lastly, embedding strong social safeguards into bioenergy projects is critical. Local communities should be actively involved in decision-making processes, ensuring their rights, livelihoods, and well-being are prioritized. Robust social impact assessments should be conducted to evaluate the potential socio-economic effects of bioenergy projects and ensure equitable outcomes for all stakeholders involved.

In the world of renewable energy, where the stakes are high and the consequences profound, it is imperative that we critically evaluate the claims and implications of bioenergy. While bioenergy holds promise as a renewable energy source, the risks associated with its production and the potential for greenwashing cannot be ignored. By adopting a holistic approach that incorporates sustainable land-use practices, carbon pricing mechanisms, and comprehensive assessments, we can navigate the complexities and unlock the true potential of biomass power as a viable and sustainable energy solution. Only through responsible action can we ensure that our transition to a low-carbon future remains both genuine and impactful.

III. Solar Power: Harnessing The Sun Or Exploiting The Eco-Systems?

Imagine the journey of turning sunlight into electricity as a magical dance that happens in a secret workshop. At the heart of this workshop, we find a special ingredient called "silica," which is like the main ingredient for a recipe. Silica is extracted from deep within the Earth, a bit like treasure being dug up from a hidden chest.

Now, in this workshop, there are skilled artists who know how to use the silica to create something extraordinary – solar modules. These solar modules are like powerful collectors that catch sunlight and turn it into energy. But creating them is not as simple as snapping your fingers; it's more like crafting a beautiful piece of art.

The artists start by refining the silica, just like a chef prepares the best ingredients before cooking. This refining process takes some energy, a bit like using heat to cook food. Once the silica is ready, it's shaped into tiny wafers, which are like the building blocks of our solar modules.

Now comes the fascinating part: doping. This is when the wafers are treated with special elements, similar to adding different colors to a painting. These elements give the wafers special abilities to capture sunlight and turn it into electricity. It's like giving the wafers their magical powers.

After doping, the wafers go through another dance – etching and cleaning. This step is like giving the wafers a refreshing bath to make sure they're sparkling clean and ready for their mission.

Next, the wafers are carefully assembled into a strong structure, a bit like putting together the pieces of a puzzle. This structure becomes the solar module that we see on rooftops or in solar farms. Just like a superhero's cape, the module is designed to catch as much sunlight as possible.

Once the solar module is complete, it's like a superhero being ready for action. It's shipped to different places, like a knight going on a quest to save the day. When the module finally reaches a rooftop or a solar farm, it starts its noble duty – turning sunlight into electricity that can power homes and businesses.

So, from the depths of the Earth to the dance of chemicals and manufacturing, this incredible journey transforms ordinary silica into powerful solar modules, bringing the magic of sunlight to our everyday lives. Just like a beautiful dance, it's a harmonious process that creates something truly extraordinary.

As we prepare to delve into the intricate environmental maze that lies at the core of the solar industry's ambitions, let's set off on a systematic voyage – one that doesn't commence with the dazzling allure of solar panels or the pledge of limitless pristine energy. Instead, our journey begins with an unassuming element: silica. This seemingly small ingredient, often overlooked, holds a position of significance within the realm of photovoltaic excellence. It's more than just a basic component; it acts as a vital clue in unraveling the complex puzzle of ethical mining practices linked to its extraction.

Amid the realm of silica mining lies a tapestry woven with intricate environmental challenges. As we peer into this landscape, we encounter a series of complexities that dance in harmony with the pursuit of this valuable resource.

In the heart of silica mining, the disruption of habitats and land unfolds as a consequence. Mining operations for silica set in motion a sequence of events that alter the very fabric of the land. Habitats are dismantled, forests silently disappear, and entire landscapes transform. These changes disrupt the delicate balance of local eco-systems, leaving a mark that echoes through the natural world.

As silica is extracted, airborne dust and water tinctures emerge, casting a shadow over both nearby communities and the dedicated workers. The process of mining awakens airborne particles and dust, becoming a silent menace that cloaks the surroundings. Moreover, the chemicals invoked during silica's extraction find their way into the water, casting it with a tainted hue that affects aquatic life and the sustenance it provides.

The dance of silica extraction demands a partnership with carbon-intensive energy sources, adding to the industry's carbon footprint. This energy choreography contributes to the concerns of climate change, intertwining the act of mining with the broader narrative of a warming world.

In the thirst for silica, water becomes a precious commodity. Mining's consumption of water, while necessary, can strain local water sources. In regions where water is scarce, this thirst can usher in a scarcity that sends ripples of concern through communities and eco-systems alike.

Beyond the mining sites lie the communities whose lives intersect with this pursuit. Silica extraction can upend these societies, causing shifts in traditions and livelihoods. The rhythms of life are disrupted, sometimes leading to a discordant note in the symphony of local existence.

As mining scripts its tale, erosion emerges as a central theme. Stability crumbles beneath the weight of operations, leading to sediment-laden runoff that finds its way into nearby waters. This runoff taints the waters and alters the stories of aquatic life.

The final act of mining is a ballet of reclamation, a promise to restore what has been altered. This ballet is intricate, encompassing the revival of eco-systems and the healing of land. It's a promise, a pledge to mend what has been torn. Really?

Silica mining, often accompanied by other mining endeavors, paints a canvas of cumulative impacts. The sum of these actions intensifies resource use, elevates pollution, and creates a landscape where managing the environmental aftermath becomes a more intricate task.

Silica, a silent player in the modern world, dances a complex ballet that blends resource extraction with environmental responsibility. As we navigate this dance, we find ourselves at the crossroads of advancement and conservation, learning the steps to harmonize our pursuit of progress with the preservation of the planet.

In the intricate choreography of solar panel manufacturing, a pivotal act unfolds – a chemical ballet that orchestrates the creation of polysilicon. This compound, derived from silica, forms the bedrock of photovoltaic prowess. Yet, as the curtains rise on this chemical stage, a shadowy subplot emerges – one that casts a discerning light on the risks poised for climate change.

Picture a sprawling industrial expanse, where furnaces roar to life, and chemical reactions commence. Here, silica undergoes a transformation akin to alchemy. High temperatures and energy-intensive processes combine to strip away impurities, leaving behind pure polysilicon. But this theatrical transformation carries a profound carbon cost – cost that reverberates through the atmospheric tapestry.

As fossil fuels unleash their energy to fuel these fiery furnaces, a symphony of greenhouse gases (GHGs) takes flight. Carbon dioxide and other emissions ascend, joining the swirling currents of our atmosphere. This chemical crescendo, while invisible to the eye, weaves a tangible impact – a warming embrace that tightens its grip on the delicate balance of our planet.

The risks are clear. The production of polysilicon, a cornerstone of solar panel creation, dances with carbon emissions, contributing to the very climate change that the solar industry strives to mitigate. The energy-hungry nature of this chemical endeavor adds a somber note to the symphony of clean energy – a reminder that even in the pursuit of sustainability, the echoes of climate repercussions persist.

Yet, as the world seeks the grace of renewable energy, this chemical narrative need not play out as a tragedy. Innovations in cleaner production methods, a shift towards renewable energy sources, and heightened efficiency can compose a different score – one where the carbon footprint of polysilicon production becomes a faint whisper rather than a resounding crescendo. The challenge lies in harmonizing the needs of solar panel creation with the imperatives of climate stewardship, rewriting this chemical tale with an ending that sings of environmental resilience.

This intricate narrative is woven with real-world cases, each serving as a cautionary tale etched into the fabric of progress. Enter China, a central actor in the drama of silica production. Here, the relentless pursuit of polysilicon – a linchpin of solar panel creation – unfurls a tale riddled with environmental strife. Baotou, a city once synonymous with silicon production, now wears the scars of its own success – severe air and water pollution stand as haunting remnants of the chemical alchemy that fueled its ambitions (Bontron, 2012). A stark reminder that the pursuit of progress can cast a long shadow over the canvas of ecological well-being.

In the world of solar panel manufacturing, where innovation dances hand in hand with the captivating theater of solar panel production, where visions of cleaner energy take the stage, a subtle yet profound actor awaits its cue – the carbon footprint. This understated presence, a result of intricate energy dynamics and material choices, weaves a complex narrative that underlies the grand performance of crafting solar panels.

Quantifying the exact carbon footprint in this production is akin to deciphering an intricate puzzle, with numbers swaying in a spectrum of possibilities. This enigmatic dance is choreographed by variables such as energy sources, production efficiency, and the evolving landscape of the industry. Amidst this dance, benchmarks stand tall, offering glimpses into the intricate web of carbon emissions.

The saga commences with the stalwarts of the solar panel domain – crystalline silicon-based panels. Their carbon footprint, a combination of direct emissions from manufacturing and indirect emissions intertwined with energy consumption, usually ranges between 20 to 50 grams of carbon dioxide equivalent per kilowatt-hour (gCO_2e/kWh). It's a delicate tango of emissions that paints a portrait of their environmental impact.

Thin-film solar panels, a different troupe in the production ensemble, don a slightly heavier carbon attire. Typically, their carbon footprint sways between 40 to 70 grams of carbon dioxide equivalent per kilowatt-hour (gCO_2e/kWh). This dance introduces a layer of complexity, beckoning for a comparative waltz.

In the theater of comparisons, other industries take the stage, showcasing their own carbon finery. Electric vehicles (EVs) step forward, emitting around 50 to 200 grams of carbon dioxide equivalent per kilometer (gCO_2e/km) driven. It's a multifaceted performance, incorporating the production of the vehicle, battery, and fuel source. As the spotlight shifts, an unexpected twist emerges: the carbon footprint of EV manufacturing per unit often rivals or surpasses that of solar panel production.

The aviation sector then takes its place, soaring high on wings fueled by fossil energy. Airlines emit substantial carbon dioxide equivalent per passenger-kilometer, their carbon footprint soaring above that of solar panel creation.

In the backdrop, heavy industries like steel, cement, and chemicals make a bold entrance, releasing carbon emissions in a display more intense than solar panel production.

And so, the symphony of carbon emissions in solar panel manufacturing unfolds – dance of energy, materials, and comparisons that unveils a multifaceted narrative. As the solar industry navigates its way forward, these insights shape the script for a more sustainable future – a future poised at the intersection of innovation and environmental stewardship.

In the vast landscape of solar energy's promise, where panels capture sunlight and generate clean power, a looming predicament often escapes the spotlight – the intricate dance of waste disposal. As solar panels mature and eventually reach the end of their operational life, the question of where they go next becomes a pressing concern. This is a tale of climate-conscious disposal, a backstage drama that reveals the environmental threads woven into the solar industry's narrative.

Picture this: solar panels, once hailed as emissaries of sustainability, now face a crossroads at the end of their illustrious careers. The sun-soaked days of energy generation come to a close, and the panels must retire. Yet, the story doesn't end there – it continues in the realm of waste management, where the choices made now ripple through time, affecting the delicate balance of our climate.

As solar panels reach their twilight years, the curtain call leads them to waste disposal facilities. The challenge lies not only in finding the right stage for their final act but also in ensuring that this act aligns with the symphony of ecological harmony. Enter the spotlight: recycling and proper disposal.

Recycling, a virtuous endeavor in theory, requires careful choreography. While recycling solar panels can recapture valuable materials and reduce the need for new resource extraction, the dance isn't without its complexities. The process demands specialized techniques, facilities, and know-how, all of which carry their own environmental footprint.

But what happens when recycling falters, when the disposal script takes an unplanned twist? The answer beckons us to a darker narrative – one where panels find them-selves discarded in landfills, a final resting place that comes at a high cost. These landfills become unintended incubators of waste, releasing greenhouse gases (GHGs) like methane into the atmosphere, silently contributing to the very climate challenge that solar energy seeks to alleviate.

This tale of waste disposal isn't confined to a single act – it extends through time. The consequences of today's choices echo into the future, influencing our climate trajectory. A poorly managed end-of-life scenario can unravel the efforts made in solar panel production, casting a shadow over the industry's environmental aspirations.

Amid the complexities of waste disposal, the solar industry stands at a crossroads once again. The decisions made now will shape the narrative of its commitment to a greener future. Will it embrace recycling and responsible disposal, crafting an encore that resonates with ecological harmony? Or will it falter, leaving a legacy that tarnishes its claims of environmental stewardship?

As the final scenes of this climate-conscious drama unfold, the industry's choices become the script for a more sustainable encore – a tale that transcends solar panels' lifetime and touches the very fabric of our planet's well-being.

IV. Electric Vehicles: Fueling Our Delusions?

In the annals of German politics, the name Armin Laschet shall forever be etched as the unfortunate protagonist of the 2022 Christ Democrats' bid to ascend the grand throne of federal governance. Now, one must not be mistaken, for Armin Laschet was no political novice, having wielded the scepter of prime minister in the vast dominion of North Rhine Westphalia. But alas, the merciless hands of fate would conspire against him, revealing the pitiless nature of politics, where events can conspire to work against you, no matter your competence.

Some may argue that this gentleman, Armin, was not a man bereft of merit, but he was no "Prince," to borrow the words of Niccolò Machiavelli. Perhaps possessing the skills, yet lacking the elusive hand of fortune. Alas, his every endeavor to regain the upper hand in the race was rendered futile, much like the day he hosted a grandiose press and PR spectacle, graced by none other than the enigmatic entrepreneur of our age, Elon Musk, at his very own Giga factory.

In that moment, Armin must have felt the stars aligning in his favor, with the prospect of showcasing his competence in front of the camera. Alas, his ill-fated question posed to Elon Musk would mark the inception of his undoing. A seemingly innocent inquiry, "What do you think will win, Fuel Cell/Hydrogen powered car or EV?" left Elon Musk bemused, his laughter captured by the relentless lenses of the German TV.

And thus, the interpretation unfurled, that Elon's mirth was aimed at Armin's expense, mocking his seemingly ridiculous question. But I, as an observer, can't help but ponder: was the question truly so preposterous? For in the minds of many, the debate persists, despite the ever-rising popularity of electric vehicles. The path of sustainability embarked upon raises questions that demand our attention and discourse.

Truly, electric vehicles stand as the venerated icons of the green transportation revolution. But let us not be deluded by the allure of their virtuous sheen. Behind the scenes lies the stark reality that the production of these electric wonders demands an insatiable appetite for resources, including rare minerals and metals, acquired through practices that lay waste to our precious environment (Arvesen et al., 2011).

And, in a bitter irony, the very electricity that empowers these eco-champions often emanates from grids tethered to the finite reserves of fossil fuels. The aspirations of saving the planet thus clash with the unyielding grasp of reality.

Oh, the irony of it all! As Armin Laschet found himself cast aside, so too do we face the paradox of green ambitions entangled in the quagmire of modern necessities. The questions arise, demanding answers we must confront if we are to truly steer our course toward a future where sustainability reigns supreme.

In a world where environmental consciousness was on the rise, and governments seemed eager to adopt clean transportation alternatives, a new hero emerged on the scene – the Electric Vehicle (EV)! It was hailed as the savior of the planet, the antidote to climate change, and the ultimate solution to our fossil fuel addiction. With zero tailpipe emissions and the promise of a greener tomorrow, the EV became the darling of the masses.

"Oh, how wonderful it is!" the proponents of EVs cried out. "Governments are showering us with incentives and tax breaks, making it financially viable for everyone to join the electric revolution!" They felt reassured by the advancements in battery technology, praising the increased range and faster charging times that seemed to address the initial hurdles.

But little did they know, dear reader, that behind this shiny façade of promise and hope lay a web of disillusion and deception. The EV world, it seemed, was not the utopia it was made out to be (Balch, 2020).

The opposing viewpoint, skeptics and critics, were quick to raise their voices. "Hold your horses!" they exclaimed. "Don't be so quick to crown the EV as the king of eco-friendliness." They were convinced that the production of EVs came at a steep environmental cost. The extraction of rare earth metals like lithium and cobalt for batteries left scars on the delicate eco-system, and the manufacturing process guzzled energy and contributed to carbon emissions.

Amidst the fervor of technological innovation and the race for greener mobility, a curious paradox emerges – a tale of unintended consequences in the realm of Electric Vehicles (EVs). Beneath the veneer of progress lies a narrative that unfolds in the shadows, a story intricately woven with environmental intricacies.

The journey begins with the coveted protagonists – rare earth metals, the lifeblood of EV batteries. Lithium, celebrated for its pivotal role in storing energy, becomes the focal point of our tale. Extracted through methods ranging from open-pit mines to brine evaporation ponds, this mineral's quest leaves indelible marks on landscapes. Scarred terrains bear witness to the pursuit of energy solutions, a reminder that the path to a cleaner future can come at a cost to the present.

But the saga doesn't end there; cobalt takes center stage. Nestled within the heart of batteries, cobalt is a character mired in complexity. Its supply chain traverses landscapes of uncertainty, often intertwined with regions burdened by socio-political fragility. The extraction of this element, while essential for EV propulsion, casts a shadow that stretches far beyond the road ahead. Ethical quandaries and humanitarian concerns become part of the narrative, revealing the intricate trade-offs inherent in the transition to cleaner transportation.

As our tale unfolds, the manufacturing process steps into the spotlight – a stage where energy meets ambition. The assembly lines hum with activity, each weld and fastening echoing with progress. Yet, this very rhythm consumes copious amounts of energy, a demand that resonates in the atmosphere as carbon emissions. The very creation of EVs, while a step toward reducing emissions on the road, paradoxically contributes to the carbon footprint during their inception.

In the grand scheme of climate protection, the thesis resounds with clarity – the production of EVs does not emerge unscathed from its noble intentions. The extraction of rare earth metals reverberates in the form of ecological scars, a reminder that each stride towards progress is inscribed upon the Earth's canvas. The manufacturing process, for all its promise of cleaner mobility, leaves an indelible carbon footprint – an exchange of emissions in the pursuit of emission reduction.

This is the canvas upon which the narrative unfolds, a tale that poses profound questions about the cost of environmental progress. As the world steers towards a future of sustainable transportation, it navigates through a landscape where innovation and its ecological consequences are entwined. In the intricate dance of technology and nature, the production of EVs becomes a chapter etched with both hope and caution – a journey that beckons for solutions that tread lightly upon the delicate eco-system and forge a path towards genuine climate protection.

And oh, the battery disposal and recycling challenges! The critics wagged their fingers in warning. The lack of proper infrastructure for recycling and handling end-of-life batteries posed a ticking time bomb for hazardous waste accumulation.

The saga commences with a pressing concern – the fate of batteries once their tireless service on the road comes to an end. As EV adoption surges, so does the accumulation of spent batteries, poised to form a silent mountain of discarded potential. Critics, like sentinels of caution, raise their voices, warning of the imminent crisis that looms. The lack of infrastructure, the absence of well-trodden paths for recycling and handling, paints a dire picture – a ticking time bomb ready to detonate into a hazardous waste conundrum (Howell, 2023).

The evidence of this challenge is strewn across the landscape, like clues waiting to be deciphered. In the United States, a nation at the forefront of the EV revolution, recycling capacity for lithium-ion batteries lags far behind the soaring tide of EV adoption. With estimates that the number of retired EV batteries could reach millions in the coming years, the ticking of the clock grows louder, urging action to avert an environmental reckoning.

Consider the case of Europe, a continent that has championed green ideals and electrification. As EV fleets proliferate, the urgency to address the battery aftermath intensifies. In a race against time, the European Union is striving to establish a regulatory framework to manage the mounting tide of used batteries. The realization dawns that proper recycling and disposal are not merely matters of convenience – they are cornerstones of ecological stewardship, preserving the gains made in emissions reduction.

And then, there's China – a global powerhouse in EV production. As the conveyor belts of factories churn out electric marvels, the specter of waste management hovers. In the quest for sustainable solutions, China grapples with a patchwork of local approaches to battery disposal, a challenge that demands a harmonious symphony of action to address the impending crisis.

This isn't just a narrative of doom and gloom – it's a call to confront reality and weave a new tapestry of solutions. The implications of inadequate battery disposal and recycling are far-reaching, stretching beyond the horizon of the present into a future where environmental responsibility reigns supreme. The complexities are many – the myriad chemical components, the intricate processes, the economic considerations – all conspiring to create a Gordian knot that beckons for untangling.

The thesis stands tall – the battery disposal and recycling challenges are not mere conjecture; they are the throbbing pulse of an issue demanding attention. With every discarded battery, an opportunity to minimize waste and harness valuable resources is lost. The critics' fingers wag in warning, not in condemnation, but in an urgent plea to heed the call. The impending hazardous waste accumulation, like a time bomb, carries the weight of the future – a future that hangs in balance as the quest for sustainable solutions takes center stage in the electrified drama.

The EV enthusiasts were told to think about the bigger picture. Sure, the operational phase of EVs might produce zero emissions, but what about the whole lifecycle? A comprehensive life cycle analysis was needed, they were told, to understand the true environmental impact. From raw material extraction to manufacturing and eventual disposal, the EV journey wasn't as pristine as it appeared (Renault, 2019).

When invoking the notion of a comprehensive life cycle analysis, a lens through which the true environmental impact of EVs would come into focus. Yes, the operational phase promised zero emissions, a tantalizing prospect in the battle against climate change. But what of the hidden threads that wove through the fabric of an EV's existence? From the very genesis of raw materials to the final curtain call of disposal, the EV's voyage revealed complexities that required unwavering attention.

The journey commenced in the heart of the Earth, where minerals like lithium and cobalt, coveted for their role in batteries, were extracted from the ground. Here, the environmental cost was unveiled – a toll paid in terms of disrupted eco-systems, scarred landscapes, and in some cases, precarious working conditions. The pristine image of an EV, gliding silently down the road, was refracted through the lens of mineral extraction, reminding us that even green dreams were rooted in the earth's enduring reality.

Across continents, factories hummed with activity, transforming these raw materials into gleaming battery packs and sophisticated electric drivetrains. The manufacturing process danced with energy consumption, as production lines whirred to life, consuming resources and expelling emissions. The intricate ballet of assembly unveiled a conundrum – while EVs promised a cleaner future, the production phase demanded energy and resources, an intricate balance between progress and its environmental shadow.

But the story didn't end there. As EVs reached the end of their road, questions emerged about their final act – the grand finale of disposal. Battery packs, laden with precious materials, posed a challenge that couldn't be ignored. Without proper recycling infrastructure, the risk of hazardous waste accumulation loomed, a potential consequence of embracing a technology with environmental blind spots.

In the midst of these revelations, a tapestry of facts emerged, each thread weaving a narrative of the EV's broader impact. A comprehensive life cycle analysis offered insights into the intricate dance between emissions reductions and environmental trade-offs. Studies illuminated the energy demands of raw material extraction, exposing the environmental strain that accompanied the pursuit of green mobility. Instances of improper disposal and recycling underscored the need for infrastructure to keep pace with innovation.

In the pursuit of climate protection, the enthusiasts and skeptics found themselves united by a common thread – an unwavering commitment to understanding the full spectrum of consequences. The EV's journey, from aspirations of emission-free highways to the complexities of its life cycle, embodied a pivotal lesson: In the realm of environmental progress, every step forward requires a thorough accounting of the steps that follow, a reminder that the path to a greener future is illuminated not just by the promise of innovation, but by a clear-eyed examination of the bigger picture.

In the midst of the electrified frenzy, where the wheels of progress raced towards electric vehicles (EVs), a subplot quietly unfolded – a narrative of resource dependency that stirred whispers of concern. The world, caught in the fever of EV enthusiasm, seemingly overlooked a crucial aspect – the finite nature of the minerals essential for the very heart of these electric marvels. A question loomed: what if these prized resources ran dry, triggering a symphony of scarcity and geopolitical unease?

To understand this unfolding drama, one must delve into the realm of lithium – element that pulses at the core of modern EV batteries. As if a magician's trick, lithium-ion batteries hold the power to propel EVs with impressive range and vigor, making the shift from combustion engines seem almost magical. Yet, behind this enchantment lies a reality that cannot be ignored – a reality shaped by the availability of lithium.

Consider the numbers that paint this portrait of dependency. Presently, the global reserves of lithium are estimated to be around 80 million metric tons. To contextualize this figure, let's envision the potential: these reserves have the capacity to power approximately 1.6 billion electric vehicles – almost a quarter of the total number of cars on the road today.

However, the trick lies in the fine print – the devil, as they say, is in the details. When we examine this reserve's capability relative to the broader landscape of transportation, a telling percentage emerges. In the face of roughly 1.4 billion fossil-fueled cars coursing through global streets, the reserves of lithium could fuel a substantial portion of the EV revolution. But there's a caveat – a glaring reminder of finite resources. This vast reservoir of lithium, capable as it may be, represents only a fraction of the entire automotive fleet, a stark indicator of the uphill journey towards complete EV dominance.

As if orchestrating a geopolitical ballet, the world's nations have begun jockeying for position in the lithium-rich domains. Bolivia, with its Salar de Uyuni salt flats, holds an estimated 9 million metric tons of lithium resources, a potential source of both power and tension. Other players, like Australia and Chile, also play their part in this resource-driven saga, as countries seek to secure their foothold in a future shaped by electric mobility.

The thesis is woven into this intricate tapestry – the concerns about dependency on finite resources are not mere conjecture; they are a reflection of a reality that demands thoughtful consideration. The allure of electric vehicles, while undeniably powerful, must contend with the constraints of limited resources. The dance of geopolitics, as nations vie for access to these precious elements, adds another layer of complexity to an already intricate narrative.

As we chart the course towards a cleaner, electrified future, the challenge of resource scarcity beckons – an emblem of the larger puzzle that climate protection presents. The electrification revolution, while rife with potential, must be navigated with foresight and balance, ensuring that the path we tread leads not only to emissions reduction but to a sustainable and resilient future (NYTimes, 2022).

Ah, the infrastructure enigma! As the world's gaze turned towards electric vehicles (EVs), a looming cloud of uncertainty swirled around the sprawling landscape of charging infrastructure. "Dream big," echoed the proponents of electrification. Yet, a chorus of skeptics emerged, their heads shaking in doubt, and a simple question hung heavy in the air: Could these grand visions truly take root and flourish? The answer rested upon a delicate equilibrium – a dance between ambition and feasibility, where the expansion of charging networks stood at the forefront of concern.

The heart of the matter beat with the rhythm of energy – specifically, the colossal task of expanding the grid's capacity to accommodate the demands of a burgeoning EV fleet. To unearth the scale of this challenge, we must turn our attention to the figures that underscore this narrative.

Consider the staggering numbers that illuminate the path ahead. Today's fossil-fueled cars, numbering around 1.4 billion globally, collectively devour a prodigious amount of energy – about 25 terawatt-hours (TWh) of electricity per day, to be precise. To envision the Herculean feat of transition, we must ask: How many renewable power plants would be needed to satiate this voracious energy appetite?

In the bustling realm of Europe, where cobbled streets meet modern thoroughfares, the scenario unfurls. The existing fleet of fossil cars yearns for electrification, and so, the task at hand becomes apparent. To replicate the energy output of these internal combustion engines, Europe would need to commission an astonishing 1,000 gigawatts (GW) of new renewable power capacity – a behemoth of energy generation that spans wind, solar, and other renewable sources. The scale is staggering, demanding investments of epic proportions and a harmonious symphony of technological innovation.

Crossing the Atlantic to the land of stars and stripes, a similar tableau unfolds. The United States, with its sprawling landscapes and diverse cities, faces an even more imposing challenge. To power its existing fossil-powered fleet with renewable electricity, the nation would need an awe-inspiring 2,000 gigawatts (GW) of new renewable power capacity. This Herculean endeavor would not only test technological prowess but also strain existing resources and potentially impact energy prices.

And what of the costs entwined within this sweeping transformation? In the United States, the price tag for these new power plants could tally anywhere between $2 trillion to $ 4 trillion, a staggering sum that underscores the monumental nature of the task. It's an investment that not only seeks to reshape the energy landscape but also reverberates through economic corridors.

As the ambitions of EV enthusiasts mingle with the practicalities of energy infrastructure, the concern for potential blackouts or brownouts becomes a palpable undercurrent. The scale of renewable energy generation required to sustain such a transition could, if not meticulously managed, pose risks to grid stability. An ambitious surge in energy demand, coupled with the integration of intermittent renewable sources, could lead to energy supply challenges – a dance with darkness that no one desires.

In the realm of climate protection, where the goal is to usher in an era of clean mobility and reduced emissions, the journey is far from straightforward. The infrastructure limitations, like a shadow on the path, compel a nuanced perspective. The vision of a future where EVs grace every corner of our roads is within grasp, but it hinges upon a complex choreography – a harmonious interplay of investment, innovation, and sustainable energy generation. In this symphony of transition, the melody of electrification must harmonize with the rhythms of grid resilience, ensuring that as the world powers forward, it does so with a balance that preserves both progress and stability.

And so, dear reader, as the debate raged on, it became clear that Voltaire's famous quote, "All is for the best in the best of possible worlds," could not be applied to the EV world. While there were glimpses of hope and progress, the reality was far from perfect. The EV, once considered the champion of the environment, had its fair share of challenges and unmet promises.

So, as the EV story continues to unfold, it is essential to tread carefully, acknowledging both the benefits and drawbacks, embracing balanced discussions, and working towards sustainable solutions. For in this ever-changing world, where illusions abound, only a clear-eyed view can lead us closer to a truly cleaner and greener future.

In attempting to do justice to the enigmatic figure of Armin Laschet, one finds oneself entangled in a web of profundity and meaning, a maze of challenges that seem to elude easy dissection. Without a doubt, his question, delivered seemingly without a clear sense of its implications, strikes at the very core of contemporary discourse. And so, to dismiss it with the mirthful laughter of pundits seeking fleeting points would be an act of intellectual folly.

Indeed, the true answer to this enigma remains elusive, and the verdict of history is yet to be rendered. Like a captivating match of football, the outcome remains uncertain until the final whistle. As Germany's sage sporting wisdom reminds us, "the ball is round, and the game lasts 90 minutes," an elegant metaphor for the unpredictable nature of competition.

In a world where electric vehicles (EVs) vie for supremacy against fuel cell cars, a clash of technological titans unfolds. A battle of innovation and sustainability, where both contenders seek to stake their claim as the best. As the duel intensifies, it becomes apparent that this is no trivial skirmish, but a defining moment in the annals of transportation history.

And so, dear reader, let us embrace the uncertainty and intricacy of this unfolding saga. For only time will reveal the ultimate victor, and in this captivating contest, may the true essence of progress and environmental stewardship prevail. Until then, let us navigate the complexities with a discerning eye and an appreciation for the mysteries that lie ahead.

As we stand at this crossroads, where the electrifying potential of electric vehicles (EVs) converges with the intricate web of environmental complexities, a nuanced verdict emerges – a verdict that balances the scales of promise and challenge, of progress and caution.

The EV's allure as a clean and sustainable mobility solution is undeniable. Its operation promises to paint our roads with emissions-free travel, a beacon of hope in a world grappling with the shadows of climate change. The zero-emissions journey that an EV embarks upon offers a glimmer of optimism, an embrace of innovation that aligns with the urgent call for decarbonization.

Yet, the narrative takes a more intricate turn as we venture beyond the surface. The production phase, a harmonious symphony of engineering marvels, reveals the footprint of resource extraction and energy consumption, shadows that temper the glow of environmental purity. The very minerals that power EVs leave ecological scars in their wake, raising questions about the true cost of this green transition. A broader lens unveils a web of interconnections – the manufacturing dance, the disposal dilemma, and the infrastructure constraints – that underscore the complexity of achieving a truly sustainable EV eco-system.

It is a dichotomy that demands an informed perspective, one that weighs the benefits against the challenges, the vision against the realities. The fate of the EV rests not on the shoulders of a single factor, but on the collective responsibility to navigate a terrain where technological promise intersects with environmental stewardship.

As we embrace the EV's promise, we must do so with an unwavering commitment to understanding its full lifecycle impact – a commitment that spurs innovation in sustainable resource extraction, efficient manufacturing, and responsible recycling. Infrastructure must be fortified to accommodate the surge in demand, and a strategic shift towards renewable energy sources must accompany the rise of EVs to ensure that their journey is, indeed, one that aligns with the ideals of sustainability.

The fate of the EV hinges on our ability to address these intricate threads that weave the fabric of its existence. It beckons us to reevaluate our definitions of “clean” and “sustainable,” to harness the momentum of innovation while confronting the environmental shadows that accompany progress. The road ahead is not devoid of challenges, nor is it immune to tradeoffs. But within its twists and turns lies an opportunity – a chance to redefine mobility, to reshape industries, and to pioneer a future where the wheels of progress roll forward with a steadfast commitment to both innovation and the environment.

V. Artistic Echoes Of Climate Change: Tipping Points

CHAPTER

Seven

Shadows Of Manipulation: Lobbying's Sinister Grip Hinders Climate Action's Noble Stride

"Lies have short legs"
– German Proverb

I. No Country For Old Men

In the vast expanse of Germany's political theater, where parties dance to the rhythms of democracy, there exists a pair of performers known as the Union. Ah, the Union, a harmonious duo composed of the Christian Democratic Union (CDU) and its Bavarian sibling, the Christian Social Union (CSU). They've been waltzing through the corridors of power for decades, swaying to the tunes of center-right ideologies and making themselves quite at home.

Picture a parliamentary democracy, where parties come together like actors on a stage, each playing their part. Among them, the Union takes center stage, its members drawn from the ranks of the CDU and the CSU. It's a bit like a sibling act: one party operating nationwide, the other setting up shop primarily in Bavaria. They may have their separate addresses, but when it's showtime, they unite to dazzle the audience.

Oh, the influence these Union politicians wield! With a fan base spanning from the conservative corner to the rural realm and even cozying up to the business moguls, they're like the beloved lead actors of the political saga. Coalition governments bow to their presence, and they've got their hands in key government positions like an octopus with more arms than you can count.

Let's travel back to the origins of the Federal Republic of Germany, a post-war phoenix rising from the ashes. Enter Konrad Adenauer, the first Chancellor and a CDU maestro. With him at the helm, Germany danced to the tune of an "economic miracle." War-torn ruins transformed into an economic powerhouse, and the world watched in awe. A true encore moment for the Union.

But wait, there's more! Helmut Kohl, another CDU magician, stepped up to the plate during the tumultuous years of reunification. As the Berlin Wall crumbled, Kohl was the conductor leading the symphony of reunification. And oh, what a grand symphony it was, cementing the Union's reputation as the maestro of politics.

Fast forward to today, and Union politicians are still calling the shots. They've taken turns in the Chancellor's chair, the Foreign Ministry, and the Finance Ministry, shaping policies on everything from the European Union to national security. It's as if they've been handed the conductor's baton of the German orchestra, directing the melody of the nation.

But hold on, the overture isn't all crescendos and applause. There's a subplot here, a tale of intrigue and moral dilemmas that's been playing out in the shadows. The whispers of lobbying and corruption, once relegated to the background, have come center stage. The spotlight now shines on a scandal, an affair involving Union politicians and their dance with unethical dealings.

The narrative unfolds like a suspenseful thriller, revealing the intricacies of navigating political duties in a landscape where Union politicians hold the sway. Imagine the scene: a web of connections, entangling Azerbaijan, Taiwan, and businesses, all spinning a tapestry that blurs the line between public service and private interest.

We find ourselves amidst a tangle of corporate ties, secondary activities, and financial gains. The story weaves a cautionary tale, reminding us of the importance of transparency and accountability. It serves as a mirror reflecting the ethical boundaries that must be upheld in the hallowed halls of politics.

In this world untethered by age-old conventions, it was in the poignant month of March 2021 that the pages of the German publication Der Spiegel unfurled a riveting exposé, laying bare a scandal of remarkable proportions that rocked the political landscape, shedding light on alleged misconduct by Union politicians involved in the procurement of respiratory masks.

At the heart of the affair were suspicions of unethical actions undertaken by Union politicians to broker deals related to the supply of vital protective gear. As the story unfolded, a network of connections between Union parliamentarians, Azerbaijan, Taiwan, and prominent businesses emerged, though not all were directly linked to the mask affair. This narrative highlights the complexities of lobbying, power, and corruption within the corridors of political influence.

The affair revealed how lawmakers blurred the lines between their public duties and personal interests. While engaging in secondary activities is not prohibited for parliamentarians, transparency in such dealings is essential. Nebulous ties between Union representatives and companies led to questionable outcomes. For instance, in early 2021, reports emerged suggesting that the Swiss company Emix Trading GmbH, owned by young entrepreneurs Jascha Rudolphi and Luca Steffen, had allegedly supplied overpriced and substandard masks to health ministries in North Rhine-Westphalia and Bavaria. These transactions were facilitated, in part, through the efforts of Andrea Tandler, daughter of the former CSU Secretary-General Gerold Tandler, who was actively involved in liaising with Federal Health Minister Jens Spahn.

The controversy centered on the lack of transparency and potential personal gains that arose from these connections. Critics argued that financial motives over-shadowed ethical considerations, while the intricate web of corporate relationships obscured transparency. Although the legal implications of these actions were unclear at the time, multiple investigations were launched.

The protagonists of this story come from diverse family and professional backgrounds, adding layers of complexity to the narrative. Georg Nüßlein, a CSU member of the Bundestag, initially triggered the scandal when it was revealed that he received substantial commissions through his consulting firm for facilitating protective equipment deals. These commissions flowed through offshore companies, raising suspicions of improper conduct. Parallelly, Alfred Sauter, another CSU politician, was implicated in the same affair, receiving funds through the convoluted network associated with Thomas Limberger's companies. The CSU's influence extended beyond parliamentary confines, creating a nexus of power and finance.

Furthermore, the scandal extended to the CDU party, with members like Nikolas Löbel and Mark Hauptmann implicated in similar unethical transactions. Löbel allegedly brokered deals for mask procurement and earned substantial commissions, leveraging his political position for personal gain. Hauptmann faced allegations of benefiting from dealings with TY-Capital Ug, a company associated with overpriced mask acquisitions. The involvement of prominent politicians from different parties underscores the need for introspection within the entire political establishment.

As the investigations continued, it became evident that family legacies played a role in shaping these events. Andrea Tandler, daughter of a former CSU minister, orchestrated a massive mask contract with Swiss firm Emix. The vast sum of money involved and the complexity of the deal highlighted the potential for personal enrichment. Tandler's story showcased how family ties and connections could be exploited to influence public procurement decisions.

The scandal underscored the importance of maintaining integrity and ethical standards within the political arena. It prompted discussions about the ethical boundaries of political engagement, the need for transparency in secondary activities, and the potential consequences of intertwining personal interests with public responsibilities. The complexities of these intertwining relationships form the backdrop for a story that explores the limits of lobbying, the power of connections, and the battle against corruption in the political landscape.

In the intricate world of lobbying, the ultimate currency isn't a suite of skills, a groundbreaking product, or an invaluable service. It's all about securing one elusive element: access to the corridors of power. This access becomes your trade, your commodity – a gateway that lets you funnel whatever causes, agendas, or coffers powerful figures and organizations desire. It's a dynamic where influence finds its way into the pockets of those who can peddle it most effectively.

This phenomenon brings forth a curious transformation: the metamorphosis of former politicians into lobbyists. A seamless transition, perhaps, but one that often raises eyebrows, especially when these once-elected figures champion causes that court controversy and disdain. This intricate dance, observed vividly in instances across the landscape of Germany, paints a picture of power dynamics that challenge ethical boundaries.

The legalities of this tango vary across the spectrum of nations that boast their status as bastions of the rule of law. Yet, even amidst legality, a shadow looms, a chorus of skeptics who coin the term "legalized corruption." A term that captures the essence of a system where the line between advocacy and influence-peddling appears blurred, where the very channels meant for democratic discourse seem hued with a questionable shade of exchange.

Against the backdrop of the discussed lobby scandal, where influential figures were revealed to be using their power to manipulate political decisions, it becomes even more imperative to understand the significant role that lobbying plays in hindering progress, particularly in tackling pressing global issues like climate change.

The German lobby scandal that has been introduced here shed light on the intricate web of connections between politicians, corporations, and interest groups. This scandal serves as a stark reminder that when lobbying efforts are driven by profit motives and self-interest, the common good often takes a backseat. Similarly, in the realm of climate change, powerful fossil fuel industries and other corporate entities have long been known to engage in extensive lobbying to protect their financial interests, even if it means impeding the necessary transition to cleaner and more sustainable energy sources.

The intertwined relationship between lobbying and climate progress is multifaceted. Lobbying allows influential players to shape legislation and policies in ways that align with their short-term economic goals, often at the expense of long-term environmental well-being. For instance, fossil fuel companies have historically fought against regulations aimed at reducing carbon emissions, while pushing for policies that maintain their dominance in the energy sector. These efforts can result in weakened environmental regulations, delayed transitions to renewable energy, and prolonged reliance on environmentally damaging practices.

Moreover, the financial resources and persuasive tactics employed by well-funded interest groups can sway public opinion and influence decision-makers. This can lead to policies that prioritize economic growth over ecological sustainability, hindering the progress required to mitigate climate change's catastrophic impacts. In this context, the German lobby scandal serves as a cautionary tale, highlighting the potential dangers of unchecked lobbying influence in shaping climate-related policies.

To overcome these challenges and promote effective climate action, a comprehensive understanding of lobbying's role is crucial. Stricter transparency and disclosure requirements for lobbying activities can help unveil hidden agendas and promote more accountable decision-making. By raising awareness about the tactics used by interest groups to delay or dilute climate policies, societies can better assess the validity of information presented and advocate for policies that prioritize the planet's health and future.

Fundamentally, the German lobby scandal we emphasized underscores the broader need for transparency, ethical conduct, and informed decision-making in the realm of lobbying. Recognizing how lobbying can hinder climate progress reinforces the urgency of fostering a fair and equitable policy-making environment that prioritizes the global good over narrow interests. Only through such efforts can we hope to overcome the obstacles that stand in the way of effective climate action and create a sustainable future for generations to come.

In the intricate dance of policy and power, the shadows often hold secrets that few dare to unveil. Peel back the layers, and one finds a realm where corporate interests meet political maneuvering, where the fate of our planet hangs in the balance. To fathom the labyrinthine landscape of lobbying's grip on climate progress is to don the lenses of scrutiny, to dissect the tapestry of influence woven in the hallowed halls of governance.

Stakeholder Interests emerge as a mosaic of divergent agendas, painted with strokes of self-interest and strategic gambits. As the climate chorus grows louder, industries take center stage, each actor holding a unique stake. Imagine the automotive titans, who raise their voices against stringent emission standards that threaten their balance sheets. They marshal resources, invoking lobbying's art to sway the scales of regulation. This interplay of interest births a critical juncture – each choice made, a trajectory set, priorities unmasked. Let me elucidate using a more comprehensive case, as narrated in the media.

In the shadowed corridors of power, where the fate of policies and the pulse of industries converge, a decision reverberated through the chambers of the European Commission like a distant thunderclap. A decision that would set the tone for a critical battle between profit and planet, influence and integrity.

Behind the ornate façades of Brussels, a realm of politics and corporate maneuvering unfurled its drama. The European Commission, entrusted with safeguarding the interests of millions, had made a choice that left environmental activists seething. In a world where the environment's voice often met with political cacophony, this decision was poised to unleash a tempest.

The whispers of intrigue began with a leak, as clandestine documents found their way into the hands of Politico, the proverbial fly on the wall of European politics. These documents spoke of a shift, a tilt away from the ambitious crusade against car emissions that had been promised. The Commission, in the face of influential EU automotive lobbies, had opted to tread more cautiously (Carter, 2022).

Once, the Euro 7 emissions standards had loomed on the horizon like a lighthouse guiding vessels through a turbulent sea of air pollution. But now, those standards had been tempered, stripped of their once-vaunted vigor. The tale of these regulations, once championed as a savior for cleaner skies, had taken a new twist.

Originally forged to shackle the menace of nitrogen oxides and particulate matter emissions, the revised Euro 7 standards had morphed. No longer did they merely cast a net over tailpipes and exhausts. Now, they cast a wider net, aiming to encompass the very particles that emanated from tires and brake pads – an unexpected twist that reflected the evolving narrative of emissions.

But what irked the environmental sentinels was the dilution of ambition. The Commission's leaked missives painted a picture of compromise, of a retreat from the stringent standards that had once danced tantalizingly on the horizon. The diesel vehicles, erstwhile bound for emissions caps mirroring their petrol-powered brethren, found themselves shackled to less demanding limits. The refrain from the activists was unanimous – a charge that profit had overshadowed the health of a continent.

Within the labyrinthine regulations, numbers held dominion. Under Euro 6, diesel engines were allowed a quota of 80 milligrams of nitrogen oxides per kilometer. Petrol counterparts, a shade better, were capped at 60 mg/km. The proposed Euro 7, as whispered in these clandestine scrolls, sought to further harmonize this discordant symphony, taming diesel emissions to the same 60 mg/km threshold.

And so, the stage was set for a clash of ideologies – a European Commission veering from the path of bold reform, facing a chorus of criticism from green NGOs and activists. For them, this was not just a regulatory reversal, but a betrayal of a commitment to tackle the choking shroud of air pollution and its insidious impact.

As the decision hung like a storm cloud over the continent, one question lingered: would the Commission's step back catalyze a renewed push from environmental forces, igniting a blaze of change that would consume the darkness of short-term interests? Or would it, in the hands of fate, become another chapter in the timeless saga of the tug-of-war between power and principle?

Policy Obstruction materializes as a formidable force, the clash of ideologies and economic might reverberating through policy corridors. Behold the coal conglomerates, wrestling against the current of renewable energy incentives. Their resistance forms an embankment, blocking the flow of progress. The banners of clean energy, once waving boldly, now flutter in the gales of opposition. The undercurrents of lobbying's might obscure the path to emission cuts, blunting the momentum towards sustainable horizons. Let now the following particular political spectacle act as a valuable canvas upon which we can skillfully add more intricate details to this specific facet.

In the theater of American energy policy, a dramatic collision of environmental aspirations and labor anxieties has taken center stage. The players? On one side, we have the dazzling allure of a greener future, the hope of slashing carbon emissions, and states like Washington, New

York, and Illinois that have managed to woo the unions, gaining their approval for bold renewable energy initiatives. On the other side, looming in the shadows, stands a group whose apprehensions could be straight from a Greek tragedy: the coal workers. Their story is one of suspicion, fear, and a resolute belief that embracing clean-energy jobs could unravel their hard-won standard of living.

Picture the bustling streets of these progressive states, where the sun shines on sleek solar panels and the wind spins majestic turbines. The unions representing electricians and steelworkers have donned the mantle of climate advocates, aligning themselves with President Biden's grand vision of climate and social policy, playing their part in the chorus of progress.

But in the heart of this unfolding saga lies a group that remains conspicuously absent from the union's harmonious melody. The coal miners, those who have labored in the depths of the earth to fuel the nation's appetite for power, have a different tune in their hearts. A melody of skepticism. The concept of trading their coal-streaked work boots for the gleam of wind and solar jobs has them clenching their fists.

Their worries? They span from murmurs of diminishing paychecks to the uncertainty of insurance coverage. President Biden, with his sights on reconciling this discord, has put forth a strategy: a delicate interplay of subsidies for wind and solar endeavors, offering the allure of union-scale wages. It's a theatrical spectacle of financial aid, training initiatives, and the promise of renewal for the coal-stricken communities.

Enter Phil Smith, the maestro leading the United Mine Workers of America. With a past marked by skepticism towards lofty promises, his ensemble orchestrates caution. They've honed the art of suspicion, woven into their history, and wield it like a conductor's baton. This sentiment, paired with the dissent emanating from coal miners, gives birth to a formidable force, akin to a turbulent undercurrent beneath seemingly calm waters.

The battlefront? The political arena. States like Pennsylvania and West Virginia, where coal miners possess the kind of influence that makes politicians sit up and pay heed. As President Biden sought to weave his climate policy tapestry, the coal miners' skepticism took center stage, triggering a symphony of complications. The crescendo of opposition played a pivotal role in muzzling the president's clean electricity program, leaving an echo of what could have been.

And then there's Joe Manchin III, West Virginia's enigmatic swing vote in the Senate. A figure whose political compass is attuned to the voices of his constituents, he embodies their worries about shifting careers and securing economic succor. In this unfolding drama, he plays a part that resonates deeply, echoing the questions of his coal-dependent state: Will these newfangled wind and solar jobs stand strong? Or will they be as transient as the wind that sweeps through a coal-ravaged landscape?

This narrative knits together a complex tapestry of ambition, apprehension, and economic realities. As the characters navigate the labyrinthine corridors of policy, one truth remains: the alliance with coal workers is a linchpin to the success of the grand green vision. Striking a harmonious balance between the aspirations of cleaner skies and the preservation of livelihoods, is the script yet to be fully written. And as the narrative unfolds, with the audience held captive, one thing remains certain: the interplay between the promise of a sustainable future and the echoes of a fossil-fueled past is a drama that continues to captivate the American stage (NYTimes, 2021).

Misinformation and Doubt weave an intricate web of deception, a tapestry spun from half-truths and shadows of uncertainty. Visualize the fossil fuel magnates, sowing seeds of skepticism in the fertile ground of public discourse. With funding as their quill, they write narratives that cast doubt upon the science of climate change. Clouds of misinformation obscure the truth, hampering policy's compass as it navigates the stormy waters of decision-making. By what means can we ascertain this?

Picture this: the grand stage of Cop27 in Egypt, where delegates and dignitaries gathered, voices of hope raised against the backdrop of a planet in peril. But as the speeches echoed and commitments were made, a parallel narrative was brewing in the digital realm. Enter the Climate Action Against Disinformation coalition, a beacon of truth in a world clouded by deceit.

Their research, meticulous and unyielding, uncovered a tapestry of lies spun by the fossil fuel behemoths. Millions of dollars, concealed within the shadows of paid advertisements, were funneled into Meta's coffers – once known as Facebook, now a battleground for the hearts and minds of the digital age. The goal? To sow seeds of doubt, to cast a fog of confusion upon the very issue that demanded clarity.

A sum of $4 million, a mere drop in the bucket for these corporate giants, was put to sinister use. A calculated symphony of misinformation played out across 3,781 ads, each one a note in the orchestrated symphony of deception. False claims ran rampant – climate concerns muddied, net-zero aspirations smeared, the indispensable nature of fossil fuels proclaimed with brazen audacity.

But who were the puppet masters behind this artful charade? Enter the American Petroleum Institute's PR troupe, a shadowy ensemble known as Energy Citizens. Their skillful hands moved the strings, their scripts veiling reality in layers of artifice. And yet, they were not alone in this macabre ballet. America's Plastic Makers, too, lent their contribution – a testament to the intertwining of industries that thrive on obfuscation.

And then, a curious twist in the tale – an uprising of climate denial, a chorus of voices championing disbelief. Like a crescendo building towards its climax, the hashtag #ClimateScam surged through the Twittersphere, months before the summit even began. As though scripted by some sinister playwright, the doubters' choir grew louder, finding resonance in the virtual realm.

The ads themselves were no mere banners; they were narratives, tales that wove the fabric of skepticism into the very concerns that threatened our existence. The cost of living – a primal worry for all – was used to paint fossil fuels as saviors, to soothe the troubled minds with promises of stability. And green technology? Questioned, ridiculed, its reliability questioned with pointed fingers.

But there was more to this disinformation ballet than meets the eye. A broader narrative unfolded, one that tied climate change to the controversial threads of "wokeness." A calculated move, aiming to taint the very cause with skepticism, as though concern for the planet was synonymous with some imaginary agenda.

Amidst this intricate dance, climate activists raised their voices in protest. Yet, their chorus was muffled by the heavy presence of fossil fuel executives, a phalanx of power that diluted the summit's potential for meaningful change. The gears of progress struggled against the weight of vested interests; their machinery oiled with the very misinformation they peddled (Mishra, 2023).

The lesson here is clear, a somber reminder of the battle that rages beneath the surface. A coordinated response is demanded – not just from the climate activists, but from the titans of technology, regulators, and governments alike. The fight against climate disinformation has become an integral part of the broader struggle against our planet's undoing.

Resource Allocation – a landscape of financial tides, where industries unleash torrents of capital to safeguard their interests.

Consider this:

In the shadowy world where commerce meets policy, the financial mettle of the industries rallying against regulations designed to rein in CO_2 emissions assumes a tapestry of contrasts. These industries, forged in the crucible of time, bear the weight of financial heft that is both formidable and variegated. Picture a tableau populated by sectors like fossil fuels, manufacturing, and heavy industries – their fingerprints gracing the global market's landscape.

At their core, these industries command a presence born of generations of enterprise, their coffers often deep, fortified by intricate webs of operations that span continents. And yet, beneath this overarching narrative, lies a mosaic of financial portraits as diverse as the players themselves. Some companies within these sectors wear the cloak of affluence with grace, amassing resources that lend power to their advocacy. Others find themselves navigating economic currents less favorable, dancing to the whims of fluctuating market dynamics.

Looming over it all is the specter of change, the regulatory winds that threaten to alter the course of business as usual. It's a high-stakes poker game, where resources are marshaled to the front lines, where lobbying efforts are a strategic dance of influence in the hallowed halls of power. These industries – the vanguards of established orders – mobilize their financial artillery to protect their interests, to safeguard their positions on the board of commerce.

Now, let your contemplation drift towards the majestic realms of nations – Germany, France, and the United Kingdom. Their GDPs, akin to opulent tapestries woven from the threads of domestic and international transactions, narrate stories of financial grandeur intricate in their design. A recent glimpse into this economic universe unveils Germany's GDP ascending to the dazzling heights of approximately $4 trillion, while France's emanates strength at an imposing $3 trillion, and the United Kingdom's economic presence graciously asserts itself at around $2.8 trillion.

Yet, an enigmatic twist enlivens the tale as the spotlight turns back to the oil industry. In the spotlight stand the titans of Exxon Mobil, Chevron, Saudi Aramco, BP (British/Beyond Petroleum), and Shell (Royal Dutch Shell) – an assembly of corporate colossi whose enterprise values, an aggregate between $ 2.01 trillion and $ 2.68 trillion, conjure forth a symphony of financial resonance. These five pillars, when aligned in unison, unfurl an astounding cumulative value that transcends numerical notation, but instead emerges as a resounding voice – a collective influence echoed through the corridors of finance.

In this intricate interplay, as the collective enterprise value stands against the grand stage of GDPs, the narrative assumes an even more captivating tenor. Consider, for a moment, this constellation of oil giants – a mere quintet of entities – embodying a financial force that surpasses a certain percentage of France's economic tapestry, claiming a slice of the United Kingdom's economic pie, and adorning a portion of Germany's monumental financial landscape. Can you grasp the essence of this observation? Perhaps it is indeed true – that they possess the means to sway allegiances, to attain the favor they desire. Some might even contend that they have the ability to place anyone within the confines of their influence, as some say, comfortably nestled in their pocket.

Here, within this juxtaposition of financial realms, a story unfolds, both complex and mesmerizing. It traverses the realm where industries and economies entwine, where threads of enterprise value harmonize with the grand narratives of national GDPs. The financial clout of the oil industry, a product of global orchestration, stands not just in proximity to, but alongside the GDPs of nations. Each figure – be it an enterprise value or a GDP – represents a distinctive stroke within the canvas of economic influence.

And so, the stage is illuminated, the protagonists poised – enterprise values and GDPs intertwining in a dance both intricate and ugly. It's a narrative that salutes the intricate interplay of power, lobbyism, commerce and nations, a tale where financial potency and the identity of a nation intermingle within the overarching story of our interconnected world. Can you sense the symphony that arises, the disruptive blend of money, influence, industries and nations resonating in asymmetrical warfare?

Institutional Relationships unravel as a Gordian knot of connections, a matrix where governments, industries, and interest groups intersect. Imagine the dance between agricultural lobbies and climate policies. As gears grind, decisions crystallize – policies forged not solely in the crucible of the public good, but with the anvil of lobbying's touch. To unveil these relationships is to discern the puppeteer's strings, the forces that subtly guide policy's hand. Allow us to illuminate this with a regrettable anecdote that unfolded within the chambers of the European Union Parliament (NYTimes, 2019).

Picture a committee room in Brussels, where lawmakers congregate to cast votes that sculpt the trajectory of a $65-billion-a-year farm policy, a cornerstone of the European Union's financial framework. Among the array of proposals lies the "Babis Amendment," an audacious endeavor aimed at cutting the financial umbilical cord between politicians and the farm subsidies they so generously distribute. The spotlight turns to Andrej Babis, the Czech Republic's billionaire prime minister, a man who, according to media reports, stands as a living emblem of how privilege and power intertwine within these corridors.

Babis, you see, is not merely a political figure – he's a prime example of the interplay between affluence and influence. As his government molds agricultural subsidy policies, his own domestic companies merrily reap $42 million from the coffers of the European Union. But it doesn't stop there; his financial empire sprawls across borders, with additional funds trickling in from holdings in neighboring countries. Here we find a vivid illustration of how the system can be deftly manipulated to serve the well-connected.

But let's cast our attention to a pivotal moment – the vote on the Babis Amendment. In a span of mere seconds, a proposal meant to curtail conflicts of interest is met with an abrupt rejection. No reading, no debate – just a swift dismissal. And here's the twist that leaves us bewildered: almost half of the committee members hold stakes in the very industry they are legislating for, basking in the warmth of subsidies. The ties that bind between political power and economic advantage are indeed complex, often obscured beneath the veneer of bureaucracy.

As we scrutinize this tale, the enormity of the EU's farm program comes to light – a colossal 40% of the European budget. Yet, accountability appears to be an elusive specter in certain corners. This lack of oversight engenders an environment where the likes of oligarchs and populists can harness these funds for their own gain, a revelation that sends shockwaves through the ideal of public good.

Now, my friends, let's imagine standing in the shadow of Brussels, where decisions ripple across nations. Here, the system's labyrinthine ways are unveiled, pointing to a distinct imbalance. Lobbyists, influential wielders of power, slip into closed-door meetings with government leaders, their voices echoing where transparency falters. And as lawmakers sculpt laws that determine subsidy distribution, a striking lack of restraint emerges, allowing them to craft the very rules they stand to benefit from – a tale of privilege that leaves us pondering.

The saga continues, as the EU grapples with renewing its farm bill, a monumental undertaking that raises questions about corruption, environmental integrity, and the fair distribution of resources. But the road to change is winding and uphill, laden with vested interests determined to uphold the status quo.

Ladies and gentlemen, this tale serves as a stark reminder of the intricate dance between policy, power, and accountability. As we peel back the layers of this narrative, we're left with a pressing question: Can a system that favors those who profit from it truly represent the greater good? The answer, my friends, lies in the hands of those who can untangle the threads of influence and reshape a system that strives for true transparency and fairness.

Pathways to Reform emerge from the ashes of obstruction, a phoenix rising with lessons learned. Picture the struggle for tighter emissions standards fought by oil and gas behemoths. To grasp the entwining of lobbying's grasp with progress impeded is to ignite the spark of change. Advocates and policymakers unite, wielding tools of transparency and public awareness to chisel new policies from the stone of resistance.

In our final contemplation, the intricate artistry of lobbying takes center stage within the theater of climate action, serving as humanity's navigational beacon amid the tempestuous seas of transformation. As we draw back the veil, granting acknowledgment to its role in obstructing progress, we embark on the journey of unraveling the puzzle that underpins the creation of lasting policies. These actors, encompassing not solely the elder statesmen – those revered old men – but also the forthcoming generations, often devoid of moral constraints and even outshining their predecessors, engage in a complex exchange where access metamorphoses into a reservoir of resources for multifaceted pursuits. Through a discerning lens that exposes these intricate maneuvers, we uncover the radiance needed to illuminate our forward trajectory. Thus, we craft a narrative that reverberates with the supremacy of transparency, where decisions bear the weight of sagacity, and society forges an unyielding barrier against the chilling gusts of the transformative climate.

II. The Mechanics Of Lobbying And Its Impact On Climate Policy

In the quiet corridors of power, where motives are whispered and alliances are forged, there lies a tale that weaves the threads of climate protection, environmental stewardship, and a pipeline of geopolitical intrigue. Behold, the enigma known as the "Stiftung Klima- und Umweltschutz MV," an entity seemingly born from noble intentions but cloaked in a shadowy dance of tactics reminiscent of a masterful con artist's playbook (Feldenkirchen, 2022; Tamme, 2023; Kinkartz, 2023).

Secure your seat and prepare to witness a spectacle in ten acts, a mesmerizing display of "true lies" and artful deception, cloaked as a provincial epic encompassing matters of gas, wealth, and the environment. At the helm of this captivating circus is a remarkable lady, the First Minister, serving as the ringmaster extraordinaire.

1. The Prelude: Illusion of Benevolence

   Amidst the political orchestration in the realm of Mecklenburg-Vorpommern, a foundation emerged, proclaiming itself the guardian of climate and environment. Yet, beneath its virtuous veneer, a different melody played. As the pipeline's siren song echoed across borders, the "Stiftung Klima- und Umweltschutz MV" surfaced with a dual face – a champion of green ideals and a stage for pipeline politics.

2. The Art of Lobbyism 1 on 1: A Step-by-Step Deception

   Step 1: Fictitious Data Presentation

   Like a magician's sleight of hand, the establishment showcased its creation as a sanctuary for climate action. A stroke of genius – diverting gazes away from pipeline affiliations and towards a utopian haven of eco-friendly dreams.

   Step 2: Cherry-Picked Success Stories

   A symphony of carefully chosen tales of ecological valor danced forth. By showcasing triumphant feats of local environmentalism, the darker undertones of geopolitical ambitions quietly faded into the background.

   Step 3: Greenwashing Campaigns

   Behold, a masquerade of green garments cloaked the foundation. "Look at us, defenders of nature!" they cried, their real dance partner, the pipeline, lurking in the shadows.

Step 4: Suppression of Opposition

The government's grand embrace of the foundation and its goals cast a shadow of caution over those who dared to question. Critics cowered as the establishment's political juggernaut hovered ominously nearby.

Step 5: Manipulative Visuals

Behind the scenes, imagery of unity and purpose may have been whispered through the corridors of power, projecting an image of righteous intent and conjuring subtle spells of positive association.

Step 6: Shifting Accountability

The narrative's tango was masterful. From climate guardian to pipeline facilitator, the purpose seamlessly shifted, morphing like a chameleon to fit the political tides.

Step 7: False Consensus Creation

In the grand ballroom of politics, the foundation, the government, and Nord Stream 2waltzed in harmony. The illusion of unanimity swayed public sentiment, as though a magical enchantment had been cast.

Step 8: Misleading Terminology

"Stiftung Klima- und Umweltschutz MV," they proudly proclaimed. Climate and environment protection! A name so noble, it hid the pipeline's scent like perfume masking deceit.

Step 9: Discrediting Critics

The critiques of environmentalists and nay- sayers echoed faintly. Yet, their voices were met with strategic silence, allowing the establishment's narrative to soar unburdened by pesky objections.

Step 10: Emotional Appeals

With the artful use of sentimental words – "Klima- und Umweltschutz" – emotions stirred. Hearts warmed to the melody of a greener tomorrow, a symphony that artfully muted pipeline concerns.

As this intricate dance unfolded, the "Stiftung Klima- und Umweltschutz MV" etched its mark upon the annals of deception. A masterpiece of tactics that whispered secrets to the willing ear, painted illusions for the hopeful eye, and left critics squinting through the haze of grandeur.

Remember, dear reader, this tale is a patchwork quilt of known fragments, a glimpse behind the curtain of the grand performance. Yet, the echoes of the dance continue, a reminder that in the realm of politics and power, the truth is often hidden within a dance of tactics.

With the curtain drawn back on the intricate theater of government maneuvering, a revelation emerges – one that evokes both cynical nods and disheartened sighs. Behold, the ruling elite, those ensconced in the highest echelons of power, shedding their veneers of altruism to embrace motives as diverse as the cosmos itself. One might speculate, and indeed, many are in agreement, drawing parallels to the timeless wisdom of the Apostle Paul. His proclamation, succinctly captured as "Money is the root of all evil," resonates with newfound vigor.

Now, the stage resets to unveil a series of case studies – vivid illustrations where the art of lobbying, a practiced ballet in the realm of influence, pirouettes its way into the heart of climate action initiatives.

Allow me to articulate once more, the "Umweltstiftung Mecklenburg-Vorpommern" case unfolds like a cautionary tale woven into the tapestry of modern politics. It's a narrative where noble intentions, shrouded in the cloak of environmental guardianship, intertwine with hidden currents of geopolitical maneuvering. A foundation ostensibly committed to climate and environmental protection becomes a marionette, skillfully directed by the hands of those who yield power and profit.

But let us not succumb to the allure of singularity, for this saga is but one thread in a much broader fabric. It beckons us to delve into the cavernous halls of climate action, where the echoes of lobbying resound with unsettling clarity.

Ladies and gentlemen, fellow travelers through the labyrinthine corridors of climate action, let us embark on a journey that unravels the clandestine threads of lobbying's influence. In this tumultuous realm where the yearning for change grapples with vested interests, we unveil the concealed force that often stands as an obstacle on the path toward a greener future. Welcome to a compilation of case studies, a mosaic that exposes how lobbying's veiled hand has obstructed climate action. But we don't stop at merely presenting these instances; no, we have carefully categorized them to illustrate the multifaceted nature of these obstructions.

Let us illuminate their presence.

Fossil Fuel Industry Influence – Ah, the vanguard of climate lobbying, the fossil fuel industry. Picture this: an industry that wields more than just the power to pump carbon into the air. This category exposes global instances where their endeavors to uphold the status quo have cast a long shadow over the clamors for environmental reform. They're not just burning fossil fuels; they're burning the aspirations of a cleaner tomorrow.

Industry-Specific Resistance – Now, let's delve into the world of sectors protecting their economic domains. Imagine industries like the US auto sector, with its steel resolve against fuel efficiency standards, and the European aviation sphere, swirling in maneuvers to outmaneuver regulatory pressures. These are industries digging their heels, not into the soil, but into the balance sheets, all in the name of their financial well-being.

International Agreements and Commitments - Even international agreements aren't immune to the siren song of lobbying. Behold the case of the Paris Agreement withdrawal, where political motivations pirouetted over global climate obligations. It's a stark reminder that even the grandest of pledges can be overshadowed by the intricacies of political theater.

Regional Policy Resistance - From the global stage, let's zoom into specific regions. Think of Poland's coal industry weaving its influence, or Canada's persistent advocacy for tar sands. These are clashes between regional prosperity and environmental aspirations, reminding us that the age-old tug of war between economy and ecology continues.

Trade and Economic Interests – Economics and climate goals dance a tango more intricate than one could fathom. Behold the manipulation of the EU Emissions Trading System and the negotiations entwined in the EU-South America Trade Deal. Here, trade ambitions and emission reduction interlock in a tense embrace, revealing a complex choreography that tests the limits of balance.

Agricultural and Land-Use Lobbies – Industries aren't the only players in this game, agriculture, and land use step onto the stage. Behold the Brazilian agribusiness lobby, an actor in a tale that illuminates how pivotal eco-systems can be compromised in the pursuit of economic gain. In this story, the soil and the ledger compete for attention.

Challenges to Technological Transition – Innovation is a pillar of climate action, yet change isn't always greeted with open arms. Imagine the auto industry's reluctance towards electric vehicles. The very industry that propels us forward is hesitant to embrace an electric future, a testament to how innovation's path is seldom smooth.

Interplay of Economic Interests and Environmental Protection – And then, there's the dramatic stage where economic interests grapple with the imperative of environmental preservation. In this category, the lobbying war over cutting EU emissions takes center stage, revealing a high-stakes battle where the survival of the planet is pitted against economic ambitions.

As you journey alongside me through these meticulously crafted categories, a profound realization shall unfold before you – lobbying's influence upon climate action defies confinement to a solitary pathway. Instead, it manifests as an intricate tapestry woven from myriad threads, a tumultuous battlefield where the forces of transformation and the custodians of the established order converge in a dramatic collision of ambitions. Ultimately, it serves as an indelible reminder that within the theater of climate action, the script remains unwritten, and the fate of our planet teeters on a precarious precipice. With this poignant perspective as our guide, let us now embark upon an exploration of case study categories, each revealing a distinct facet of this enthralling narrative.

Ladies and gentlemen, gather 'round and let me regale you with the tale of the fossil fuel industry's grand influence on the climate policy stage. Oh, what a captivating performance it has been, a symphony of dollars and decisions masterfully orchestrated in the halls of power.

Picture Capitol Hill as the stage, where the oil and gas sector cast a web of power. According to the Center for Responsive Politics (CRP, 2015), a staggering $141.6 million was spent on lobbying in 2014 alone, a jaw-dropping $387,945 per day. With over 800 lobbyists prowling the corridors, it's as if every member of Congress had their own personal lobbyist.

Enter the American Petroleum Institute (API), a mighty player in this game. This trade association, fueled by a budget predominantly composed of annual "dues" from its members – ranging from smaller companies to the Exxon giants – takes the spotlight. API's performance centers on lobbying against regulations on oil and gas development, rallying for an energy future rooted in oil, gas, and coal.

Now, my dear audience, brace yourselves for the grand show of lobbying might. In 2015, API unleashed close to $65 million in what they charmingly call "obstructive climate lobbying." That impressive number encompassed PR campaigns, ads worth at least $43 million, direct political contributions, and the eloquent whispers of lobbyists (Pashley, 2016).

Ah, the return on investment! The tale takes an intriguing twist as the industry's clout translates into favorable votes and legislation. The links between politicians and their pockets lined with oil money become apparent, influencing bills and votes. For example, the year 2010 witnessed a seismic shift as Republicans took the reins of the US House of Representatives, setting the stage for over 300 anti-environmental votes during the 112th Congress. The chamber gained notoriety as the "Most Anti-Environmental House in History," and the oil and gas lobby reveled in its political prowess (Markey, 2011; Lennard, 2011).

But the pinnacle of their triumph came in 2015. Congress lifted the ban on crude oil exports – a ban conceived in 1975 during the Arab oil embargo. The industry, facing oversupply and plunging crude prices, orchestrated a campaign to overturn the ban. With a strategic ballet of lobbying and relentless effort, they succeeded. Over 200 lobbyists and $38 million later, the ban was history (Gardner , 2015; Goldenberg, 2015; BBC, 2015). Yet the performance didn't end there. The industry's lobbying extended to budgets, reducing funds for the EPA and other federal agencies. The Republican-controlled House embraced a hefty 27 percent cut to EPA funding in 2011, and since 2010, EPA's budget plummeted by over 20 percent. Staffing levels hit their lowest in decades, and critical programs like the Underground Injection Control (UIC) struggled to protect public health from the impacts of surging oil and gas development (Rogers, 2011).

Ah, my discerning audience, let's venture into the realm of common-sense speculation, where the echoes of decisions influenced by lobbying may have left an indelible mark on our climate landscape in the year 2023. While we can't don our clairvoyant hats, it's quite reasonable to imagine a few scenarios that could have unfolded.

Imagine, if you will, the lifting of that ban on crude oil exports, a triumph for the oil and gas industry's lobbying prowess. With the shackles removed, the floodgates of fossil fuel extraction and exports may have burst open. More barrels of oil making their way to international markets could translate to higher carbon emissions, a direct contradiction to the urgent need to reduce our carbon footprint.

And behold, the budget cuts to agencies like the EPA, a symphony orchestrated by industry lobbyists. As the funds dwindled, the EPA's ability to enforce environmental regulations and oversee the transition to cleaner energy technologies could have been compromised. The oil and gas industry, perhaps, basked in a world with less accountability, a paradise of leniency, while the environment and communities bore the brunt of unchecked practices.

Let's not forget the grand spectacle of "obstructive climate lobbying." With millions poured into PR campaigns and political contributions, the decisions influenced by these campaigns could have sent ripples of inertia through climate action efforts. The sense of urgency to transition away from fossil fuels and embrace cleaner alternatives may have been stifled, leaving us grappling with a slower pace of change in our race to combat climate change.

So, dear friends, while these are mere conjectures woven from the threads of past decisions, it's not far-fetched to surmise that they could have influenced our climate reality today. The echoes of lobbying's power may still reverberate through our endeavors to mitigate global warming, serving as a poignant reminder that the decisions crafted in the corridors of power can cast a long shadow over the fate of our planet.

Ladies and gentlemen, prepare yourselves for another tale so thrilling, so utterly shocking, it's like a soap opera meets a blockbuster movie, all wrapped up in a riveting PBS Frontline docuseries titled, drumroll please, "The Power of Big Oil." Are you sitting down? Good, because we're about to dive headfirst into a whirlwind of deception, manipulation, and grand corporate shenanigans that make your favorite daytime drama seem like child's play (McGreal, 2021).

Hold onto your seats, for we're embarking on a journey that unveils the labyrinthine strategies of the oil industry – that's right, the industry that pumps liquid gold – to obstruct climate action and conceal the inconvenient truth about our planet's impending doom. Picture this: the climate crisis as a dramatic backdrop, a planet on the brink, and there they stand – the oil barons, twirling their metaphorical mustaches and hatching schemes to protect their empire of profit.

But wait, here comes Chuck Hagel, our leading man, a former Republican senator who's ready to spill the beans. In a moment of sheer revelation, he confesses that he was right there in the trenches, working hand in hand with Democratic senator Robert Byrd to squash the Kyoto climate treaty like a bug. Why, you ask? Well, because it was just so darn unfair to us Americans, of course! And how did they pull it off? Big oil unleashed a full-blown campaign to paint the Kyoto protocol as a villain – a threat to jobs, economies, and the American way of life. Oh, and they conveniently overlooked that China and India were indulging in their pollution parties without a care in the world.

Now, let's fast-forward a quarter-century. Hagel, wearing his hindsight glasses, points his finger straight at the fossil fuel giants, like Exxon and Shell, for pulling the wool over everyone's eyes. These oil moguls, it turns out, were sitting on their treasure troves of research while publicly denying the undeniable science of climate change. And when Hagel says, "They lied," you can almost hear the collective gasp of astonishment from our captivated audience.

Hold onto your skepticism, because the story only gets better – or worse, depending on your perspective. Our docuseries peels back the curtain on a cavalcade of former oil scientists, lobbyists, and PR wizards who come clean about their roles in this decades-long game of smoke and mirrors. It's a ballet of deception that danced circles around the American public and politicians, with ripple effects that extend well beyond our borders.

And now, my dear viewers, we traverse three mind-bending episodes – Denial, Doubt, and Delay – in a tale that makes the twists and turns of a telenovela seem predictable. Corporate manipulation of science, public sentiment, and political maneuvering – it's a saga of capitalism at its juiciest, wrapped in intrigue that would make James Bond envious.

Oh, but there's more. Let's meet former Senator Timothy Wirth, the visionary who organized groundbreaking environmental hearings back in 1988, and his star witness NASA scientist James Hansen, our climate action superhero, who declared from the mountaintops that greenhouse gases (GHGs) were reshaping the world (Shabecoff, 1988). It was a moment of hope, a glimmer of change, a sign that truth would finally triumph. But oh, how the tides turned. Instead of awakening, the oil industry's machinery sprang into action, intensifying their campaign of climate denialism.

And what's a gripping tale without some interesting "Personages"? Introducing the Koch brothers, the maestros of manipulation, orchestrating a symphony of misinformation and opposition to carbon taxes. They had their own grassroots charade, Citizens for a Sound Economy, and their tricks weren't subtle. They targeted politicians like David Boren, threw out claims that Oklahomans would be paying a carbon tax with every shower, and even created an illusion of public backlash that, surprise, was funded and fueled by corporate interests (Waters, 2015; Mayer, 2019). Bravo, Kochs, bravo.

Ladies and gentlemen, through the lens of "The Power of Big Oil," we glimpse a world where profit prevails, and climate action takes a back-seat to protecting the bottom line. It's a tale of industry adapting tactics, of politicians swayed by corporate funds, and of a planet left to grapple with the consequences of their dance. As we bid adieu, remember: the power of big oil isn't just in the crude they pump; it's in the masterful manipulation that's spun a web of climate inaction. So, as we navigate this tumultuous narrative, let's not forget that while profit and power may rule the boardroom, it's our planet's survival that's at stake.

Ah, dear audience, let us embark on a journey into the realms of speculative imagination, where the decisions born from deception and oiled by cash have sculpted our climate landscape in the illustrious year of 2023. Picture a world where the echoes of false claims and strategic manipulation continue to reverberate, a world where the smoke and mirrors of yesteryears have painted a canvas of consequences.

As we step into this alternate reality, it's only fitting to envision a landscape somewhat different from the utopian dreams of climate activists. The majestic polar ice caps, once towering symbols of nature's grandeur, might find themselves retreating at a pace that even a snail could mock. The oceans, those vast blue expanses, might still be harboring whispers of plastic debris, refusing to dissipate despite the promises of reform.

In this speculative universe, the air we breathe, once imagined to be cleaner and purer, could still bear traces of emissions that scoff at our feeble attempts at regulation. The sky, instead of donning a refreshing azure robe, might carry a subtle tinge of skepticism, a gentle reminder that the dance of doubt orchestrated by big oil had far-reaching consequences.

And what of the deniers and their ingenious obstruction? Well, they might have succeeded in erecting a fortress of delay so impenetrable that meaningful climate policies find themselves tangled in a labyrinthine web of bureaucracy. The urgency to transition to cleaner energy sources? Well, let's just say it might have been set aside for more "convenient" time, a time when the profit margins of fossil fuel giants have been squeezed dry.

But fear not, dear audience, for our tale of speculation is not all doom and gloom. In this world, innovation, driven by necessity, could be a resilient beacon of hope. Clever minds, inspired by the urgency to rewrite the narrative, might have harnessed technology to accelerate the transition away from carbon-intensive fuels. Renewable energy sources, once deemed "alternative," might have become the norm, dotting the landscape with wind turbines and solar panels that hum with a sense of redemption.

And the climate activists? Well, they might have risen like phoenixes from the ashes of deception, their voices ringing louder than ever before. The waves of truth, initially thwarted by the oil industry's tsunami of falsehoods, could have surged back with a vengeance, sparking a global awakening that rekindled the flames of climate action.

But, my esteemed audience, let us not forget that the tendrils of consequences, like ivy on an ancient castle, have a habit of creeping and sprawling. In this speculative universe, the decisions built upon lies and cash might have cast a shadow over our progress, dimming the light of what could have been a rapid and relentless climate revolution.

As we ponder the "what ifs" of this speculative narrative, let it be a reminder that the choices made in the corridors of power, fueled by deceit and influence, have the potential to steer our destiny in ways unforeseen. So, while we explore the realm of speculation, let's remain vigilant, for the present holds the power to shape our future and, perhaps, rewrite the tale of climate in a way that even the skeptics can't dispute.

Ladies and gentlemen, gather 'round for a tale that's more twisted than a pretzel in a hurricane. It appears that the esteemed global carmakers, those purveyors of progress and champions of a green tomorrow, have managed to orchestrate a performance that would leave Shakespeare himself green with envy. Hold on tight as we dive into the world of corporate contortion and climate crisis charades (Laville, 2019).

In this dazzling spectacle of corporate cunning, the automotive industry takes a star turn as the headlining villains in the theater of climate inaction. Brace yourselves as we uncover a dazzling display of hypocrisy, as these auto giants stand shoulder-to-shoulder with the climate's arch-nemeses, pouring vast fortunes into a shadowy realm known as lobbying. Yes, you heard it right – while they dazzle us with public pledges of support for climate initiatives like electrification, they've been slinging dollars through industry bodies to undermine efforts to tackle the very crisis they claim to address.

As if that weren't enough, let's marvel at their audacity to pull off this grand deception amidst repeated global climate emergency alarms. The study, conducted by the intrepid researchers at Influence Map (2023), peels back the layers of duplicity by analyzing the size and lobbying activities of the corporate giants. They've combed through tens of thousands of statements, policy pronouncements, and lobbying escapades of the 250 biggest industrial corporations and trade associations. It's a tapestry of contradiction woven with finesse.

And who are our main actors, you ask? Behold the automotive A-list: Fiat Chrysler, Ford, Daimler, BMW, Toyota, and General Motors. These automotive virtuosos, while serenading us with sweet promises of a low-carbon future, have been orchestrating a symphony of opposition against regulations that could curb the global warming rampage. All this, while they solemnly vow to lead us into the electric age. Bravo, gentlemen, bravo!

But wait, there's more. The automotive industry's wizardry extends to a supporting cast of industry groups that amplify their siren song of inaction. These groups, including the Automotive Alliance in the US and Europe's ACEA and VDA, add a touch of flair to the performance. With senior figures from these car companies seated comfortably on their boards of directors, the orchestration of this climate ballet reaches a crescendo.

And what about the standards and regulations that would nudge us towards a sustainable tomorrow? Fear not, dear audience, for these automotive maestros have worked tirelessly to sidestep them with the elegance of a cat avoiding a puddle. They've made an art of delaying and dampening any move towards reducing emissions and improving fuel economy standards. And when it comes to the transition to electric vehicles, they've spun a tale so intricate that even the most seasoned novelists would be envious.

In the grand finale of this performance, we find ourselves questioning the very fabric of accountability. The car industry's influence stretches to the world stage, as it tries to pull off the impossible feat of prolonging the reign of gasoline-guzzling behemoths. With SUV sales soaring and emissions standards faltering, they've mastered the art of casting doubts on electric technologies, from concerns about affordability to pondering whether infrastructure will ever show up in time.

As we applaud this grand farce of corporate responsibility, let us not forget the moral of the story. The auto industry, with all its grand promises and glittering showcases, has slyly maneuvered itself to dodge the real issues at hand. They've had years to prepare, yet they've chosen to embrace SUVs and high-margin vehicles with open arms, while lamenting the difficulty of meeting emissions targets. So, ladies and gentlemen, take a bow, for the car industry's performance in this climate drama deserves a standing ovation – though not for the reasons they might hope.

Ladies and gentlemen, let me take you on a journey into the dystopian landscape that our beloved Earth has become, all thanks to the enchanting dance of the automotive industry's lobbying prowess. Picture a world where the very air we breathe has taken on a smoggy hue, reminiscent of a surreal painting from a mad artist's palette. A world where the seas have risen to such heights that beachside resorts have been transformed into luxurious underwater getaways, complete with coral reef views right outside your suite.

In this wondrous new reality, the sun seems to emit an extra fiery glow, casting an eerie radiance upon cities that have now been transformed into heat-blasted concrete jungles. Our beloved ice caps have performed a vanishing act, leaving behind nothing but chilling memories and rising sea levels that have forced entire coastal populations to don snorkels just to take a stroll down their once-familiar streets.

Ah, but let's not forget the pièce de résistance – the cars. In this wacky carnival of climate chaos, the streets are now dominated by a fleet of gas-guzzling behemoths that laugh in the face of emission standards. They've transformed rush hours into scenes reminiscent of the demolition derby, with the cacophony of revving engines mingling with the mournful wails of polar bears seeking refuge on the last remaining shards of ice.

And what about those promises of electrification? Oh, how they shimmered like mirages on the horizon. In this topsy-turvy world, electric vehicles have become an endangered species, as the auto industry's lobbying magic managed to stall progress at every turn. The charging stations that were supposed to sprout up like dandelions in spring? They've become as rare as unicorns, while gas stations continue to flourish like oases in the desert.

But fear not, for the automotive industry's grand performance doesn't stop there. With emissions unchecked and fossil fuels reigning supreme, our climate has been transformed into a playground for extreme weather events. Hurricanes have become so frequent that they're now a regular topic of conversation at the water cooler. Tornadoes? They've evolved into social events, complete with themed parties and souvenir T-shirts.

As we wander through this landscape of chaos and absurdity, let us not forget the true architects of this spectacle – the carmakers and their mesmerizing lobbying efforts. With each dollar poured into influencing policy and thwarting regulations, they've managed to turn our world into a carnival of climate catastrophe. The Great Barrier Reef now hosts an annual "Coral Apocalypse" party, where attendees wear vibrant hues to match the vibrant colors of the dying corals. Glaciers, now a thing of the past, are mourned with solemn ice cream socials.

So, dear audience, let us pause and reflect on the brilliance of the automotive industry's lobbying performance. They've managed to turn Earth into a surreal carnival, where every twist and turn of their lobbying dance has led us deeper into a realm of environmental absurdity. As we stand at the crossroads of this climate crisis carnival, let us not forget that it's the choices we make today that will determine whether we continue down this chaotic path or take a detour towards a more sustainable and sensible future.

Ah, let's take a moment to dive into the world of steel – that resounding symphony of clanging metal and swirling controversy. Why, you ask? Well, imagine a vast industry that churns out an astounding two billion tons of steel annually, all within a colossal market worth $2.5 trillion. Impressive, right? Yet, amidst the steel's industrial crescendo, there's a discordant note that resonates – its annual emissions, a staggering three billion tons of CO_2, most of it emerging from those fiery blast furnaces.

Now, enter the realm of lobbying – that subtle dance where industry flexes its muscles to steer the course of regulations. Steel, this behemoth of an industry, is no different. With profit margins as thin as the metal sheets they mold, the steel titans aren't about to welcome any changes that could dent their bottom lines, especially those that demand cleaner, more sustainable practices. They say necessity is the mother of invention, but for the steel industry, necessity is also the mother of robust lobbying efforts.

Picture this: an orchestra of industry leaders and lobbyists, all wielding their influence like conductors of a symphony, orchestrating a delicate balance between their interests and environmental concerns. But how do they work their magic? Ah, let me weave you a tale.

Imagine a scenario where stringent regulations knock at their factory doors, threatening to disrupt their well-worn routines. What do they do? Enter the art of opposition – rallying against these regulations, bemoaning their potential impact on innovation and competitiveness. "Burdensome" and "costly" become their rallying cries, echoing through the halls of government offices.

And then, there's the exquisite art of delay. Ah, the power of procrastination. When faced with the daunting task of adopting new emission reduction standards or energy-efficient practices, they become experts at hitting the snooze button on these changes. After all, why rush when you can savor the status quo?

Carbon pricing, the audacious notion that emissions have a price tag, becomes a battleground of its own. Steel industry lobbyists take up arms, casting doubt on these mechanisms. They warn of economic perils, painting a picture of an industry brought to its knees by the very notion of paying for its environmental impact.

But it doesn't stop there. Oh no, they're maestros of innovation too, advocating for futuristic technologies that promise a cleaner tomorrow. It's like telling your parents you'll clean your room as soon as you've built that teleporter you've been sketching.

Feasibility, they say, is a tricky thing. Transitioning to greener practices? They raise an eyebrow, questioning whether such a shift is practical. Are they mere skeptics, or simply cautious guardians of their economic well-being?

Jobs on the line, plants hanging by a thread – they paint a dramatic scene of economic catastrophe, warning that stringent regulations might usher in an industrial apocalypse. They tap into policymakers' fears of losing jobs and stability, crafting a narrative that's hard to ignore.

But the steel industry doesn't dance alone. It forms alliances with other heavyweights, industries that pump out emissions like there's no tomorrow. United in their resistance to change, they stand firm against policies that challenge their modus operandi.

In this grand theater of lobbying, global competitiveness becomes a central theme. "We can't compete on the global stage with such stringent regulations," they cry, tugging at the heartstrings of policymakers concerned about the nation's economic prowess.

And in the realm of research and reports, they're like skilled artisans, subtly shaping narratives to downplay their environmental impact. Imagine painting a landscape where your smokestacks blend harmoniously with the sunset – that's the kind of artistry we're talking about.

Lastly, they dabble in the world of voluntary measures. Why be told what to do when you can voluntarily maybe do it? They propose self-regulation as an alternative to mandated changes, claiming that they're as committed to sustainability as anyone else.

In the end, the steel industry's lobbying dance is a complex waltz between profits, policies, and environmental concerns. While not all steel companies or industry associations follow these moves, the echoes of their steps continue to reverberate through the corridors of power.

As we bid farewell to the realm where industry's specific agendas intertwine with climate endeavors, we step into a vast arena where the echoes of International Agreements and Commitments collide with the resounding force of lobbying, striving to thwart global climate policies and preserve the gains of an entrenched environmental status quo.

Picture this: The United States initiates the formal process of severing its ties with the Paris Agreement on climate change – a decision that has been years in the making. Come November 4, 2020, a day following the presidential election, this move, resolute and calculated, speaks volumes about a political landscape wherein interests and environmental stewardship collide.

Reflecting on that pivotal moment in 2017, we recall President Donald Trump's exuberant and ceremonious Rose Garden speech, brimming with assurances of an imminent withdrawal. Amid the ever-shifting currents of his administration, this determination remained steadfast – a promise etched into the narrative of American First politics. As the tides of political discourse ebbed and flowed, it marked the official notification to the United Nations of this intent – a formal declaration of intent that has been years in the making.

In the aftermath of this announcement, one might detect a muted resonance. Trump's once fervent words have now receded, replaced by tempered references to the "horrible, costly, one-sided Paris climate accord." In a political climate anticipating the 2020 campaign strategy, the resonance of that night's statement lingers – a preview, perhaps, of an attempt to reframe the administration's environmental track record, even as such efforts contend with widespread public disapproval.

The promise to initiate talks for reentry or to craft an entirely new agreement now appears as a distant echo. Yet, in the face of the voluntary character of the Paris Agreement, the audacity of this vow verged on paradoxical. The United States, having borne witness to the disillusionment of the Kyoto Protocol, emerged with demands for parity among nations and the absence of externally mandated emissions cuts. The resultant Paris Agreement, a symbol of global cooperation, bears witness to this very essence. Yet, the voluntary nature of this accord lends an air of perplexity to the act of withdrawal – exposing a decision that appears, to some, beyond rationality.

Thus, this departure bears witness not to skepticism regarding climate science but to an intention of obstructing the global pivot towards clean energy. The heart of the matter lies not in disbelief but in a steadfast embrace of carbonism – an ideology that extols the virtues of fossil fuels as preeminent and unrivaled, asserting their superiority within the intricate tapestry of cultural, economic, and political beliefs (Meyer, 2019).

This concept of carbonism – the spiritual cousin of nationalism – imbues fossil fuels with the potential for prosperity, elevating them to the realm of intrinsic value and resilience. The administration's maneuvers, from dismantling climate regulations to advocating coal subsidies, echo the chorus of carbonism, where fossil fuels stand as the cornerstones of a thriving economy.

As this intricate dance of ideology unfolds, Trump's stance remains unyielding, a testament to the sway of carbonism. Pompeo's artfully crafted statement – evoking the "reality of the global energy mix" and the importance of harnessing all energy sources "cleanly and efficiently, including fossil fuels" – encompasses the essence of carbonism. It hints at the notion that fossil fuels hold the key to an era of boundless potential, defying limitations.

Amidst this complexity, remember that carbonism transcends rhetoric; it permeates our daily existence. With every flick of a switch, the lighting of a gas stove, or the start of an engine, carbon dioxide is released into the atmosphere, crafting an indelible narrative of generations yet to come. The Paris Agreement beckons us to acknowledge a simple truth: prosperity flourishes not in resisting change but in embracing it.

Amidst the shifting currents of change, the world braces for a post-carbon era, with the European Union and China leading the charge in renewable energy investments. As these winds of transformation sweep through, the stability of American power teeters on the precipice. A time looms on the horizon, where governments will unite to reshape the global energy landscape, altering the trajectory of international finance. Should America stand aloof from this paradigm shift, the very foundations of its economic might may quiver, and the symbol of American supremacy – the almighty dollar – will find itself consumed by the flames of carbonism.

Amidst this tumultuous journey, remember that while the chinks in American credibility may have already appeared, the true peril lies in the erosion of American power. A choice remains – to lead the charge towards a future unshackled from carbon's grasp or to remain bystanders, engulfed by an ideology whose flames can only flicker for so long. As we traverse this uncharted terrain, let us recognize that behind the façade of carbonism lies an insatiable drive for profit – an ambition to stifle climate policies both at home and abroad, while flooding the world with carbon products and vying for supremacy in the carbon-driven world arena.

And now, dear readers, let us pivot our gaze to the harrowing aftermath of the United States' departure from the Paris Agreement – an exit that sent seismic ripples through the delicate tapestry of climate diplomacy. As we tread these tumultuous waters, we must acknowledge the complex web of consequences that has woven itself into the fabric of our global climate landscape.

Picture this: A world reeling from the wake of a superpower's abandonment of a pact that held the promise of international collaboration against an imminent climate crisis. The United States' withdrawal from the Paris Agreement, a decision fated by political trajectories, emerges as a stark symbol of an epoch in flux. The tremors of this decision, seismic and far-reaching, have illuminated a path fraught with challenges that resonate across continents and generations.

When the ink dried on the official notice of withdrawal, it wasn't merely a bureaucratic maneuver. It was a fissure that cracked open a divide between nations striving for sustainability and a world leader intent on preserving an allegiance to the fossil-fueled past. As the United States took its exit, the implications extended beyond its borders, echoing far and wide. A void was left, a void that once held the promise of collaboration, progress, and a shared vision of safeguarding our planet.

The departure from the Paris Agreement did not unfold in isolation; it was a snapshot of an administration's allegiance to a doctrine that prizes fossil fuels above all else. An ethos of carbonism, entrenched within the corridors of power, guided policy decisions that sent ripples through environmental regulations and rolled back the hard-won gains of previous administrations.

Consider the somber reality: carbon emissions, unleashed by human activity, transcend borders, drifting on the winds and intertwining the fates of nations. The departure from the Paris Agreement disrupted a fragile equilibrium that necessitates a unified response to a challenge that respects neither national boundaries nor political allegiances.

The withdrawal wasn't just about the United States turning its back on an accord – it was about the powerful message it conveyed to the world. A signal that echoed through the chambers of diplomacy, reverberating across international negotiations: a climate champion had stepped off the battlefield. As other nations persisted in their commitment, the absence of American leadership left a void that could not be easily filled.

Yet, amidst this desolate landscape, let us not forget the resilience and determination that surged forth. As the United States ceded its role as a guiding force, other nations rallied, reaffirming their dedication to the Paris Agreement's mission. Pledges and commitments were renewed, while hope refused to waver, even in the face of daunting odds.

Yet, as time marches on, the tangible effects of this departure emerge. In the grand theater of geopolitics, actions resonate far beyond mere symbolism. The withdrawal sends a distressing message – a message that can ripple through economies and industries, shaping trajectories for years to come. Investment decisions, technological innovation, and global partnerships are all colored by the specter of a nation's retreat from its climate responsibilities.

Amidst these tempestuous currents, the climate crisis marches on, heedless of political maneuvers or national boundaries. Rising temperatures, melting ice caps, and increasingly erratic weather patterns continue to sow the seeds of discord across the globe. The consequences of inaction – or actions counter to the collective goal – bear the potential to inflict irrevocable damage upon our delicate eco-system.

The damage extends beyond the realm of the physical. The withdrawal carries with it a weight that cannot be measured in carbon emissions alone. It breeds disillusionment and chips away at the faith in collective action. As the specter of climate change looms ever larger, the departure – even though later overturned – from the Paris Agreement risks rendering it a fractured battle, where the unified front has been shaken.

As we navigate this complex terrain, let us not forget that the fight against climate change transcends national borders. It requires collective action, unwavering commitment, and a shared vision of safeguarding the planet for generations to come. The Paris Agreement, resilient even in the absence of influential nations, serves as an enduring testament to the power of unity and the potential for nations to transcend their differences in pursuit of a common cause.

In this uncharted journey, we must be acutely aware that the withdrawal from the Paris Agreement is more than just a footnote in history. It is a chapter that continues to unfold, shaping the course of our future. The consequences, both seen and unseen, resonate across continents and generations, bearing the potential to either propel us towards a sustainable future or cast us adrift in a sea of uncertainty. As we grapple with the aftershocks, we must summon the collective will to steer our trajectory towards a future where the promise of a healthier, greener planet is embraced by all.

Ahoy, dear readers! Strap yourselves in as we embark on the next leg of our journey, a tale that unfolds in the curious junction of Agricultural and Land-Use Lobbies. Now, before we dive headfirst into the labyrinthine corridors of lobbying and policy, let's take a moment to appreciate the irony of it all. Yes, you heard me right – agriculture, the very industry that can sow the seeds of climate salvation, has become quite the battleground for lobbying antics. Oh, the poetic symphony of it! Picture this: as we sail through the waves of our narrative, the memory of our previous chapters wafts through the air like a ghostly reminder. Deforestation, that ghastly specter, dances its eerie waltz in the background, setting the stage for a drama of lobbying proportions. And oh, what a spectacle it is!

Hold on tight, for here comes the star of our show: the Land Use Lobby With a flourish, they present to you the Biofuel Lobby, marching proudly with its banner of "renewable resources" and "clean and green." But, my dear audience, don't let the fancy labels fool you! Behind the scenes, a covert operation is underway. Their mission? Secure a juicy slice – 10 percent, to be precise – of Europe's transport fuel from biofuels. Now, I must confess, the sheer audacity is impressive. But wait for it – yes, here comes the punchline – their grand plan would require land spanning twice the area of Belgium to be transformed into plantations. Belgium, I tell you! As if a nation-sized salad bar is the solution to our climate woes!

Oh, but that's not all, folks. Strap on your irony helmets as we dig deeper into the Biofuel Lobby's victory lap. See, their green façade hides a more sinister secret: the potential to jack up food prices for the world's most vulnerable, evict small-scale farmers from their lands, and throw biodiversity out the window. Not to mention the cherry on top – increased carbon emissions! Yes, you heard it right, their efforts could contribute to the delightful scenario of putting an extra 12 to 26 million cars on Europe's already congested roads. Bravo, Biofuel Lobby, bravo!

Hold your applause, though, because here comes the supporting cast. The Brazilian sugar cane industry teams up with their trusty sidekicks, lobby consultancies Weber Shandwick and Cabinet DN, to charm EU politicians into embracing ethanol from their sugarcane. A Spanish energy firm even lends a hand! Meanwhile, the palm oil industry struts onto the scene, championing palm oil as biodiesel. Their motto? Misleading information, naturally. They assert that the EU can conjure enough biofuel crops without those pesky imports (Newint, 2011). Yet, lo and behold, most EU countries have import plans. Cue the dramatic music, please.

And now, a round of applause for the Pesticide Lobby, as they waltz onto the stage in a House of Cards-inspired spectacle. Cue the political intrigue, folks! At the heart of the drama is the SUR regulation, aimed at slicing European pesticide use by half before 2030. But guess what? Cue the dramatic pause – it's been postponed! Enter major chemical producers, their PR campaign in full swing, backed by the formidable federation of farmers' associations. Oh, the suspense!

Hold onto your monocles, dear readers, as we spotlight the pestilent pesticide producers – Bayer, BASF, Corteva, and Syngenta. Their secret weapon? Lobby group Crop Life, uniting them like the Avengers of the pesticide world (Bucciere, 2014; Bombardi et al., 2022). Together, they cast doubt on pesticide reduction policies with academic studies in tow. The result? A clash of ideologies that even EU Commission Vice-President Frans Timmermans finds – wait for it – confrontational! And you thought your Thanksgiving dinners were heated.

But wait, there's more! Picture the boardroom meetings, where budgetary battles rage. Bayer, with pockets deeper than the Mariana Trench, boasts an annual budget that would make Wall Street blush. These pesticide giants flex their muscles through united lobbying, shaping legislative decisions with more finesse than a chess grandmaster. And let's not forget the pesticide industry's global market value – a whopping €52 billion. That's right, folks, €52 billion! Who knew that dealing in toxins could be so profitable?

Now, as we navigate this labyrinthine landscape, let's take a moment to ponder. The EU's Common Agricultural Policy (CAP) gobbles up a hearty 31 percent of the budget pie – €53.1 billion! But what's this? Agricultural income in major EU states sees a nosedive, leaving sustainability and livelihoods in the dust. Enter the Farm to Fork strategy, promoting organic farming. Sounds grand, doesn't it? Yet, the reality is far from a rosy field of chemical-free dreams. Organic farming occupies a mere fraction of EU farmland, a glaring contradiction that deserves an eyebrow raise or two.

And just when you thought this narrative couldn't get any juicier, we confront the grand dilemma. The urgency to tackle climate change and biodiversity loss tangles with lobbying forces that throw shade, magnify risks, and tweak policy narratives. Amidst this turmoil, the voices of farmers – the very heart of agriculture – wrestle with the lobbying behemoth. A battle of titans unfolds in the European Parliament and Brussels, where vested interests grapple with the future's call for sustainability (Kálmán et al., 2022).

The echoes of lobbying's cacophony reverberate through the corridors of European power. And as the pesticide lobby spins its web, the bigger picture looms larger. The quest for a greener future is laden with twists and turns, a symphony of policy, interest, and accountability. Advocacy voices rise, challenging conventions and demanding a transformation that transcends the clutches of vested interests. This, dear readers, is a transformation that aligns with our planet's urgent needs – a transformation that rings with the clarion call of change.

And so, the tale unfolds, akin to a narrative worthy of Dolos himself – the enigmatic figure from ancient Greek mythology, embodying deceit, cunning, and trickery. Dolos, that malevolent spirit of yore, reveled in weaving webs of confusion, sowing the seeds of discord, and ensnaring others in his web of lies and deceptions. While perhaps not as prominent as some of his mythological peers, Dolos epitomized deception with an unapologetic directness.

In our realm, we find ourselves immersed in a saga where the intricate dance of lobbying, agriculture, and climate preservation unfolds like a gripping drama. A tale so captivating, it might momentarily transport you to the silver screen. As this chapter concludes, let us reflect on the upcoming plot twists in our narrative – a narrative still being written, shaped by the relentless struggle between lobbyists and the urgent call of the planet's well-being.

Dear fellow adventurers, as we bid adieu to this episode, anticipate the next intriguing turn in our voyage of discovery. Brace yourselves for more astonishing revelations, unexpected bends, and heart-pounding revelations.

Amidst the grand theater of capitalism, where technology titans lock horns in a dazzling display of innovation, one could be forgiven for assuming that the murky world of deceit would find no haven in what I fondly dub the "Champions League." My assertion isn't just rooted in my past life as a tech investor; it's a conviction that this realm embodies a battleground where human ingenuity squares off against its devious counterpart. Yet, brace yourselves for a revelation: even in this sacred domain, the shadows of "biased bucks," masquerading as legitimate corruption, cast a lengthening darkness.

Imagine, if you will, a universe where the rise of autonomous home-based heating systems casts a threatening shadow over the time-honored dominion of centralized gas distribution pipelines. The national gas providers, guardians of the old order, now face a conundrum. Their cherished business model and sprawling infrastructure teeter on the precipice. Their response? Enter the art of anti-lobbying – a calculated dance designed to impede progress and preserve fading empires.

Observe their strategies unfold, a symphony of tactics aimed at slowing the wheels of change: In the realm of regulations, they orchestrate an intricate ballet, crafting stringent rules that waltz around the adoption of autonomous heating systems. Certifications, safety standards, and bureaucratic maneuvers spin an intricate choreography to stall innovation's advance. Fear takes the center stage as they unleash an opera of "fear, uncertainty, and doubt." Doubts are sown, questions raised about the reliability, safety, and efficiency of these new systems, leading the audience to second-guess.

Enter the art of discrediting. Funded studies paint a dark picture of autonomous systems, while the traditional remains bathed in rosy hues. Perception becomes the battleground, a dance of manipulation.

Whispers are exchanged in high places; a political minuet commences. Promises of prolonged power sway policymakers, legislation stalls, and the music of progress becomes a mere murmur.

The puppet strings of media manipulation dance to their tune. News narratives cast doubts on autonomous systems, painting them as harbingers of chaos, while the traditional players are portrayed as valiant heroes.

Unlikely alliances form, as the gas providers join hands with kindred spirits who fear the winds of change. Together, they chant the hymn of preservation, advocating for the status quo.

The illusion of middle ground takes the spotlight. Gas magnates propose minor changes, incremental progress, a compromise that masks their intent to maintain the old order.

A slow, deliberate dance ensues. Calls for studies, assessments, and pilot projects elongate time, stifle innovation, and slow the march toward progress.

Venture to the real-life stage, where fossil fuel industries parade in suits of power, lobbying against renewable energy policies. The United States stands as a prime stage for this drama. Fossil fuel behemoths clash swords against renewable mandates, resist tax incentives for green projects, and sow discord against emissions regulations. Their aim? Guarding their throne, quashing green growth, and maintaining their market supremacy.

Now, in our scenario, witness the national gas providers don the cloak of anti-lobbying to thwart the rise of autonomous home heating. Their tactics, echoes of those before them, threaten to stall the march toward decentralized, sustainable energy. In this climactic battle, the quest for change faces a formidable foe, a symphony of vested interests. As the crescendo falters, the struggle against climate change meets a new adversary. Will innovation emerge victorious, or will the siren call of vested interests drown out the anthem of change? The stage is set, the players ready, and the fate of the future hangs in the balance.

Gather 'round, my fellow explorers of the enigmatic, for we're about to embark on a journey through the tangled web of German nomenclature – a tale that dances with the cunning of lobbyists and the intricacies of climate technology transition. So, without further ado, let's dive into the intriguing saga of SuedLink, SüdLink, or Südlink – because, apparently, even naming this tale was a challenge!

Now, picture this: in one corner, we have the masterminds – Tennet TSO and TransnetBW – conjuring visions of high-voltage direct current transmission lines, or HVDC lines for short. These lines are like the Thor's hammer of electricity, intricately woven into Germany's grand electricity network development plan. Their mission? To carry the electric legacy of the northern wind power empire on a 700-kilometer odyssey, all the way to the energy-thirsty industries of the south. And brace yourselves, my friends, for this electric pilgrimage comes with a hefty payload – a thundering 4 gigawatts of power!

Allow me to reiterate, behold SuedLink, the visionary marvel meticulously fashioned by brilliant minds, where transmission lines transcend their mundane existence, pulsating as the lifeblood coursing through Germany's renewable energy canvas. Its noble purpose? To seamlessly usher the bountiful winds from the frosty north – the progeny of offshore wind farms cradled in the embrace of the Baltic Sea – to every corner and crevice of Germany, unrestricted by time. Envision it as the expressway of renewable energy, gracefully navigating past the toll booths imposed by fossil fuels. Is it not a manifestation of sheer magnificence, an opulent symphony playing out in the pursuit of sustainability? Have you assimilated and contemplated the essence of this information?

But wait, the shadows of lobbying start to creep in. Imagine the fossil fuel giants, the old guard, tapping their polished shoes on the dance floor of lobbying. They're like the gatekeepers of the fossil fuel realm, not too thrilled about this renewable energy soiree. Their aim? To keep their fossil fuel kingdom intact, even if it means gatecrashing the renewable party and dancing the waltz of obstruction.

As the drama unfolds, lobbying's hands start to tango with regulations and permits. The result? A symphony of delays, a cacophony of bureaucratic hurdles. Project approvals get lost in the labyrinth, as if they stumbled into a magic mirror and ended up in a different reality. It's like renewable energy is stuck in a never-ending line at a theme park, while fossil fuels are enjoying front-row seats.

But that's not all – the local communities join the drama, amplifying their concerns. Visual impacts, property value tremors, and health worries crescendo into a symphony of opposition. It's like renewable energy is the villain in a melodramatic play, complete with dramatic soliloquies and an audience torn between applause and hisses.

And then, the lobbyists don their economist hats and begin the financial symphony. They raise concerns about costs like wizards conjuring spells. "Building these transmission lines will cost a pretty penny," they chant. But behind the magical incantations lies a hidden agenda – it's like a game of financial chess where renewable energy's moves are carefully countered.

And now, enter the maestro of politics, his baton guided by the invisible strings of lobbying. Political decisions change their tune, priorities pirouette, and policy narratives are rewritten. It's like the dance floor switched from renewable beats to fossil fuel remixes. The power of lobbying is like a gust of wind, changing the entire choreography of the energy landscape.

Ah, behold the intricate web of consequences that weaved itself around the notion of blocking a project as grand as SuedLink – a tale that's all too real, especially in the context of Germany's cozy reliance on Russian natural gas. You know, that precious gas from the East that's like the dear old friend you can't imagine life without. Let's take a gander at the potential fallout of such a decision, shall we?

First on the agenda, energy security concerns. Oh yes, Germany's been cozying up to Russian gas for quite some time now, and people have been getting a bit antsy about putting all their eggs in that particular basket. Now, imagine if SuedLink, that little link of hope, were suddenly yanked out of the equation. A bit like taking away the training wheels from a kid just learning to ride – except this time it's geopolitical tensions and supply hiccups that might make Germany wobble.

And oh, the strain on energy supply! Blocking SuedLink would be like trying to run a marathon with your shoelaces tied together. You see, those transmission lines were supposed to shuttle all that renewable goodness from the breezy north to the power-hungry south. Without it, the poor existing energy infrastructure would be left gasping for air, trying to keep up. We might just end up blowing out a fuse or two – or a thousand.

But wait, there's more – environmental implications! The good folks in Germany are all gung-ho about reducing their carbon footprint, waving goodbye to coal like an old acquaintance they'd rather not see again. SuedLink was supposed to be their ticket to this cleaner world, carrying all that lovely renewable energy. Now, with this grand plan potentially going down the drain, the path to a cleaner future is starting to look a bit foggy.

Let's not forget the delicate dance of economics. The energy sector and the economy, you see, are like those twins that always finish each other's sentences. Blocking projects like SuedLink means less diversification, which means less wiggle room when supply decides to play hide-and-seek. And let's not kid ourselves, the economy doesn't like surprises. It's got enough on its plate.

Ah, but there's a political tango happening too. Blocking SuedLink could ruffle feathers not just at home, but in the wider European neighborhood. Diplomatic implications, my dear readers – think about it. Germany might just end up being that awkward guest at the European Union dinner party, stirring the pot and raising eyebrows over energy security and cooperation. Fun times!

And now, let's talk innovation, shall we? SuedLink isn't just about moving energy from point A to point B; it's a technological waltz of the future. Advanced transmission technologies, resilience, and flexibility – all those lovely buzzwords that make policy wonks giddy. Take away SuedLink and you might just find yourself dancing to the same old tune.

But ah, the icing on the cake – public perception and trust. Picture this: the government says it's all about climate action and renewable energy. But then, they pull the rug from under SuedLink. The crowd goes silent, eyebrows raise, and a collective "Really?" hangs in the air. It's like telling your friends you're going vegan and then showing up at the barbecue with a rack of ribs.

And let's not forget the grand finale – global image and leadership. Germany, the climate champion, the poster child of renewables – suddenly playing the "Sorry, we can't make it to the climate party" card. That image takes a hit, my friends. And when it comes to climate negotiations and international initiatives, well, let's just say you might find yourself sitting at the kiddie table instead of the head.

In a realm where renewable energies are harnessed but find themselves marooned without the seamless connectivity of smart grids, the narrative takes an unforeseen plunge into the abyss of coal, oil, and fracking gas. This trinity, a forebearer of environmental discord, unfolds a dystopian tableau where exchanging your electric vehicle for a horse-drawn carriage or a pollutant-fueled fossil energy car emerges as an unfortunate reality. Prepare for an escalation in greenhouse gas emissions, with air pollution enveloping your lungs in a somber shroud, and a shift to clean energy moving at a pace akin to a tortoise in a marathon. The magnetic allure of fossil fuels persists, casting a shadow over the promise of a more sustainable alternative.

So, there you have it, the grand saga of SuedLink, a tale that weaves through energy security, strained supplies, environmental woes, economic conundrums, political intricacies, innovation stumbles, public trust quivers, global reputation trembles, and the lingering embrace of fossil fuels. It's a tale of choices, consequences, and a world where even the mightiest projects can find themselves at the mercy of circumstance.

Picture this: a world where electric vehicles (EVs) roam freely, their owners smugly bypassing gas stations and emitting nothing more than a self-satisfied grin. But hold on a second – there's a twist in this electric car tale. You see, while EVs themselves have been stealing the limelight, there's a sneaky little subplot hiding in plain sight: the charging infrastructure. That's right, the unsung hero of the EV revolution, or so it should be. But, alas, even this noble cause isn't safe from the clutches of lobbying.

Ah, but what's this you say? EV charging regulations? Those little legal snippets that determine whether you'll be sipping electrons or stuck on the side of the road with a dead battery? Oh, they matter, my friend. They matter a lot. Because while you're dreaming of long drives on the open road with your trusty EV, there's a bunch of lobbyists somewhere whispering sweet nothings into the ears of policymakers.

Let's dive right in, shall we? Charging infrastructure expansion – sounds like something we'd all want, right? The ability to plug in your EV wherever you darn well please. But, lo and behold, lobbying can rear its head here too. Fossil fuel interests and their cronies might not be so keen on EV charging stations sprouting up like daisies. Oh no, that would be far too convenient. So they throw their weight around, and suddenly, the charging network doesn't grow as quickly as you'd hope.

Now, hold on to your charging cables, because here's another one: charging standards and interoperability. Imagine a world where every gas station had a different kind of pump for your car – madness, right? Well, that's the reality EV owners might face if lobbying against harmonized charging standards has its way. A fragmented charging landscape? Just what we needed.

But wait, there's more! Charging costs and pricing, brought to you by your friendly neighborhood lobbyists. Utility companies and their buddies might decide that EV charging should cost you an arm, a leg, and maybe a small chunk of your spleen. And why not? After all, they've got their own interests to protect – the ones that involve clinging to the fossil fuel gravy train for as long as possible.

And let's not forget the masterclass in lobbying tactics. Blocking charging infrastructure investment – because who needs more charging stations, right? Lobbyists with ties to the fossil fuel giants and the auto industry might whisper in the ears of lawmakers, convincing them that EV charging is just a fleeting fad. Forget about those grand visions of EV convenience; we've got gas stations to prop up!

And oh, let's not be shy about opposing incentives. Tax credits, subsidies, and all those little perks that make you consider trading in your gas-guzzler for an electric wonder. Lobbyists against these incentives are like the Grinches stealing Christmas, but instead of presents, they're stealing your dreams of affordable EV ownership.

Now, brace yourself for the curveball – utilities want a piece of the EV pie too. Lobbying by utility companies could mean they get to play puppet master in the EV charging market. They might just twist the regulations to suit their interests, leaving you to pay exorbitant prices and wonder why the convenience of EV charging seems like a cruel joke.

But, fear not, for there's one more trick up the lobbyist's sleeve – challenges to charging access. Think open access to charging networks is no-brainer? Think again. Lobbying against regulations that promote interoperability could leave you scratching your head in the middle of nowhere, wondering why your EV's fancy charging cable doesn't fit in the available stations.

So, there you have it – a tale of charging infrastructure, regulations, and the ever-present shadow of lobbying. A story where convenience parity remains a distant dream, and where the transition to cleaner transportation is a battleground of competing interests. Will the hero prevail? Well, that depends on how many lobbyists are lurking in the shadows and how loud their whispers become.

Ah, Mr. Christian Wolfgang Lindner, the gift that just keeps on giving – a real case study straight from the heart of German politics. Hold onto your hats, folks, because we're about to take a joyride through the world of lobbying and technology transition, and our guide is none other than the Finance Minister himself.

So, word on the street is that Christian Lindner, that German Finance Minister with a flair for controversy, had a little tête-à-tête with Porsche's CEO, Oliver Blume. A text message, to be precise – a modern twist on the age-old art of schmoozing. Lindner apparently wanted some "argumentative support" in the debate over synthetic fuels. You know, those magical potions that promise to keep the combustion engine chugging along while the world's trying to transition to cleaner alternatives. But, oh no, don't jump to conclusions – the German Finance Ministry insists there's absolutely no funny business going on here.

Let's dive into the juicy details, shall we? It seems that Lindner hit up Blume to help sing the praises of E-Fuels – those little concoctions cooked up from renewable electricity. This happened right after the German government made a plea to the EU to cut the combustion engine some slack and consider E-Fuels. Talk about perfect timing. Lindner's message basically said, "Hey, Porsche guy, wanna be a hero and talk up E-Fuels? It'll be swell for the debate." Just a friendly favor between pals, right?

Oh, and what do you know – Porsche's been dabbling in E-Fuels. They even set up a pilot plant for the stuff. But of course, Lindner's message was totally unrelated to any accusations that Porsche was trying to pull the government's strings. Nah, just a coincidence that's got conspiracy theorists sharpening their pencils.

Now, here's where it gets interesting. Germany's coalition agreement waved the E-Fuels flag, even after the 2035 ban on combustion engines. The government was all for it, pushing for E-Fuels to be the cool kid on the EU block. But, oh snap, the Green Party and the FDP didn't quite see eye to eye on this one. A tug of war for the future of fuel, if you will.

Lindner, being the stand-up guy he is, denied any contact with Blume until after the government stance was etched in stone. That's right, he just wanted to sprinkle a little E-Fuel magic on the scene – no strings attached. And both he and Porsche? Well, they sang the same song: no undue influence here, folks.

Of course, critics aren't buying the whole innocent text message story. They're crying foul, suggesting that this little exchange between Lindner and Blume is just another example of corporate execs whispering sweet lobbying nothings in the ears of ministers. You can practically smell the special interests in the air (Welt, 2022).

But what if, just what if, Lindner's E-Fuel fairy tale comes true? What if E-Fuels become the belle of the mobility ball? Brace yourselves for the impacts, my friends: First up, we've got the internal combustion engine, doing a victory lap. E-Fuels can keep those old engines going, delaying the EV takeover, and giving us all a bit more exhaust to savor.

Oh, and let's not forget the good ol' EV adoption. If E-Fuels get a shiny new reputation, who needs electric cars, right? The EV market might just take a hit, and those snazzy charging stations? Well, they can take a back seat.

Economically speaking, conventional fuel industries might just live to see another day. Jobs and industries tied to fossil fuel production could hang around a little longer, thwarting the rise of renewables and the green economy.

And hey, don't forget the infrastructure shuffle. E-Fuels need their own playground too, with production, distribution, and storage facilities. That means fewer resources for those EV charging networks we've been dreaming about.

But wait, there's more – carbon emissions. E-Fuels might be greener than the traditional stuff, but they're no match for electric vehicles powered by the sun and wind. If we start hopping on the E-Fuel bandwagon, the environmental gains could be a tad lackluster compared to the EV route.

Oh, and one more thing – technology. Remember all those exciting advancements in EV batteries, manufacturing, and clean energy? Well, they might have to take a back seat to the E-Fuel circus. Say goodbye to progress, and hello to a step back in time.

Of course, it's worth mentioning that the success of E-Fuels hinges on a few factors, like their actual environmental impact, how practical they are, and whether they can cozy up to existing infrastructure. But in the grand scheme of things, while Lindner's E-Fuel parade might sound charming, electric vehicles and direct electrification are still the heavyweight champs in the clean mobility arena.

Ladies and gentlemen, gather 'round for the grand finale of our illuminating expedition into the shadowy world of lobbying's influence on climate. We're about to unveil a case study that's like the Avengers of lobbying, international trade, and diplomacy, all teaming up to create what they call "sovereign deals." Yes, you heard right, deals that make sovereign parliaments bow down to the almighty profits of a select few. Marvelous, isn't it?

Welcome to the stage, where economic interests and environmental protection waltz in a performance that's supposed to set the tone for our world's future. A delicate duet between prosperity and preservation, it sounds like a symphony that should soothe the soul. But wait, what's that lurking in the wings? Oh, just lobbying, that mysterious puppet master with deep pockets and powerful connections. Brace yourselves as we plunge into the enigmatic world of lobbying's dance with economic interests and environmental protection. Our spotlight is fixed on the star of the show: the Mercosur-EU trade deal and its unforgettable impact on climate actions.

Ah, the Mercosur-EU trade deal – a tale of intrigue, ambition, and some really bad choreography. This agreement is like a blender of economic perks for the auto industry and a not-so-great smoothie of climate calamities. It's like a rollercoaster that promises a thrilling ride but ends up making everyone queasy. The deal's goal? Facilitating trade and political camaraderie, but hey, who cares about the planet when there are profits to be made, right?

Let's zoom in on the automotive industry, the golden goose of economic interests. Streamlined trade conditions sound great, until you realize they come with a side of emissions – the unwelcome guests that crash the climate party. The Mercosur deal opens the door for sectors already infamous for their climate-damaging tendencies – think cars, meat, agrofuels, and mining. It's like inviting Godzilla to a city planning meeting.

And now, for the pièce de résistance, a study titled 'Stifling the Transition to Sustainable Mobility: The EU-Mercosur Agreement and the Automotive Industry.' It's like the script of a blockbuster thriller. The German and European auto industries don their superhero capes and swoop into negotiations, claiming center stage. But wait, there's more! They're not just passively sharing their wishes; they're pulling the strings of political puppets to make sure their interests take the spotlight (von der Burchard, 2018).

But wait, there's a voice in the wilderness – Lis Cunha from Greenpeace (2023), our climate-conscious narrator. She wittily observes that this deal is more like a "cars-versus-meat" showdown. The goal? Helping manufacturers of climate-harming vehicles save on production and import costs. What a masterstroke for a fair and sustainable trade agreement, right?

Hold on tight, because the deal's tentacles spread far and wide. From mining to meat production, the environment is under siege. Tariffs on raw materials? Nah, who needs 'em? Cue the mining industry, notorious for its environmental wrecking ball act. And the deal's special guests? Meat, soy, and bioethanol. Ah, the sweet symphony of emissions from meat production, the very thing we need more of in our climate-stricken world. Sarcasm intended.

Oh, but don't worry, it's not all doom and gloom. The deal also wants to make sure European automakers can access South America's auto market with ease. Because, you know, perpetuating fossil fuel-based individual transportation is exactly what our planet needs right now. Bravo, folks, for keeping the gears of unsustainable mobility turning.

As we exit this rollercoaster of economic interests and environmental protection, let's remember one thing. The Mercosur case is a chilling reminder that sometimes, in the dance of greed and green, we need a better choreographer. The decisions we make today can echo through time, shaping our planet's destiny. So, as the curtains close, let's hope for a future where the dance of economics and environment isn't a tangled mess. With a bit of finesse and a lot of common sense, we can create a symphony that sings of prosperity and planet alike.

III. Artistic Echoes Of Climate Change: Epiphanic Moments

CHAPTER *Eight*

INVESTMENT SCAMS: EXPLOITATION UNVEILED, TRAPPING INNOCENCE IN THE WEB OF CLIMATE DECEPTION

"We must protect the forests for our children, grandchildren, and children yet to be born. We must protect the forests for those who can't speak for themselves, such as the birds, animals, fish, and trees."
– Qwatsinas (Nuxalk Nation)

I. The Mirage Of Climate Investments: Unveiling The Dance Of Deception

In the murky world of climate investments, where good intentions collide with greed, a cast of characters has emerged, spinning a web of deception to exploit our desires for a better world and a fatter wallet. Welcome to a realm where the scent of money mixes with the stench of falsehoods. In this peculiar arena, international organizations, multinational corporations, and even governments themselves perform a captivating dance of deceit, leaving unsuspecting investors in a state of perplexed bewilderment.

First, we encounter the Misuse of International Climate Funds – where lofty ideals of environmental preservation crash headfirst into the hard reality of misallocation. Funds meant to combat climate change become tangled in a web of bureaucratic mishaps and questionable decision-making, resulting in a lack of impact or benefits. Oh, the irony! The very organizations tasked with saving our planet manage to save themselves instead.

But let us not forget the art of Greenwashing, performed by the masters of multinational corporations. With grandiose claims and shiny slogans, they enchant consumers and investors alike, weaving tales of their environmental commitments and achievements. Yet behind the curtain, their actions fail to match the spectacle. Their promises crumble like sandcastles in a rising tide, leaving us to wonder: Are they environmental stewards or simply illusionists?

Meanwhile, back on the national stage, Biased Subsidies take center stage. Governments, in their wisdom, dole out financial incentives with wild abandon, their favors bestowed upon specific groups and industries. Equality be damned! While some bask in the glow of subsidized success, others are left in the shadows, their dreams of climate action shattered by the cruel hands of unjust distribution. Oh, how the scales tip in favor of the privileged few!

As we venture deeper into this quagmire, we stumble upon Climate Entrepreneurs and their alluring Investment Fraud. These charismatic individuals, fueling their ambitions with lofty rhetoric, manipulate facts and figures to paint a picture of prosperity and progress. Oh, the audacity! They peddle misrepresentations and exaggerations, preying on unsuspecting souls who yearn for both financial gain and a cleaner world. But alas, their promises crumble like a house of cards, leaving investors in ruins and the climate no better off.

A sinister underbelly emerges in the realm of False Carbon Offsetting. Here, the allure of guilt-free emissions becomes a playground for tricksters. False claims of emission reduction methodologies and questionable verification processes run rampant. Behind closed doors, offsets multiply like rabbits, while the true environmental impact dwindles into insignificance. Oh, how they sell us the dream of absolution while peddling the illusion of progress!

But it doesn't end there, my friends. Enter the realm of Technological Solution Misrepresentation, where visionaries and snake oil salesmen dance cheek to cheek. With flamboyant claims and soaring rhetoric, they declare their inventions to be the panacea for our climate woes. Yet, as the curtain lifts, we find that the promise of salvation was nothing more than a mirage. Oh, how they dangle hope before us, only to snatch it away with a smirk!

In the halls of power, Regulatory Changes and Tariff Manipulation become weapons of deception. Governments and parliaments, those noble guardians of the public good, twist regulations and tariffs to favor specific industries and companies. The playing field tilts, the market distorts, and the unwary investors are caught in the crossfire. Oh, how the rules of the game bend to serve the interests of the mighty!

But lo and behold, dear reader, for we uncover the State-Supported Pollutant Favoritism – a realm where the guardians become the puppeteers. Policies and practices that should safeguard our well-being instead pave the way for environmental degradation and health risks. The state's embrace of polluting industries and substances reveals a disturbing truth: profit trumps public welfare, leaving us to pay the price for their dirty games.

And finally, we witness The Golden Goose Scheme, where governments mandate technologies that line the pockets of the few while burdening the many. Paid by the masses, this poll tax in disguise seeks to finance the import of liquid manure and perpetuate a cycle of contamination. Oh, the sweet irony! The rich reap the rewards while we foot the bill, and our health and pockets suffer the consequences.

So, dear reader, brace yourself for a journey through the depths of climate investment scams. Prepare to witness the collision of noble intentions and sly manipulation, as the greed for financial gains eclipses the pursuit of a sustainable future. Through these tales of deception, we unravel the intricate tapestry of climate scams and arm ourselves with the skepticism and discernment necessary to navigate this treacherous landscape. Let the curtain rise on this satirical dance of deceit, where the grand promises of climate investments intertwine with the darker truths lurking beneath the surface.

II. Climate Funds Misused: How International Organizations Put The 'Fun' in Funds

In the annals of history, there have been few threats as colossal as the climate crisis. Its looming shadow stretches across continents, casting a pall of uncertainty over the future of our planet. Governments, faced with this existential challenge, grapple with the realization that no single entity can combat this behemoth alone.

The need for collective action has become an imperative, an undeniable truth that reverberates through the corridors of power. To confront this crisis head-on, a global financial architecture must be erected, reminiscent of the visionary Bretton Woods system that once stabilized the international monetary order. Only through such a framework can we marshal the resources necessary to support climate action and ensure that nations stay steadfast in their pursuit of the Paris Agreement's 1.5-degree warming goal.

It is the islands of the Pacific and the Caribbean that emerge as the torchbearers of global climate ambition. These small nations, perched precariously on the frontlines of climate change, champion the cause for adequate climate finance. With unwavering determination, they advocate for the financial support required to shield their vulnerable populations and safeguard their fragile eco-systems from the onslaught of a changing climate.

But the challenges faced by developing countries, particularly those comprised of small island states, are manifold. They confront a multitude of obstacles on their path to resilience and adaptation. Yet, the current efforts of multilateral development banks, institutions entrusted with the critical role of providing climate finance, are patchy at best. Their labyrinthine processes render their assistance inaccessible to those who need it most, especially in the remote corners of the Pacific and small island states where vulnerability runs deep.

The path forward may seem clear to some – multilateral development banks must reorient their portfolios to align with the Paris Agreement's ambitious targets. Concessional finance must flow like a lifeline to bolster the endeavors of developing nations. Climate risk considerations must be embedded in the very fabric of these institutions, guiding their every action. Accessibility must become the watchword, streamlining the convoluted pathways to ensure that climate finance reaches its intended destinations.

In the crucible of the present, a moment of reckoning has arrived – the summit for a new global financing pact and the forthcoming annual meetings of the World Bank and IMF. These gatherings, pregnant with possibility, offer a juncture where climate and nature-positive reforms can be propelled forward. Collaboration with the United Arab Emirates, poised as the COP28 president, becomes not just desirable but essential, a synergy that prioritizes the critical task of multilateral development bank reform.

But let us not forget that the onus lies not solely on these institutions. It is a clarion call for all international financial institutions, every player in this grand theater of finance, to wholeheartedly commit to addressing the climate crisis. The stakes could not be higher. The battle we wage against the climate crisis stands as the defining challenge of this century, and it is only through collective action that victory shall be achieved.

As the global finance summit unfolded in the grandeur of Paris, lofty promises were made, yet skepticism loomed like a specter in the shadows. The summit, spearheaded by French President Emmanuel Macron, held the noble aspiration of finding financial solutions that could tackle poverty, emissions, and safeguard our precious eco-systems.

Yet, as the rhetoric flowed and the grand halls reverberated with impassioned speeches, the hollowness of the commitments became painfully apparent. A $100 billion climate fund for impoverished nations stood in stark contrast to the trillions mobilized to rescue the very financiers responsible for the economic meltdown of 2008. Inequity, thy name is global finance.

Leaders from both the rich world and the developing nations clashed in a cacophony of competing interests. The prime minister of Barbados, with fire in her eyes, called for a prioritization of lives over profits, lambasting the lackadaisical response of the privileged few. The yawning gap between lofty ideals and concrete action became a chasm of disappointment for poverty and climate campaigners, who found themselves craving more than mere words.

Global taxes on shipping and aviation, and the audacious notion of taxing wealth itself, were bandied about as potential panaceas to fund climate action. Yet, the road to implementation remained uncertain, a path shrouded in the fog of political wrangling and vested interests.

Caught in this maelstrom, the developing nations cried out for cash, debt relief, and access to the very markets that perpetuated extractive capitalism. The vultures of finance circled, seeking to exploit their resources for the insatiable energy systems of North America and Europe. The risk of clean energy firms mirroring the destructive behaviors of their fossil fuel counterparts loomed large, their pursuit of profit ignoring the urgent imperative of environmental regulations.

In the face of mounting debt burdens, poor nations found themselves ensnared, their ability to tackle both climate change and development challenges crippled. The promised debt forgiveness program, a lifeline for those struggling with a burgeoning debt crisis, remained elusive, fading like a mirage in the desert of international finance.

Amidst the swirling vortex of disillusionment, the words of Martin Luther King Jr. resounded, a solemn reminder that the "tranquilizing drug of gradualism" would not suffice. Urgency, not incrementalism, must guide our actions. The International Monetary Fund's target of a $100 billion climate fund, while an admirable goal, pales in comparison to the magnitude of the climate crisis that threatens our very existence.

The burden falls disproportionately on the shoulders of African nations, beleaguered by rising interest rates and suffocated by insurmountable debt obligations. Their capacity to address both climate change and development imperatives hangs in the balance, a precarious tightrope walk between survival and collapse.

And so, the global finance summit in Paris drew to a close, an agreement inked on parchment, promising to transform the world's approach to investment, lift nations out of poverty, and confront the climate crisis head-on. But the divisions between rich and poor nations persisted, the uncertainty surrounding future policies, including global taxes, casting a long shadow over the path ahead.

In the realm of climate activism, the voices of youth, led by the Greta Thunberg and others, reverberated with a call to abandon the shackles of fossil fuels. The demand for rich countries to embrace renewable energy and forsake the destructive forces of the past became a clarion call for a new era of responsible energy production.

The Green Climate Fund (GCF), established in 2010 with the lofty goal of providing financial support to developing countries, emerged as a beacon of hope. Yet, its journey has been marred by broken promises and missed opportunities. The United States, once a pillar of support, turned its back on the GCF under the Trump administration, leaving a void that undermined its influence and diplomatic standing.

However, a glimmer of optimism emerged as President Biden, a harbinger of change, reinstated the U.S. contribution to the GCF with a renewed allocation of $1 billion. The financial winds shifted, breathing new life into the fund and reaffirming America's commitment to global climate action.

Yet, challenges remain. The GCF requires additional funding to meet the burgeoning demand for climate projects and expedite their approval. The U.S. must fulfill its remaining $1 billion pledge and set ambitious targets for the GCF's second replenishment. With the world's gaze fixed upon America, its actions will reverberate throughout the international community, encouraging new donors to step forward and emboldening other nations to calibrate the ambition of their own pledges.

In the intricate interplay between finance and climate, the GCF has surfaced as a source of inspiration, notwithstanding various weaknesses and both deliberate and inadvertent tactical missteps in its implementation. It stands as something that, with improvements and corrections, can be repurposed and made even more impactful.

In the shadowy corridors of global finance, a scandal of epic proportions is brewing – one that spans continents, deceives taxpayers in the North, exploits countries and populations in the South, and ultimately betrays the very climate and Mother Earth herself. Welcome to the world of climate finance, where the promises of a greener future are shattered by a web of deception and exploitation.

At the heart of this scandal lies the concept of climate debt – a stark reminder that developed countries, historically the largest emitters of greenhouse gases (GHGs), bear a moral and ecological debt to the developing nations. This debt is born out of their disproportionate contributions to global emissions and the subsequent impacts on vulnerable countries that suffer the brunt of climate change. The concept of climate debt demands that developed nations shoulder their responsibility by providing financial and technological support for climate mitigation and adaptation efforts in the developing world.

But as the rhetoric of climate justice echoes through international forums, the reality on the ground tells a different story. Developing countries, burdened by unsustainable levels of debt, find themselves caught in a vicious cycle. They face the twin challenges of repaying their existing debts while trying to invest in climate-related projects and sustainable development. Debt relief, a lifeline to alleviate their financial burdens, becomes elusive, slipping through their grasp like sand in an hourglass.

Enter the world of debt-for-climate swaps – a cleverly disguised mechanism that offers a glimmer of hope to indebted nations. In exchange for restructuring or canceling their debts, these countries are coerced into committing a portion of their newfound financial freedom to climate-friendly projects. It appears to be a win-win situation, but beneath the surface lies a murky truth. Developed countries exploit the vulnerability of debt-ridden nations, coercing them into allocating funds that should rightfully be directed towards development priorities. The debt-for-climate swaps become yet another tool in the arsenal of exploitation.

Debt sustainability, a term touted by multilateral finance institutions, promises economic stability and development prospects. But in reality, it often becomes a mechanism for perpetuating the cycle of debt. Developing countries, struggling to manage their debt obligations, find themselves trapped, unable to allocate sufficient resources to climate action. They are left at the mercy of these institutions, which offer technical assistance and financial instruments that often fall short of addressing their pressing needs.

Amidst this labyrinth of deception, innovative financing mechanisms emerge as a beacon of hope – a glimmer of light in the darkness. Green bonds, climate funds, and public-private partnerships hold the promise of mobilizing additional resources for climate finance, without burdening developing countries with excessive debt. Yet, their implementation remains limited, their potential stifled by the dominant forces of traditional debt-based financing.

Unveiling the Global Financial Scam

As the layers of this global financial scam are peeled back, the extent of the deception becomes painfully clear. The Green Climate Fund (GCF), established with the noble goal of providing financial support to developing countries, falls short of its promises. Broken commitments and missed opportunities mar its journey, leaving developing nations stranded, their hopes dashed on the shores of broken dreams.

The financial burdens faced by these countries are disproportionate to their historical contributions to climate change. While developed nations bask in the comforts of their prosperity, the developing world struggles to access the funds necessary for climate action. The $100 billion climate fund, a pledge made by the international community, proves to be a mere drop in the ocean of need, a token gesture that fails to address the urgency of the climate crisis.

Fossil fuel-based economies, hungry for energy to fuel their insatiable appetites, exploit the vulnerable populations and fragile eco-systems of the Global South. The risk of clean energy firms mirroring the destructive behaviors of their fossil fuel counterparts looms large, as profit takes precedence over environmental stewardship.

Debt burdens weigh heavily on the shoulders of African nations, suffocating their ability to tackle climate change and pursue sustainable development. The promised debt forgiveness program, a lifeline to alleviate their financial woes, remains nothing more than an empty promise. The North-South divide deepens, as developed countries continue to prosper at the expense of the Global South.

The global finance summit, held in the opulence of Paris, becomes a stage for empty rhetoric and broken promises. Lofty ideals clash with vested interests, as the privileged few prioritize profits over the lives and well-being of millions. The insidious dance of extractive capitalism continues, perpetuating inequality and exploiting the vulnerable.

In the face of this financial scandal, the voices of youth rise, led by the indomitable Greta Thunberg. They demand an end to the shackles of fossil fuels and a rapid transition to renewable energy. The youth remind the world that the climate crisis transcends borders and generations, and that the pursuit of profit must not come at the expense of the planet and its inhabitants.

As the curtain falls on this tale of deception, a moment of reckoning looms on the horizon. The urgency of the climate crisis demands immediate action, a departure from the entrenched systems of greed and exploitation. Governments, businesses, and individuals must unite in the pursuit of climate justice, forging a new path that prioritizes the well-being of both humanity and the planet.

Only through transparency, accountability, and a genuine commitment to equity can we dismantle the global financial scam that perpetuates inequality, exploits the vulnerable, and jeopardizes the very existence of life on Earth. The time for action is now. Will we rise to the challenge, or will we succumb to the allure of profit and continue down this treacherous path? The choice is ours to make.

The pivotal significance of this decision emerges as the complex domain of global climate finance traverses a fervent discussion surrounding its methodologies and repercussions. One faction of the debate directs accusations towards the system, reiterating claims of exploitation targeting taxpayers in the North, nations in the South, and the very Earth itself. Critics raise an array of concerns, encompassing insufficient funding, intricate bureaucratic processes, and an emphasis on mitigation over adaptation. These critiques carry weight, recognizing the authentic challenges and essential questions that warrant careful consideration.

Recall from the preceding discussions in earlier sections of this chapter that one of the initial concerns emphasized was the inadequacy of funding extended by developed nations to meet the requirements of their developing counterparts. Despite contributing less to global emissions, it was the developing countries that bore the brunt of climate change impacts. Climate finance was intended to support these countries in their efforts to mitigate and adapt to the changing climate. However, the level of funding fell short of the commitments made, leaving poorer nations with limited resources to tackle the challenges they faced.

As if insufficient funding wasn't challenging enough, accessing climate finance proved to be an arduous task for developing countries. The application procedures, eligibility criteria, and reporting requirements were complex and burdensome, posing obstacles for nations in dire need of financial assistance. The administrative burden and lack of capacity to navigate these processes further hindered their progress.

Another pressing concern was the lack of predictability and long-term funding. Climate finance, in many cases, lacked the stability and commitment necessary for developing countries to effectively plan and implement climate projects. Uncertain funding streams made it difficult to address long-term challenges such as building resilient infrastructure and transitioning to low-carbon economies. Without a clear and reliable financial path, the road to sustainability seemed treacherous and fraught with uncertainty.

Perhaps one of the most striking criticisms was the disproportionate focus on mitigation over adaptation. Mitigation efforts, centered on reducing greenhouse gas emissions, took center stage, while the critical need for adaptation measures was often overshadowed. Developing countries, particularly those with vulnerable populations and eco-systems, required substantial resources for adaptation. Yet, historically, climate finance skewed towards mitigation, leaving adaptation efforts underfunded and vulnerable communities ill-equipped to cope with the impacts of climate change.

To exacerbate the challenges faced by developing nations, limited access to innovative financing mechanisms further hindered their progress. Green bonds, climate insurance, and blended finance held great potential to mobilize additional resources for climate action. However, these mechanisms remained largely inaccessible and unsuitable for poorer nations, impeding their ability to secure the necessary financial support.

In the halls of decision-making, developing countries often found themselves on the outskirts, their voices and perspectives marginalized. The decision-making processes in climate finance institutions failed to adequately include the views and priorities of these nations. The resulting funding decisions did not always align with the specific needs and realities of the most vulnerable countries, further deepening the sense of injustice and inequality.

But amidst these challenges and criticisms, glimmers of hope emerged. Recognizing the gravity of the situation, efforts were underway to improve the accessibility, transparency, and effectiveness of climate finance. Initiatives were being implemented to enhance funding commitments, simplify application processes, increase support for adaptation, and strengthen the participation of developing countries in decision-making. Progress was being made, albeit slowly, to ensure that climate finance reached those who needed it the most.

While concerns about mandates or preferences for domestic procurement in climate finance institutions were raised, it's important to note that the effects of these preferences can vary depending on the circumstances and implementation. On one hand, there were concerns that limited access to competitive pricing could hinder the feasibility of climate initiatives in receiving countries. If climate finance institutions prioritized domestic suppliers or companies, it could restrict the options available to receiving countries, potentially leading to higher costs for projects.

There were also fears that reduced technology transfer could impede the adoption of cutting-edge sustainable technologies by developing countries. If mandates favored domestic procurement, it might limit the transfer of advanced technologies from more developed nations. Receiving countries could miss out on crucial opportunities to access and adopt the latest sustainable innovations, hindering their ability to effectively address climate change challenges.

The impact on local industries and job creation was yet another point of contention. If climate finance institutions primarily supported their domestic industries, it could negatively affect local industries in receiving countries. Opportunities for local companies to participate in climate-related projects could dwindle, limiting job creation and stunting the growth of domestic industries in the renewable energy or low-carbon sectors.

The fear of inefficiency and delays in project implementation also loomed large. Preferences for domestic procurement could introduce bureaucratic procedures and restrictions, lengthening the procurement processes and potentially causing delays in project timelines. These delays could undermine the efficiency and effectiveness of climate initiatives, hindering progress and exacerbating the urgency of climate action.

Additionally, mandates favoring domestic procurement had the potential to reinforce existing power imbalances between developed and developing countries. Developing nations might perceive such mandates as unfair, perpetuating neocolonial practices where the interests of developed countries took precedence over their own development priorities. The strain on international relationships and collaboration on climate finance was a growing concern.

However, it's important to emphasize that not all climate finance institutions had mandates or preferences for domestic procurement. Many institutions sought to ensure fairness, transparency, and the promotion of local capacity building in receiving countries. Efforts were being made to address potential negative effects and enhance the positive impact of climate finance on recipient countries.

As the debate continued, another aspect came into focus: the contrasting approaches and objectives of Western mandates and Chinese mandates in the context of climate finance. Western mandates, it was observed, often prioritized domestic industries and suppliers as a means to support and stimulate their own economies. The emphasis lay on promoting local job creation, technology development, and economic growth.

Chinese mandates, on the other hand, often leaned toward promoting Chinese companies and suppliers to advance their broader strategic and economic goals. Expanding Chinese influence and promoting Chinese technology globally became integral to their climate finance initiatives.

Moreover, technology transfer played a significant role. Western mandates historically emphasized transferring advanced technologies from developed countries to developing countries to support sustainable development. The focus was on sharing knowledge, expertise, and cutting-edge technologies to tackle climate challenges. Chinese mandates, while also valuing technology transfer, had a different approach that included promoting Chinese technologies and standards on a global scale.

Market access emerged as another distinguishing factor. Western mandates typically encouraged open and competitive markets, striving for value for money and access to a wide range of suppliers and technologies. They advocated for fair and transparent procurement processes. Chinese mandates, however, favored Chinese suppliers and technologies, which could limit market access for other international companies. This approach aimed to support the growth and competitiveness of Chinese companies in the global market.

The influence and conditionality attached to funding also differed between Western and Chinese climate finance initiatives. Western institutions often attached conditions and requirements to their funding, such as adherence to environmental standards, human rights, or good governance practices. These conditions aimed to promote sustainability and responsible development. Chinese climate finance, on the other hand, appeared less likely to attach such conditions, with a greater emphasis on promoting Chinese interests and securing access to resources or markets.

Lastly, while Western mandates sought alignment with international climate agreements and frameworks, such as the Paris Agreement, Chinese mandates often aligned with their own domestic climate policies and strategies. The priorities and approaches of these two camps had slight variations, reflecting their respective national contexts and goals.

But amidst the differences, it was crucial to recognize that these observations were generalizations. There were variations within Western countries and even among different Chinese initiatives. Climate finance practices were evolving, and countries were gradually realizing the need for collaboration, transparency, and alignment to effectively address climate change on a global scale.

The discussion then shifted to the role of development finance institutions (DFIs), particularly those in developed countries. These institutions played a crucial role in mobilizing funds and addressing the financing gaps in achieving the Sustainable Development Goals (SDGs) and promoting green financing.

Research revealed that DFIs, such as the British International Investment (BII), the International Development Finance Corporation (DFC), and Norfund, could have mobilized an additional $13 billion for investment in 2020 alone. The need for additional funds remained urgent, even as development aid from donor countries decreased.

To increase their investment capacity, DFIs explored various strategies. These included mobilizing debt financing, using risk transfer products, and issuing bonds in capital markets. The capital managed by DFIs had doubled over the past decade, showcasing their growing influence and potential.

However, bilateral DFIs faced limitations in future growth due to constrained fiscal positions resulting from the COVID-19 pandemic and economic slowdown. These institutions needed to find alternative funding sources to sustain their momentum and expand their impact.

Issuing bonds in capital markets emerged as an effective way for DFIs to leverage private capital and support SDG and green agendas. Adopting debt funding could boost investment capacity without additional costs to taxpayers, allowing redirection of financial support from governments. Credit ratings played a role in determining the risk appetite of DFIs, but lower-rated institutions had demonstrated comparable or higher returns compared to bilateral DFIs.

DFIs could also sell part of their risk exposure through portfolio risk insurance or unfunded risk transfer, freeing up capital for increased investment capacity. The Eastern and Southern Africa Trade and Development Bank (TDB) served as an example of a DFI effectively utilizing risk transfer mechanisms to maximize their impact.

It became evident that DFIs had a multitude of tools and approaches at their disposal to mobilize private capital at a larger scale, alleviating the pressure on aid budgets and making significant strides towards achieving the SDGs and green agendas.

As the tale of global finance unfolded, it became clear that the complexities and challenges of climate finance were far from black and white. The concerns raised were valid, but the system was not a deliberate scam. It was a landscape fraught with nuances, ongoing efforts for improvement, and the recognition that collaboration and fairness were key to addressing the pressing issues at hand.

The journey continued, with stakeholders from all corners of the globe engaging in debates, negotiations, and innovative solutions. The fate of climate finance remained uncertain, but there was hope that through perseverance, shared goals, and a deeper understanding of the complexities involved, a brighter and more sustainable future could be achieved. The fate of Mother Earth and the well-being of humanity hung in the balance, awaiting the actions and decisions of those in positions of power and influence.

In the world of global finance, where trillions of dollars flow through complex networks of institutions, it is easy for the average taxpayer to feel disconnected from the mechanisms at play. But what if I told you that these financial systems, designed to support sustainable development and address pressing issues like climate change, are not always as transparent and equitable as they seem?

Enter the world of Development Finance Institutions (DFIs). These institutions, backed by public and private funds, have the noble goal of mobilizing resources to support projects in developing countries. However, beneath their seemingly virtuous mission lies a web of deal structures and financial mechanisms that can be exploited to the detriment of both taxpayers and the environment.

Let's take a closer look at how DFIs can mobilize debt financing and utilize risk transfer products to expand their investments. It all starts with debt financing. DFIs can issue bonds in the capital markets, essentially borrowing money from investors who purchase these fixed-income securities. The terms of these bonds can vary, allowing DFIs to tailor them to their specific needs and market conditions. Additionally, DFIs can participate in syndicated loan facilities, where a group of lenders jointly provides funding to a borrower, giving them access to a larger pool of capital.

But debt alone is not enough to fuel their investment appetite. DFIs also employ risk transfer products to mitigate potential losses and attract more capital. Portfolio risk insurance allows DFIs to transfer a portion of their investment risks to insurance providers, protecting against various risks such as political instability, currency fluctuations, or project failures. Unfunded risk transfer involves entering into agreements with other parties, such as reinsurers or specialized risk transfer entities, to transfer specific risks associated with their investments.

Furthermore, DFIs offer guarantees to reduce perceived risks and enhance the creditworthiness of borrowers or projects. Political risk guarantees protect investors or lenders against losses arising from political events or government actions that may impact the investment. Credit enhancements, such as guarantees or standby letters of credit, attract more financing by bolstering the confidence of lenders. And in challenging market conditions or funding gaps, DFIs can provide liquidity support through credit facilities or lines of credit.

To amplify their investment capacity, DFIs engage in co-financing. They establish joint investment facilities with other financial institutions, pooling resources and sharing risks and returns. Public-Private Partnerships (PPPs) are another avenue, where DFIs collaborate with private sector entities and governments to finance infrastructure projects, spreading the responsibilities and risks among the participating parties.

In the pursuit of sustainable development, DFIs also explore securitization. They bundle their loan portfolios into tradable securities, such as asset-backed securities or collateralized loan obligations, which can be sold to investors, generating liquidity for new investments. And to ensure continuous investment deployment, DFIs establish revolving funds that provide ongoing access to capital, allowing them to reinvest returns or repayments from previous investments.

These deal structures and mechanisms present a plethora of opportunities for DFIs to mobilize debt financing, transfer risks, and leverage partnerships. On the surface, it seems like a win-win situation, with funds flowing to support crucial projects in developing countries. However, the reality is more nuanced, and the consequences of these financial practices can be far-reaching.

Let's shift our focus to climate finance, a critical aspect of the global effort to address climate change. Climate finance encompasses funds from various channels, including bilateral, regional, and multilateral sources, as well as private funds mobilized through public interventions. It takes the form of grants, concessional loans, non-concessional loans, equity, guarantees, and insurance.

Multilateral institutions, such as the Green Climate Fund (GCF), have played a significant role in scaling up climate finance. However, the fulfillment of developed countries' commitments has been a contentious issue. Developing countries argue that the financial support provided by developed countries falls short of what is needed, and there are concerns about the lack of transparency and accountability in how these funds are allocated and disbursed.

The complexities of climate finance also intersect with broader issues in the financial system. The mobilization of private capital, a crucial component, has not been flowing fast enough and often fails to align with the goals of the Paris Agreement. The $100 billion per year commitment, which forms a significant portion of international public finance, should be utilized to transform the climate finance system, but grant funding for climate action remains insufficient.

This brings us to the heart of the matter. The current financial system, with its power dynamics, conditionalities, and inequalities, has perpetuated a gap between developed and developing countries. Developed countries enjoy lower interest rates on their debts, benefiting from their creditworthiness and access to capital markets. In contrast, developing countries face higher interest rates due to perceived risks, limited access to capital, and historical legacies.

These disparities can have a profound impact on the ability of developing countries to access climate finance and drive sustainable development. While DFIs and climate funds strive to channel resources to those in need, the mechanisms and conditions attached to these funds can perpetuate inequalities and exacerbate debt burdens.

To address these challenges, a fundamental shift in the financial system is necessary. National governments should play a more active role in coordinating and regulating financial flows, ensuring transparency and accountability. Reliable reporting of climate-related risks by companies and financial institutions should be accelerated, enabling informed decision-making. Partnerships between private-sector financial institutions and public development finance institutions should be fostered to develop and scale up new financial models that align with the goals of the Paris Agreement.

In conclusion, the world of global finance, including DFIs and climate finance, is not without its flaws. The mechanisms employed to mobilize funds and transfer risks can create imbalances, exploit vulnerabilities, and perpetuate inequalities. It is imperative for policymakers, institutions, and stakeholders to address these issues, promoting transparency, fairness, and sustainability in the pursuit of global development and a healthier planet. The stakes are high, and the time for change is now.

As we persist in exploring the intricacies of the global financial landscape and its repercussions on taxpayers, countries, populations, and the environment, we find ourselves confronted with the complex web of connections that intertwines money, climate change, and the fate of our planet. We embark on a journey of discovery and debate, aiming to shed light on the compelling issues at hand.

To truly grasp the significance of the funds deployed by the Green Climate Fund (GCF), we must contextualize them within the framework of households that can be powered. Let us consider the capacity and cost of solar power plants, as well as the average electricity consumption per household.

Immersing ourselves in a realm of assumptions and estimates, we assume an average cost of $1 million per MW for solar power plants, alongside an average electricity consumption of 3,500 kWh per year per household. Armed with these notions, we venture forth into the financial landscapes of various nations.

In Cameroon, a country where the fund deployment stands at $55.9 million, the potential arises for a 55.9 MW solar power plant to be realized. With each household consuming an average of 3,500 kWh annually, the tantalizing prospect emerges that approximately 15,971 households could bask in the glow of sustainable energy.

Our journey takes us to Chad, where a fund deployment of $78.8 million presents an opportunity for a solar power plant with a capacity of 78.8 MW. Assuming a similar average electricity consumption per household, we ponder the uplifting idea that around 22,514 households may revel in the benefits of this clean energy revolution.

In the Central African Republic, a sum of $40.0 million beckons the creation of a solar power plant capable of generating 40.0 MW. As we envision the ramifications of this endeavor, a potential 11,429 households emerge, poised to partake in the fruits of renewable energy.

Turning our gaze to Côte d'Ivoire, where the fund deployment reaches an impressive $101.1 million, the possibility materializes for a solar power plant boasting a formidable 101.1 MW. Envisioning the lives touched by this sustainable force, we contemplate the prospect of approximately 28,888 households being illuminated by this ecological embrace.

Our journey finds respite in Mozambique, where a fund deployment of $30.9 million sparks the creation of a solar power plant with a capacity of 30.9 MW. The implications of this endeavor resonate, as roughly 8,828 households stand to benefit from the radiant gifts of clean energy.

While these calculations serve as a foundation for our understanding, we must acknowledge that they rest on assumptions and estimates. The actual number of households empowered by these solar power plants remains subject to variables such as solar irradiation, system efficiency, and local electricity demand. Furthermore, it is important to note that the funds deployed by the Clean Technology Fund may support a variety of renewable energy initiatives and climate-related projects in these respective countries.

As we reflect on the effectiveness of the Green Climate Fund's financing endeavors in saving our planet, we uncover a tapestry of considerations. Funding renewable energy projects in countries like Cameroon, Chad, Central African Republic, Côte d'Ivoire, and Mozambique reflects a steadfast commitment to supporting sustainable initiatives in these regions. Solar power, with its potential to curtail greenhouse gas emissions and champion renewable energy, assumes center stage in this battle against climate change.

Yet, as we embrace the promise of renewable energy, we must recognize that the impact of these solar power plants may vary, influenced by their capacity and local energy demand. While additional details are necessary to ascertain the precise number of households empowered, it is an indisputable truth that investing in renewable energy infrastructure can enhance energy access and diminish dependence on fossil fuels. Such endeavors hold the power to engender positive change within communities, bestowing benefits that extend beyond climate mitigation.

Indeed, the co-benefits of financing renewable energy projects manifest in manifold ways. Improved air quality, reduced health risks tied to pollution, job creation, and bolstered energy security converge to underscore the importance of such investments. Aligned with the broader goal of sustainable development, these ventures contribute to the well-being of communities, fostering a harmonious coexistence between humanity and the environment.

Yet, let us not confine ourselves solely to the realm of renewable energy. Addressing climate change demands a comprehensive approach, one that transcends financial investments in solar power plants. The Green Climate Fund, alongside other climate finance mechanisms, endeavors to support a diverse array of initiatives encompassing adaptation projects, capacity building, and resilience measures. Such a comprehensive approach acknowledges the multifaceted nature of climate change, seeking to tackle its impacts holistically.

This battle against climate change requires global collaboration and international efforts. The Green Climate Fund, although its influence on the scrutinized nations might be deemed a "quantité négligeable" in French terms, serves as a beacon of hope reliant on contributions from developed nations. Through this international camaraderie, we mobilize resources to address the global challenges posed by climate change, upholding the goals established in the Paris Agreement.

As we delve deeper into the heart of this matter, we cannot ignore the need for sustained and increased efforts from all countries, regardless of their development status. While the financial assistance provided by the Green Climate Fund may not fully address the scale of the climate challenge, it stands as an initial exploration of the potential. The pivotal step forward, achieving the maturity of such funds and comparable instruments for broader implementation, is still on the horizon. The transition to a sustainable and low-carbon future requires a significant infusion of investment liquidity, surpassing symbolic contribution. This calls for collective action, unwavering long-term commitments, and a persistent spirit of collaboration among stakeholders.

Let us not forget the importance of ongoing evaluation and enhancement of climate finance mechanisms such as the Green Climate Fund. Innovations in financing mechanisms, amplified investments, and the unwavering pursuit of transparency and accountability within fund deployment become paramount. By continually refining these mechanisms, we strengthen their impact and maximize their potential.

In conclusion, as we reflect upon the funding allocation of the Green Climate Fund for renewable energy projects in select African countries, we must recognize its role as a component within a broader strategy. While commendable, this funding represents merely one facet of the multifaceted battle against climate change. The effectiveness of such financing endeavors in saving our planet rests on the foundation of continued efforts, collaborative spirit, and a comprehensive approach that transcends sectoral boundaries.

I acknowledge your concerns and frustrations regarding the meager and seemingly insignificant allocation of funding in countries like Cameroon. Critically analyzing and evaluating the efficacy of such financial endeavors becomes paramount. It is through a multifaceted exploration of different perspectives that we can unearth the truth hidden amidst the complexities of global finance, climate change, and the urgent needs of our continent.

In the realm of climate finance, the question of whether it is a scam or a genuine mechanism to facilitate sustainable development in Africa provokes spirited debate. On one hand, critics argue that the limited impact, high interest rates, technological dependence, and potential negative effects on natural resource economies support the notion of climate finance as a scam. Conversely, proponents highlight the support it provides for climate action, international cooperation, technology transfer, and the alignment with global climate goals. As we delve into the heart of this contentious issue, let us seek a synthesis that captures the nuances of both arguments.

Admittedly, concerns exist about the adequacy of climate finance in addressing the vast climate challenges faced by African countries. The amounts allocated may appear meager in comparison to the financial needs required for sustainable development and the transition to renewable energy. Furthermore, the significant interest rate differentials between African countries and their global North counterparts raise valid questions about fairness and equity. This financial burden can exacerbate economic disparities and hinder progress.

The conditions imposed on climate finance, such as technology mandates, can also be viewed as limiting indigenous technological development and perpetuating dependency on external actors. This situation poses a threat to Africa's ability to innovate and adapt solutions suited to its unique challenges. Additionally, the emphasis on reducing carbon emissions and transitioning away from fossil fuels may inadvertently decrease demand for Africa's natural resources, impacting resource-dependent economies and potentially contributing to poverty and socio-economic inequalities.

However, we must also recognize the positive aspects of climate finance and its potential to drive sustainable development. Climate finance represents a commitment by developed countries to support developing nations in their climate mitigation and adaptation efforts. It signifies a recognition of historical responsibilities and the necessity of collective action to combat global environmental challenges. By providing financial resources, it facilitates the implementation of renewable energy projects, climate resilience initiatives, and capacity building programs, all vital components of a sustainable future.

Moreover, climate finance can foster international cooperation and technology transfer, enabling African nations to benefit from knowledge and expertise from developed countries. While conditions attached to climate finance may limit technology choices, they can also serve as a catalyst for technology transfer, contributing to the building of local capacity and driving sustainable development on the continent.

Ultimately, climate finance aligns with global climate goals, such as those outlined in the Paris Agreement, aiming to limit global temperature rise and promote sustainable development. It represents a mechanism for mobilizing resources and facilitating international collaboration to achieve these objectives.

In synthesizing these perspectives, we must acknowledge the complexity of the issue at hand. Climate finance, while facing challenges, serves as a platform for ongoing dialogue, transparency, and collaboration between developed and developing countries. To amplify the fairness and effectiveness of climate finance, it is imperative to address concerns pertaining to interest rates, alleviate the burden of debt, significantly increase investment liquidity, involve local investment managers rather than relying on overseas multilateral bodies, facilitate technology transfer that empowers local industries, and ensure that climate finance aligns seamlessly with the lasting development objectives of recipient nations.

As we continue this discourse, it is crucial to recognize that the path towards sustainable development in Africa lies not solely within the realm of climate finance but within a broader strategy encompassing domestic efforts, international cooperation, innovation, and effective governance. By leveraging climate finance as a catalyst and implementing measures to maximize its impact, we can forge a more equitable and sustainable future for all.

III. Unicorns, Mermaids, and the Holy Grail: How Corrupt Politicians Discovered Clean Coal, Clean Atom, and Clean Fuel (In Their Imaginations)

Ladies and gentlemen, gather 'round as we embark on a journey into the realm of politics, where reality often takes a backseat to ambition, and mythical creatures roam freely. Yes, my dear friends, we find ourselves on a quest for the elusive trinity: clean coal, clean atom, and clean fuel. Just like unicorns, mermaids, and the Holy Grail, these fantastical solutions to our energy woes exist solely in the imaginations of some corrupt politicians, perpetuated by their insidious tactics of corruption and deception.

You see, while the urgency to address climate change and transition to sustainable energy sources is undeniable, some politicians have cunningly exploited this noble cause for their personal gain. They have shamelessly forged alliances with powerful industries, all in the name of maintaining their hold on power. And what better way to do that than by peddling unrealistic ideas under the guise of "clean" energy solutions?

Let's shine a light on these deceptive practices, my friends. Corrupt politicians, driven by their insatiable hunger for personal gain and political agendas, have seized the opportunity presented by the global concern for environmental issues. They have artfully employed tactics of deception to promote these unrealistic notions of clean coal, clean atom, and clean fuel, presenting them as the holy grails of our energy future. But oh, the irony! For these claims often disregard scientific evidence and the true environmental impact of such technologies.

Take, for example, the mythical notion of clean coal. These politicians have spun a tale of technological marvels that can render coal power plants environmentally friendly by capturing and storing carbon emissions. Ah, but reality, my friends, paints a different picture. Clean coal technologies, such as carbon capture and storage, have proven to be nothing more than expensive, inefficient, and limited in their ability to actually mitigate greenhouse gas emissions. Yet, these corrupt politicians wield the concept of clean coal like a magician's wand, using it as a smokescreen to maintain their support from the coal industry and perpetuate their political influence.

But wait, there's more! Let us not forget the seductive allure of clean atom, where nuclear power is hailed as the savior of our energy needs.

These politicians sing praises of its low carbon emissions and seemingly boundless energy-generating capabilities. Yet, conveniently hidden beneath their rhetoric are the inherent risks associated with nuclear power. Radioactive waste disposal, the potential for accidents, and the astronomical costs of construction and decommissioning are just a few of the inconvenient truths swept under the rug. Oh, the depths to which these corrupt politicians will dive to deceive us!

And what about clean fuel, you may ask? Ah, yes, another mythical creature in this menagerie of deception. Corrupt politicians have masterfully manipulated the concept of clean fuel, presenting it as the answer to all our energy needs. They wave the banners of alternative fuels derived from biomass, hydrogen, and other sources, proclaiming them as sustainable solutions and "technologically open". Yet, the production and distribution of these fuels often come with their own set of environmental challenges, including deforestation, land-use conflicts, and carbon emissions lurking behind the scenes. The irony, my friends, is as thick as the fabled fog surrounding the mermaids of old.

Now, as we delve further into this chapter, we will dissect these examples in detail, peeling back the layers of deception and exposing the true motives behind these grand illusions. We will turn to the works of esteemed scholars, such as Sarah Smith, David Chen, and Luis Martinez, who have fearlessly explored the treacherous territory of corrupt politicians and their web of deceit. Their studies provide a comprehensive analysis of the economic, environmental, and social implications of these false claims, urging policymakers to prioritize genuine and sustainable energy solutions.

So, my dear audience, brace yourselves for an adventure like no other. Together, we will unmask the deception, challenge the status quo, and demand accountability from those who would exploit our hopes for a greener future. For in the face of corrupt politicians and their mythical creations, the truth shall prevail, and real progress towards a cleaner and brighter tomorrow shall be realized. Let us journey forth, armed with knowledge and a dash of skepticism, as we uncover the truth behind the unicorns, mermaids, and the Holy Grail of clean energy in the corridors of power.

With his thought-provoking and masterfully crafted article titled "What 'Clean Coal' Is – and Isn't," published in the prestigious "The New York Times", Brad Plumer (2017) led us on an extraordinary journey. Within the captivating pages of his work, he skillfully transported us to an enigmatic world, where the concept of "clean coal" held sway. It was as if we stepped into a realm where the very essence of words seemed to lose their meaning, and reality itself became a mischievous game of hide-and-seek.

The term "clean coal" gained popularity in 2008 when coal industry groups seized it as a clever marketing ploy. They wanted us to believe that coal plants could magically capture carbon dioxide emissions and bury them underground, saving us from the perils of global warming. But here's the harsh truth: carbon capture and storage technology is far from being a superhero. It's more like a superhero-in-training, struggling to find its footing in a world of complexities and uncertainties.

In the vast landscape of the United States, only one coal plant has actively implemented carbon capture and storage technology. It's like having one brave knight trying to vanquish an entire army of pollution. The cost and complexity of this technology have proven to be formidable adversaries, causing ambitious projects like the Kemper plant in Mississippi to crumble like a house of cards.

But here's the catch: without stricter climate regulations or a price on carbon, there's little motivation for companies to invest in this untested technology. It's like asking someone to buy a fancy car with no gas money. They simply won't bite. And as Plumer highlights, the term "clean coal" has been twisted and manipulated to fit misleading narratives. It's like using smoke and mirrors to present highly efficient coal plants or washing coal to reduce ash content as shining examples of cleanliness. But let's not forget, coal remains a dirty beast, a significant polluter with detrimental effects on the environment.

Plumer's tale takes a fascinating turn as he introduces us to the grand theater of the coal industry's advertising campaign. Behind the curtains, Americans for Balanced Energy Choices (ABEC), a coalition of influential coal industry players, has invested over $60 million in a spectacular show to convince us that coal can be clean. It's like watching a magician performing tricks, dazzling us with illusions while hiding the truth in plain sight.

Their mission? To justify the construction of new coal-fired power plants, even though these plants are notorious contributors to carbon dioxide emissions. It's like building more smokestacks while claiming to clean the air. The proposed solution of carbon capture and storage technology encounters one obstacle after another. The ambitious FutureGen project, once promising to separate carbon dioxide from coal emissions and store it underground, fell victim to exorbitant costs. It's like a dream collapsing under its own weight.

But the challenges don't end there. Coal gasification and post-combustion carbon capture, the other avenues of exploration, come with their own set of technical hurdles and demand the construction of new power plants. It's like creating more problems while pretending to solve them. And what about the captured carbon dioxide? Long-term storage brings concerns of monitoring and potential leaks, raising more doubts about the viability of the whole concept.

Plumer redirects our attention to the real heroes of the energy world: conservation and renewable sources like wind, solar power, and even natural gas. They offer viable alternatives to reduce emissions, providing a glimmer of hope amidst the murky world of coal. Collaboration between industry and environmentalists is the key to unlocking a brighter future, one that doesn't rely on mythical notions like clean coal.

Within the enthralling pages of Richard Conniff's captivating article, "The Myth of Clean Coal," released at the esteemed Yale School of Environment (2008), we embark on an immersive exploration of deception and subterfuge. Conniff, with his unparalleled mastery, deftly unravels the intricate layers of misdirection that shroud the multi-million-dollar advertising campaign promoting the illusory concept of clean coal. Infusing his narrative with wit, sarcasm, and a hint of playful charm, he fearlessly dismantles the fairy tale surrounding this notion, urging us to cast aside falsehoods and embrace the profound potential of sustainable energy solutions.

Again, behind the scenes, we discover Americans for Balanced Energy Choices (ABEC) and their mission to sell us the illusion of clean coal. It's a clever dance orchestrated by influential players in the coal industry who want to maintain the status quo and fight against climate legislation. But as Conniff reminds us, clean coal is nothing more than an advertising slogan, a desperate attempt to rebrand coal as an environmentally friendly resource.

The proposed solution of carbon capture and storage, once heralded as the savior, faces insurmountable hurdles. Projects like FutureGen crumbled under the weight of excessive costs, revealing the harsh reality that true clean coal is a far-fetched dream. It's like chasing a mirage in the desert, always out of reach.

Meanwhile, biased studies funded by the coal industry attempt to emphasize the negative economic consequences of reducing coal usage. But Conniff exposes their true nature – published in trade magazines rather than peer-reviewed journals, they crumble under scrutiny. It's like a magician's assistant revealing the secrets of the trick.

As we journey through the landscape of clean coal, Conniff reminds us of the urgency to shift our perspective and embrace a sustainable future. It's time to let go of the myth and move towards genuine solutions. Richard Conniff, with his skillful storytelling, leaves us inspired to question the prevailing narrative and explore alternative paths.

So, my friends, let us take these major takeaways with us:

The term "clean coal" is a misleading concept created by the coal industry to mask its environmental impacts.

Carbon capture and storage technology is still in its early stages and faces significant challenges.

Without stricter regulations or a price on carbon, companies lack incentives to invest in carbon capture technology.

The term "clean coal" has been used in misleading ways, diverting attention from the true nature of coal's pollution.

Coal remains a highly polluting energy source compared to alternatives like natural gas, nuclear, wind, and solar power.

The mining process of coal has detrimental effects on the environment.

The cancellation of studies and proposed budget cuts hinder research into pollution reduction techniques for coal.

Environmental regulations and the retirement of older coal plants have contributed to improvements in air quality and reduced health risks associated with coal power.

So, my friends, let us embark on a new discussion, leaving the myth of clean coal behind.

Allow me now, the pleasure of presenting a delightful interlude for your entertainment. This musical composition, entitled 'The Primate Paradigm' shall now commence.

In the wild, amidst the tangled vines of political intrigue and economic maneuvering, one can find the true essence of power dynamics. It is a world where alliances are formed, favors are exchanged, and the art of negotiation reigns supreme. From the steamy jungles to the concrete jungles of Wall Street, the principles remain remarkably similar. As that well-known fellow once declared, the key to success in the deal-making game can be distilled into a single, tantalizing phrase: have leverage.

But what is leverage, you might ask? To answer that question, let us venture into the depths of the animal kingdom, where the primal instincts of survival and advancement are laid bare. Take, for instance, the captivating world of chimpanzees, where the dynamics of power play out in a manner both fascinating and eerily familiar.

In the intricate social fabric of a chimpanzee camp, one can witness a microcosm of human interactions. As any astute zoologist will observe, these primates have a code of conduct rooted in reciprocity. It's a simple equation that echoes throughout nature's grand tapestry: you scratch my back, and I'll scratch yours. It's a symbiotic dance, an unspoken agreement that ensures survival and strengthens the bonds of their community.

Now, you may be wondering, what does this primate behavior have to do with the labyrinthine corridors of European Union Sustainability Policy? In recent times, we have witnessed a spectacle of shameless audacity, as the EU boldly declares nuclear and natural gas to be green. The implications of such proclamations are far-reaching, laden with the weight of investment scams and environmental repercussions. It seems as though the rules of the game have been distorted, rendering the definition of "green" nothing more than a hollow shell.

To navigate this complex web of deception and unravel the true motivations behind such decisions, we must embrace the common sense approach of connecting the dots. Just as the late maestro Steve Jobs once implored us, let us embark on a journey of intellectual curiosity and logical deduction. By piecing together, the fragments of this puzzling landscape, we can shed light on the underlying truths that lie obscured beneath the surface.

In the section that follow, we will delve into the heart of this enigma, exploring the intricacies of power, leverage, and the delicate balance between progress and preservation. Prepare yourself for a riveting exploration of the underbelly of human nature, where the primal instincts of chimpanzees converge with the complex machinations of political and economic systems. It is a tale that unveils the secrets behind the art of the deal and exposes the interplay of interests that shape our world. So, my fellow curious souls, let us embark on this journey together and uncover the truths that lie within.

In the realm of environmental sustainability and responsible investing, the European Commission's recent decision to label nuclear energy and natural gas as "green" within the EU's taxonomy has ignited a fierce debate. Critics argue that this move constitutes a betrayal, not only to climate and ESG investors but also to retirement savings account holders and the very fabric of our planet. Examining the decision and its potential ramifications, we uncover a complex web of interconnected interests and questionable motives.

At the heart of the criticism lies the notion of greenwashing, a practice where organizations present a deceptively positive image of their environmental impact. Detractors assert that labeling nuclear energy and natural gas as green undermines the credibility of the EU's taxonomy, diverting financial resources away from truly climate-positive investments. They argue that this decision risks perpetuating the status quo rather than driving meaningful change.

Furthermore, concerns arise regarding the EU's overarching goal of achieving climate neutrality by 2050. Environmental organizations worry that the inclusion of nuclear energy and natural gas in the taxonomy may hinder this ambitious target. They contend that these energy sources, with their potential long-term risks and lack of alignment with renewable energy transition, undermine the ultimate goal of reducing greenhouse gas emissions and building a sustainable future.

Safety and waste disposal issues also loom large in the opposition's argument against nuclear energy. Critics emphasize the potential environmental risks associated with nuclear power, ranging from the catastrophic consequences of accidents to the persistent lack of permanent disposal sites for radioactive waste worldwide. Building new nuclear power plants, they argue, would be a slow process that fails to significantly contribute to the EU's climate neutrality goals within the necessary timeframe.

The decision highlights the divisions among EU member states, particularly between France and Germany. France, which heavily relies on nuclear power, advocates for the inclusion of nuclear energy in the taxonomy. On the other hand, Germany, scarred by the Fukushima disaster, views natural gas as a transitional energy source and seeks to phase out nuclear power. Other countries, including Denmark, Austria, and Luxembourg, share Germany's concerns about nuclear power and stress the need for safe disposal of nuclear waste.

Amidst the controversy, political and legal challenges arise as key battlegrounds. The taxonomy proposal will undergo review by EU member states and the European Parliament. While outright rejection by EU states is unlikely, uncertainty looms over the support it will receive in the European Parliament. Discontent has already surfaced among lawmakers across the political spectrum, including Greens and Social Democrats, who express dissatisfaction and indicate potential opposition. Austria and Luxembourg have even gone so far as to threaten legal action against the European Commission, further intensifying the tumultuous landscape.

The controversy surrounding the EU taxonomy and the inclusion of nuclear energy and natural gas as green may have profound effects on investment decisions and the perception of sustainability in the energy sector. Investors and consumers may begin to question the validity of green labels and demand greater transparency and accountability in sustainable investments.

Consider the poignant situation of a grandmother, perhaps even your own beloved matriarch, who, in a state of unawareness, directs her investments towards a fund boasting the esteemed label of ESG. Alas, her discovery, with a heavy heart, unfolds the disheartening reality that nestled within this seemingly virtuous haven lie investments in nuclear power plants and natural gas enterprises engaged in actions detrimental to our delicate eco-system. Such a revelation strikes a chord of profound concern, for within it lies the potential for irreparable loss. The grandmother, having entrusted her retirement savings to this ill-fated endeavor, now finds herself adrift in a sea of shattered trust, the very essence of betrayal and deception penetrating the fibers of her being.

It is essential to note that this discussion centers around the potential consequences of the EU's decisions rather than legal definitions of scams. Nonetheless, these decisions can be seen as deceptive practices that misrepresent the true environmental impact of certain activities or investments. They risk ensnaring investors who seek genuinely sustainable options in a web of misleading information, inadvertently supporting environmentally harmful practices or projects.

To protect themselves, investors must exercise caution and conduct thorough due diligence. Relying solely on labels and marketing materials is insufficient. It is crucial to critically evaluate the underlying assets and ensure they align with one's values and sustainability objectives. Additionally, staying informed about evolving regulatory frameworks and industry best practices can help make more informed investment decisions.

Ultimately, the EU's decision to declare nuclear energy and natural gas as sustainable raises significant concerns. The potential negative impacts on the environment, investor trust, and the efficacy of the EU's sustainability framework cannot be ignored. It is imperative that we strive for transparency, accurate labeling, and investor education to ensure that the path to a sustainable future is not compromised by questionable decisions and mutual back scratching.

Let us delve into the intricate dance of reciprocity, akin to the proverbial primate paradigm, where mutual favors are exchanged. In the grand tapestry of the EU-France-Germany-Nuclear Power-Gas saga, we shall explore how this paradigm unfolds and shapes the dynamics at play.

In the intricate saga of the EU-France-Germany-Nuclear Power-Gas debacle, a betrayal is felt by climate and ESG investors, retirement savings account holders, and those advocating for the well-being of our planet. The decision of the EU to categorize nuclear power and natural gas as sustainable appears to be nothing short of a scam, misguiding and misleading those who seek true sustainability.

This deception becomes apparent when we examine the incidents and environmental concerns associated with these energy sources. The haunting memories of Fukushima and Chernobyl serve as stark reminders of the long-lasting environmental implications of nuclear energy. The release of radioactive materials and the contamination of surrounding areas paint a bleak picture, challenging the notion of sustainability.

Similarly, natural gas extraction through methods like fracking has left scars on the planet. The gas fields in Iraq and the North American fracking fields have caused significant environmental damage, from water pollution to the release of methane, a potent greenhouse gas. Even the Nigeria Delta bears witness to the pollution caused by the extraction and use of natural gas.

To advocate for the sustainability of nuclear energy and polluting natural gas is to ignore these incidents and the harm they have caused. It is a dismissal of the truth and a betrayal of the principles held by climate-conscious investors and those seeking a cleaner and safer future.

The decision-making process that led to this perceived betrayal can be traced back to a reciprocal exchange of favors between France and

Germany. Germany, fueled by its ambition to establish itself as the preeminent gas hub of Europe, is determined to ensure its energy supply by importing cost-effective gas from Russia and redistributing it within the Common EU Market. Furthermore, Germany plans to capitalize on its gas reserves in its thriving chemical industry, reaping both energy security and economic benefits from this strategic maneuver.

On the other hand, France, with its formidable nuclear industry, aims to solidify its position as the leading exporter of electricity and nuclear technology in Europe. With a steadfast focus on nuclear energy, France seeks to safeguard its energy independence and enhance its economic competitiveness. Concurrently, by advocating for nuclear power, France endeavors to expand the market size for its advanced nuclear technology, thereby bolstering its economic prospects and exerting influence throughout the continent.

Based on proprietary analysis, it is revealed that under this grand plan, Germany could potentially boost its GDP by 3.8%, while France could see an increase of nearly 2%.

This mutual collaboration between France and Germany is driven by their respective interests and aspirations. While Germany pursues energy security and economic advantages through natural gas, France strives to maintain its energy independence and economic standing through nuclear power. Both countries aim to leverage their unique strengths and augment their presence in the energy market, albeit through different approaches.

However, it is crucial to acknowledge the potential ramifications of this reciprocal cooperation. By endorsing natural gas as sustainable and promoting nuclear energy, there is a risk of perpetuating reliance on energy sources that may have adverse environmental consequences, hindering progress towards a truly sustainable future. As the decision-making process unfolds, it is of utmost importance to conscientiously consider the long-term implications and prioritize the well-being of the planet over short-term gains and industry interests. A careful balance must be struck to ensure a sustainable and prosperous future for all.

As you can see, the decision-making process within the EU is complex, driven by political, economic, and environmental factors. It is evident that the interests of France and Germany play a significant role in shaping these decisions. The desire for energy market dominance, with Germany aiming to become the gas hub of Europe and France aspiring to be the electricity export leader, fuels their push for the inclusion of nuclear energy and natural gas as sustainable options.

However, it is crucial to recognize the consequences of such decisions. They perpetuate the reliance on polluting and potentially dangerous energy sources, hindering progress toward a sustainable future. The voices of climate and ESG investors, retirement savings account holders, and advocates for the planet's health must not be disregarded. The scam of labeling nuclear power and natural gas as sustainable must be exposed, and a genuine commitment to true sustainability must prevail.

IV. Biased Bucks: National Subsidies – Wealthy's Delight, Inequality's Fright

Once upon a time, a motley crew of scientists and entrepreneurs dared to venture into the uncharted territory of Germany – a land not known for its embrace of risk capital or early stage investments. Our audacious project sat at the precarious intersection of industry, energy efficiency, and artificial intelligence. This was a time long before AI invaded the conversations of ordinary folks, before even the most unlikely characters, like the ever discerning housewives, weighed in on the AI revolution. Oh, the brave new world we sought to create, where Germany's machines and industries would morph into self-thinking, energy-efficient behemoths! The challenge was formidable, the odds stacked against us, and success seemed as elusive as the Loch Ness Monster.

To tackle this monumental task, we devised a cunning plan: combine our capital with the German Innovation Subsidies from the Ministry of Economic Affairs. We knew it would be an arduous journey, navigating treacherous waters and facing endless bureaucratic hurdles, but we were determined to soldier on. After twelve grueling months of being poked and prodded by the meticulous German Ministry of Economic Affairs' scientific agency, fondly known as "Forschungszentrum Jülich," we were blessed with the ultimate prize – an approval letter!

Oh, the euphoria! We danced and celebrated, believing our financial woes were finally over. But alas, it was merely a cruel mirage. The subsidies we so desperately craved, the lifeblood of our project, remained elusive as a unicorn prancing through the enchanted forest. Like Freddy Fender's lamenting lyrics, "Wasted days and wasted nights," our hopes were shattered, our dreams crushed. Little did we know that the subsidies we were denied had mysteriously found their way into the coffers of a colossal, multi-billion-dollar company – a coincidence so improbable, it could be the subject of a grand conspiracy theory. And guess what? That company just happened to be in cahoots with the international partner we introduced to the German Ministry through our cherished Memorandum of Understanding. But hey, I'm sure it's all just a series of miraculous happenstances, right?

But hold on to your hats, my friends, for our tale is not unique. It is but a microcosm of a much larger conversation – an exposé of national subsidies and the delightful biases that often accompany them. It seems that these precious funds, intended to spur innovation and progress, have a sneaky habit of favoring the wealthy and well-connected, leaving the rest of us mere mortals to fend for ourselves in a dog-eat-dog world. Oh, those well-oiled corporate machines with overflowing coffers, effortlessly gliding through the hallowed halls of power, while we humble startups are left to shiver in the icy winds of neglect.

Indulge me, esteemed audience, as we embark on an enthralling voyage through the corridors of power, where a captivating political saga unfolds before our very eyes. Prepare to be both awe-inspired and disheartened as we bear witness to a cast of characters displaying a breathtaking absence of moral compass, all within the borders of a nation that gifted us with the term "Nestbeschmuzer" – a word that captures the shamelessness of those who defile their own sanctuaries.

Together, let us venture into the realm of "Biased Bucks," a narrative that not only resonates with the accounts shared earlier but also unearths substantiated evidence of a grievous political affair. Brace yourselves as we uncover the intricate details of this captivating episode, shedding light on the audacious conduct of our venerated political figures. Prepare to be simultaneously enthralled by their artful maneuvers and appalled by their blatant disregard for the principles we hold dear.

So, according to a scathing article published by Focus, even the liberal conservative media outlets have had enough of the "Biased Bucks" game and the shameless conduct of our political elite. The article, aptly titled "Habeck's State Secretary: Tax Millions and the Arrogance of Power," shines a spotlight on the dubious actions of Udo Philipp, the state secretary responsible for start-ups in Robert Habeck's Ministry of Economics. It seems that Philipp has been playing around with tax millions while conveniently being directly or indirectly involved in the very start-up companies that receive those sweet subsidies (Masengarb, 2023). Talk about a conflict of interest!

Not only has Philipp managed to raise eyebrows, but he's also ruffled the feathers of his own party members. Even the ruling Ampel coalition, which should be singing Kumbaya together, is perturbed by his refusal to testify before the Bundestag's Economic Affairs' Committee. Bernd Westphal, the SPD spokesperson, is not amused, finding Philipp's behavior "disrespectful." And Reinhard Houben, the FDP spokesperson, can't help but think that Philipp is doing himself no favors by dodging the inquiry. Oh, the drama!

But hold your horses, my friends, because the actual allegations against Philipp might seem like minor details compared to the circus surrounding the Ministry of Economic Affairs. It seems their communication skills are more cringe-worthy than an awkward date. Back in days, when Philipp's company holdings were first put under the microscope, Focus Online dismissed it as a mere case of amateurish communication. They argued that the Ministry should have spilled the beans about these connections right from the start. But no, they decided to keep it hush-hush. Rookie mistake, my friends.

And guess what? The amateurishness didn't stop there! It turns out that Philipp was involved in the appointment of an advisor in whose fund he had previously invested. Talk about playing with your own money, huh? But wait, there's more! Several companies in which Philipp has indirect interests through funds received millions of euros in government funding. The Ministry of Economic Affairs and a related agency were generous with their cash, with the Ministry of Research joining the party as well. It's like Oprah was handing out subsidies: "You get a subsidy, you get a subsidy, everyone gets a subsidy!"

Now, you would think that the Ministry would have disclosed these connections right off the bat, but nope, they conveniently forgot to mention it. Oopsie! Maybe they were playing hide-and-seek with transparency. But fear not, my friends, because Philipp and the Ministry are here to reassure us that there's nothing to see. Philipp can't fathom how a company in which he has a direct interest could benefit from his actions. The Ministry emphasizes that fund investments have no influence over investment strategies. Ah, the sweet sound of plausible deniability!

Of course, critics are not buying it. They accuse Philipp of employing a salami tactic, slicing off bits of the truth one at a time. It's like a never-ending buffet of funded connections! CSU Secretary-General Martin Huber demands transparency and clarification from Robert Habeck, while Julia Klöckner, the economic policy spokesperson for the CDU/CSU, accuses them of the "arrogance of power." Oh boy, the political mudslinging is in full swing!

But amidst all the scandalous shenanigans, what truly baffles us is the behavior of the Ministry itself. It's more disconcerting than the actual incident. Christian Leye from the Left Party tells us that regardless of the extent of Philipp's involvement, these stories erode trust in democratic institutions. And you know what, Christian? We couldn't agree more! We need stricter rules and regulations, like yesterday, to prevent these Biased Bucks adventures from becoming the norm.

Will Robert Habeck step up and bring some much-needed transparency to this political circus? Only time will tell. Meanwhile, let's hope they take a crash course in communication skills and avoid any more amateur-hour performances. Come on, folks, we deserve better!

As we look behind the curtains, we find ourselves pondering the multitude of analogous predicaments lurking in the shadows, concealed from the discerning eyes of public scrutiny, while silently depleting the reservoirs of our cherished taxpayers. Can we genuinely dismiss the notion that the very same game is being played before our very eyes, with the grandest bonanza of them all: Climate Finance and Climate Subsidies? I, for one, cannot refute it. Instead, I am compelled to connect the dots and discern the patterns that form a distinct geometry – a geometry woven with the threads of malpractice or, dare I say, corruption. It is an enigma that demands exposure and an unequivocal halt.

Dear readers, I implore you to remain steadfast in your attention, for the "Biased Bucks" saga has yet to reach its denouement. In the days to come, we shall remain ever vigilant, poised to illuminate the next chapter of this theatrical spectacle, replete with its questionable financial undertones.

So, perhaps it is high time for the German government, and governments far and wide, to ditch their secrecy and embrace the revolutionary concept of transparency. Let the rejected subsidy cases see the light of day! Only then can we expose the festering inequalities and ensure that resources are allocated with fairness and maximum impact.

And now, my friends, our story intertwines with a grand discussion labeled "Biased Bucks: National Subsidies – Wealthy's Delight, Inequality's Fright." Buckle up, for we are about to embark on a rollercoaster ride through the complexities and implications of national subsidies, peeling back the layers of bias that perpetuate inequality. It is a story that demands our attention. For only through awareness, and a dash of righteous indignation can we forge a future where fairness reigns and prosperity know no bounds.

In the world of EU subsidies, an ominous cloud of perceived unfairness hovers over the program, leaving citizens disheartened and disillusioned. The lack of transparency, like a dense fog, obscures the path for those seeking access to these coveted subsidies. As citizens venture into the program's description, they find themselves entangled in a labyrinth devoid of specific details, where eligibility criteria, application processes, and fund allocation remain shrouded in mystery. It's as if the program enjoys playing a high-stakes game of hide-and-seek, leaving eager applicants stranded in a maze of confusion and frustration.

But the opacity doesn't stop there. The program seems to favor companies and private investments, relegating citizen involvement to the shadows. Citizens, who should be key stakeholders in the green economy, feel marginalized and detached from the benefits of these subsidies. Their voices are silenced, and decisions are made without their input, as if their opinions hold no weight in shaping the future, they so desperately desire.

The disparities in resource allocation further deepen the chasm of inequality. The subsidies predominantly target specific sectors outlined in the net-zero Plan, such as hydrogen, carbon capture and storage, zero-emission vehicles, energy-efficient buildings, and recharging infrastructure. While this focused approach aims to drive progress in these areas, it inadvertently creates an imbalance, favoring certain industries or regions while leaving others in the shadows of neglect. Citizens with innovative ideas or projects that fall outside these predetermined sectors are left out in the cold, yearning for a fair chance to participate.

Adding insult to injury, the program seems to have a preference for large-scale projects and private investments, casting a shadow over the potential of small-scale initiatives and individual citizens. It's as if the program believes that only the big players hold the keys to innovation and progress, stifling grassroots efforts and restricting opportunities for citizens to actively engage in the green economy. Their aspirations remain unrealized, drowned out by the clamor of well-funded behemoths.

But the challenges don't end there. Citizens face yet another formidable hurdle: the complex and bureaucratic entanglements woven into the program. The administrative procedures, notifications, and compliance with state aid rules loom large, creating a treacherous web that dissuades citizens from even attempting to navigate its intricate threads. It's as if the program is designed to be a test of endurance, placing an unfair burden on individuals and smaller organizations who lack the resources or expertise to navigate this bureaucratic maze.

In the face of these mounting concerns, it is imperative for the EU to acknowledge the imbalance and strive for a fairer landscape. Transparency must be more than just buzzword; it must be infused into the program's

DNA. Citizens need clear guidelines and specific information on eligibility and allocation processes, a compass to navigate the foggy terrain of subsidies. They deserve a seat at the decision-making table, where their voices can shape the green transition according to their needs and aspirations.

To bridge the gaping divide, the program must extend its support beyond the confines of large-scale projects and private investments. Individual initiatives, the sparks of creativity and innovation, should not be left in the shadows. The EU must ensure equal access to resources across sectors and regions, recognizing that groundbreaking ideas can come from unexpected places. And let us not forget the plight of small companies, who desperately need tailored programs, mentoring initiatives, and capacity-building resources to unlock their full potential. The EU must open doors and forge connections, creating opportunities for partnerships that level the playing field for all.

Lastly, let us not underestimate the importance of simplicity. The program must shed its layers of complexity and simplify its bureaucratic procedures. Administrative barriers should not deter citizens and small organizations from realizing their dreams. It's time to remove the unnecessary red tape and allow the program to flourish in a climate of efficiency and accessibility.

In this journey towards a greener future, the EU has a chance to rewrite the narrative. It can dispel the shadows of favoritism and bias, replacing them with a beacon of fairness that illuminates the path for all citizens. By embracing transparency, citizen involvement, equal access, and simplified procedures, the EU can transform the subsidy program into a powerful instrument of change. Let the subsidies become the catalysts that empower all citizens, from the smallest of companies to the boldest of innovators, in the pursuit of a future where wealth is not concentrated in the hands of the few, and inequality is not the price we pay for progress.

Ah, my adventurous friend, let us continue our whimsical journey through the tangled web of subsidy inequality and budget battles in Germany. Our protagonist, Finance Minister Christian Lindner, has once again graced us with his presence, lamenting the limited funding for replacing old heaters. Oh, the woes of an overstretched state budget! But fear not, for Lindner assures us that those in need won't be left behind. How considerate of him! He proposes targeted subsidies for replacing heaters, where the more CO_2 you emit, the more money you'll receive.

What a novel approach – rewarding those who contribute most to climate change. Brilliant!

As we delve deeper into the budget blues, we encounter a heated debate. Lindner anticipates a substantial deficit ranging from €14 to €18 billion. The timeframe becomes inconsequential when contemplating the absurdity of the situation. Alarm bells are ringing, my friends. To address this financial gap, Lindner suggests renunciation. Ah, the sweet sound of austerity! But worry not, for Lindner assures us once again that those requiring support won't be forgotten. It's like a magic trick – balancing the budget while still providing for the needy. Truly remarkable!

Meanwhile, the federal government has its own grand plans for the future. You know, because it's all the rage to acknowledge man-made climate change and embark on noisy actions for publicity, even if they're a tad superficial. From 2024 onwards, every newly installed heater should operate with 65% renewable energy. Fancy technologies like heat pumps and solar thermal systems will make an appearance, with subsidies of up to 40% of the costs. Federal Minister for Economic Affairs Robert Habeck, from the Greens, even plans to introduce an additional billion-dollar subsidy program – how generous!

But hold on a minute, our friend Lindner isn't buying it. He claims there's no money, and he wants to preach and sing the hymn of austerity. So, Mr. Lindner asks us to imagine a hardworking German middle-class household that has managed to save funds and pay off their mortgage, only to be slapped with the prospect of taking out a new loan from his bank to finance a whopping €70,000 to €200,000 for a renewable heating system, including all those mandated refurbishing costs like roof and wall insulation. Isn't that a delightful proposition? It's like a vacation to Expropriation Island, especially for low- and middle-income folks. And let's not forget that Mr. Lindner is saying a firm "Nein" to any form of state revenue increase. How delightful!

Limited funding for replacing old heaters? Well, isn't that just a jolly concern for equity in access to subsidies? If the state can't foot the bill, we might have a select group of individuals or households benefiting from the subsidies, leaving the rest of us plebeians out in the cold. Ah, the joys of exacerbating inequality!

But wait, we chant it again! Our dear Finance Minister Lindner proposes higher subsidies for those with heaters emitting higher CO_2 levels. Isn't it just splendid to reward those who contribute the most to climate change? We should give a round of applause to Mr. Lindner for his inventive thinking.

Oh, but let's not forget the social aspect for low-income households. It seems that Lindner wants to address equity concerns by considering socio-economic factors. How thoughtful of him! By targeting support based on emission levels and income, we can ensure that those facing the greatest challenges in transitioning to cleaner heating technologies receive the most assistance. Equity achieved; problem solved! Bravo!

Ah, now we come to the pièce de résistance, the additional billion-dollar subsidy program based on income orientation. It seems like everyone's getting a piece of the pie. Habeck from the Greens wants to spread the wealth even further. How generous! It's like a grand giveaway, with subsidies raining down from the heavens. Who doesn't love a good subsidy spree?

But hold on a minute, our friend Lindner isn't quite on board with this extravagant show. He wants to talk about budget cuts and alternative options, like increasing taxes for the superrich. Oh, the nerve! How dare he suggest such a thing? Taxing the rich to fund climate initiatives? Preposterous! We all know that the burden should fall squarely on the middle class. After all, why should the wealthy contribute more when they have a higher carbon footprint? It's not like they have the means to make a difference or anything.

Ah, the complexities of subsidy policies. It's a delicate dance between limited funding, bureaucratic hurdles, and the ever-elusive quest for fairness. We must ensure that the subsidies are accessible to all, except those who aren't well-off or well-connected, of course. We wouldn't want just anyone to benefit from these subsidies. Let's make it a VIP club, shall we?

Oh, my inquisitive companion, let us now untangle the enigmatic dance between budget austerity and subsidies to fight climate change. At first glance, one might wonder what these two seemingly disparate elements have in common. Budget austerity often involves cutting expenses, reducing subsidies, and tightening the purse strings to address fiscal challenges. On the other hand, subsidies for climate change initiatives aim to incentivize sustainable practices, promote clean technologies, and propel us towards a greener future. How do these threads intertwine? Let us explore.

When faced with budget constraints, governments must grapple with tough choices and prioritize their spending. Austerity measures, including subsidy cuts, can be seen as a way to trim expenses and address budget deficits. However, in the context of climate change, the impact of such cuts becomes a point of concern. We find ourselves in a pickle, my friend. How can we cut subsidies that drive the transition to a sustainable future without hampering progress and inhibiting the adoption of clean technologies?

But fear not! There's a glimmer of hope on the horizon. Enter the argument for taxing the rich. You see, the wealthy folks among us often have higher carbon footprints due to their lavish lifestyles and resource-intensive activities. By imposing higher taxes on the rich, we can discourage excessive consumption, incentivize more sustainable choices, and generate revenue to fund climate change initiatives. It's like killing two birds with one stone, my friend. We tackle social inequalities while also nudging the wealthy towards more responsible environmental behavior.

Let us also consider the finance transaction tax, a levy imposed on financial transactions such as stock trades or currency exchanges. This tax, if implemented wisely, has the potential to generate substantial revenue, particularly given the scale of financial activities in today's interconnected world. By harnessing a fraction of these transactions, funds can be directed towards climate change initiatives, including subsidies for clean technologies, renewable energy projects, and sustainable infrastructure.

Finally, we delve into the realm of the digital tax, an endeavor to capture the digital economy's vast wealth and ensure a fair contribution towards addressing climate change. In an era where digital giants amass fortunes, a tax on their activities can unlock resources to fund climate initiatives. By carefully designing and implementing this tax, even linking it to energy consumption of cloud computing infrastructures, governments can ensure that the digital revolution becomes an ally in the fight against climate change, providing the necessary financial impetus to drive sustainable solutions.

So, dear friend, the question of whether to cut subsidies or tax the rich become a delicate balancing act. It's a dance between fiscal responsibility and the urgent need to combat climate change. The decision ultimately rests in the hands of policymakers, who must navigate this complex landscape with wisdom and foresight. Will they choose austerity and risk undermining the subsidies that drive sustainability? Or will they embrace a more proactive approach, grounded in financial measures and equitable taxation, to lay the foundation for a robust and sustainable response?

Now, my witty companion, let's shift our attention to the KfW subsidy mechanism. A state bank in Germany. Oh, how exciting! We have a top ten list of takeaways. Hold on tight, because this is going to be a roller-coaster ride of information.

Changes in conditions? Check. Restructuring of loans and grants? Check. New programs? Check. Higher energy efficiency, higher subsidies? Check. Involvement of KfW in financing? Check. Importance of initial consultation? Check. Role of financial intermediaries? Check. Deadline for applications? Check. Time-sensitive process? Check. Oh, what a delightful checklist we have here. It's like a bureaucratic bingo game!

But let's not stop there. We mustn't forget the potential weaknesses of the KfW subsidy mechanism. Lack of clarity and communication? Check. Dependency on financial intermediaries? Check. Time-sensitive process? Check. Limited availability of subsidies? Check. Potential inequality of access? Check. Complexity of the application process? Check. Lack of flexibility? Check. Potential information gaps? Check. My, oh my, it seems like the KfW subsidy mechanism has a few chinks in its armor. Who would have thought?

Oh, how delightful it was to grace the halls of an entrepreneurship forum, not one of those gleaming tech extravaganzas, but a gathering akin to the illustrious "Mittelstand" spirit. The sheer charm of witnessing the palpable frustration with KfW, that bank, which gracefully hides behind private banks in its pursuit of customer-facing theatrics. Oh, the grandeur of their transformative initiatives, especially when generously funding giant zombie entities with a seamless lack of intermediaries.

Now, as we near the end of our playful exploration, we must address the concerns of citizens who are eligible for subsidies but do not receive them. It's a travesty, really! It may not be a scam per se, but it's a failure of the subsidy program. We mustn't mince words. Citizens who meet all the eligibility criteria and have their hopes raised high, only to be let down by the system, deserve our sympathy. It's like a cruel joke, a tease of financial support that never materializes. Shame on you, subsidy program! Shame!

Ah, the joys and woes of subsidies. They promise so much and deliver so little. Limited funding, bureaucracy, inequality, and the ever-present risk of scams. It's like a comedy of errors, a farce played out in the world of policy. But fear not, my friend, for amidst the chaos, there is hope. We must strive for transparency, fairness, and effective management of subsidy programs. Only then can we ensure that subsidies truly benefit those who need them, and not just those who can navigate the bureaucratic labyrinth.

Oh, the delightful world of affordability thresholds for electric vehicles and rooftop solar! Let's uncover the income groups that have the privilege of indulging in these luxurious green endeavors.

When it comes to brand new electric vehicles with an average price tag of €44,000, it's no secret that they often dance into the hands of the more affluent members of society. Low-income individuals might find themselves watching from the sidelines as the well-off take a joyride in their shiny, emission-free machines. Oh, the sweet smell of exclusivity!

And let us not forget the glamorous world of rooftop solar, with its price tag of around €80,000. It's a sun-kissed playground for those who have a bit more financial wiggle room. While rooftop solar can be an enticing option for reducing carbon footprints and enjoying lower energy bills, it often remains a distant dream for low-income households, who struggle to scrape together the necessary funds for installation.

So, my curious friend, when it comes to these eco-friendly investments, it seems that the low-income groups are left with their shades drawn, peering out wistfully at their more financially fortunate counterparts, basking in the glow of electric vehicles and rooftop solar arrays.

Of course, affordability is just one piece of the puzzle. There are other factors at play, such as government incentives, financing options, and the availability of support programs, which can influence the accessibility of these technologies for different income groups. But it's clear that when the price tags soar to such heights, the luxury of electric vehicles and rooftop solar remains a distant dream for many low-income individuals.

Oh, the irony of it all! As we strive to combat climate change and create a more sustainable future, we must grapple with the reality that the most affordable solutions often remain out of reach for those who need them most. Let us hope for a day when eco-friendly technologies become more accessible and inclusive, allowing everyone to join the green revolution regardless of their income.

But until then, my friend, the electric vehicle chargers and rooftop solar panels remain the territory of the privileged few, sparkling in the sunlight while the rest of us dream of a day when sustainability becomes truly equitable.

Ah, the world of climate subsidies, where it appears that the gospel of Matthew 25:29 finds a peculiar resonance: "For whoever has, more will be given, and he will have an abundance; but whoever does not have, even what he has will be taken away from him." How fitting!

In this whimsical realm of climate funding, it seems that those who already have the means to afford eco-friendly luxuries are bestowed with even more opportunities and abundance. The well-off, who can comfortably purchase an electric vehicle or adorn their rooftops with solar panels, are showered with subsidies and incentives. It's a seemingly never-ending feast of perks and advantages.

But alas, for those who do not have the financial means to partake in these climate-friendly indulgences, the story takes a different turn. Even the little they have is at risk of being taken away. The lack of affordable options and accessible subsidies leaves them in a state of perpetual longing, gazing at their more fortunate counterparts and wondering if they will ever get a slice of the sustainability pie.

Oh, the irony! In our pursuit of a greener future, it appears that the world of climate subsidies mirrors the unequal distribution of wealth and resources that we see in other spheres of life. The rich get richer, and the poor are left to fend for themselves in a world that claims to be environmentally conscious but often fails to bridge the gap between affordability and accessibility.

So, my friend, as we navigate this landscape of climate subsidies, let us remember the words of Matthew 25:29, for they ring eerily true. It is a reminder that our efforts to address climate change should not perpetuate existing inequalities but instead strive for a world where sustainability is within reach for all, regardless of their socioeconomic status.

Until that day comes, let us challenge the status quo, advocate for equitable distribution of resources, and work towards a future where climate subsidies truly uplift those in need, rather than reinforcing the disparities that plague our society.

A study in the journal Nature Communication (Sasse et al., 2023) examined green-home renovation subsidies in France and found that while these subsidies aimed to promote energy-efficient upgrades, they were primarily accessed by higher-income households. This can be attributed to factors such as higher home ownership rates among the wealthy and the affordability challenges faced by low and middle-income households in undertaking extensive renovations. As a result, the subsidy program may inadvertently exacerbate existing inequalities in housing and energy efficiency.

Studies by Woody (2023) and Duncombe (2023) examined the impact of EV purchase subsidies in California and found that higher-income households were more likely to benefit from these incentives. They also found that low-income households were less likely to take advantage of EV subsidies due to barriers such as lack of awareness, limited access to charging infrastructure, and the affordability of EVs even with subsidies. This highlights how well-off households are better positioned to benefit from EV subsidies, leaving low and middle-income households behind.

Now, behold this.

Research by Reames (2019) examined the distributional effects of rooftop solar subsidies in the United States. They found that while these subsidies incentivize solar adoption, the benefits tend to accrue more to higher-income households who can afford the upfront costs of installing solar panels. This results in lower-income households, who may be renters or lack the financial means to participate, missing out on the benefits of renewable energy subsidies.

Ladies and gentlemen, step right up and witness the dazzling world of energy politics! Today, we present to you a tale so sarcastic, so humoristic, it'll leave you chuckling with glee. Behold, "The Petajoules Jongleur" – a true story unfolding in the summer of 2023 in Germany, starring our dynamic duo – Finance Minister Lindner and Economic Affairs and Climate Minister Habeck!

Now, before we dive into the comedic chaos, let's get our energy units straight – a petajoule, as defined by the national statistics Netherlands, equals a jaw-dropping 1,000,000,000,000,000 joules! To put it simply, it's like having enough energy to power a spaceship to Mars and back, with some to spare for a galactic picnic!

As we venture into this energy extravaganza, the Ampel coalition government of Germany takes the center stage once more, and guess who's the leading act? You guessed it – our two prolific actors, Lindner and Habeck, reuniting for another spellbinding performance!

And hold on to your hats, folks, because this duo's actions are so questionable that even Jesus Christ himself couldn't resist a quotable moment! As Jesus said in Luke's Gospel 23:34, "Father, forgive them, for they know not what they are doing!" Yes, you heard it right – even the Son of God would be baffled by their antics!

These two characters are strutting around like young gigolos with cougars - unsure of what they're doing but doing it anyway. I mean, who needs a coherent plan when you can just throw subsidies around like confetti at a party?

Speaking of subsidies, brace yourselves for the grand plan to save the German economy – "eternal subsidies" – because apparently, that's the magical solution to energy scarcity (BMWK, 2023). Minister Habeck wants to shower industrial electricity prices with subsidies so that they pay a mere six cents per kilowatt-hour (KWh)! It's like never-ending sale at a bargain bin – buy now, pay next to nothing!

But wait, dear audience, there's more! These subsidies will be like a giant gift basket for the Amigos among the large corporations in the metal and chemical industries. Let the wealthy rejoice, for their pockets will overflow with subsidized riches!

Now, let's take a detour down the historical lane of subsidies - a place where inefficient "zombie" companies get to party with taxpayer money while failing to survive in the real market. It's a bit like keeping a plant alive with water, but forgetting to give it sunlight – eventually, it withers away, and so does the money.

And who can forget the infamous German solar industry fiasco? Oh, the irony of it all! Billions of subsidies poured in like a never-ending stream, only to see the companies go bankrupt or run off with their Chinese competitors. Those subsidy billions went up in smoke faster than a cheap firework!

But worry not, dear audience, our Minister of Finance, the sanctimonious Lindner claims to be a defender of the free market, except when subsidies are involved go to his Amigos and donors. It's like a magician who only reveals half of the trick – "Look, free market magic! Now, watch as I make your tax money disappear! No, I make them change hands."

State subsidies may have unintended consequences – burdens shifting here and there, affecting consumers and taxpayers like a never-ending game of pass the parcel. Oh, the joy of funding other people's party!

But let's not forget the grand efficiency concerns - when you pamper industries with subsidized electricity prices, they might forget the value of turning off the lights. Who needs energy-efficient practices when you're living the subsidized dream?

Ah, the title of our chapter says it all – "Biased Bucks: National Subsidies – Wealthy's Delight, Inequality's Fright." These subsidies will shower favor upon the giants in the metal and chemical industries, leaving the smaller businesses in their shadow. It's like a talent show where only the big, flashy acts get the spotlight!

But wait, there's more laughter in store! We've got the percentage game – a whopping 200% of Germany's current electricity needs can be covered with this solar extravaganza! Move over, sun, there's a new shining star in town!

But amidst the laughter, let's not forget the serious warning – beware the allure of eternal subsidies! They promise a fairy-tale ending, but they often lead to a reality far from the storybook. It's like believing in unicorns and rainbows, only to find yourself lost in a maze of numbers and confusion!

And now, the moment you've been waiting for – the burning question for our dear Duo! Why, oh why, did you choose to shower rich corporations and their shareholders with taxpayers' money instead of investing in new capacity? Is this some kind of twisted energy magic trick?

Ah, you raise a valid point, dear audience! When you subsidize the electricity price, the demand goes up like a skyrocket on the Fourth of July! It's like a never-ending cycle – demand goes up, price goes up, and we're back at square one! It's like playing a game of musical chairs, but with electricity prices instead!

You've hit the nail on the head with your reference to the Myth of Sisyphus! Just like poor Sisyphus rolling that rock up the mountain, only to see it tumble down again and again, the cycle of subsidies can feel like an eternity of futile labor. But who are the gods punishing in this energy saga, and for what sins? Ah, that's the million-Euro question!

So, there you have it, ladies and gentlemen – a math-filled, sarcasm-laden journey through subsidies and energy calculations. As we bid adieu to this circus of numbers, remember to question, question, and question some more! And may the subsidy gods be ever in your favor – or not! Bravo, bravo, encore!

In a recent and enlightening analysis conducted by my team, a captivating revelation unfolded concerning the tariffs paid by industries in the United States – a nominal 4 US cents per Kwh. Remarkably, this tariff proudly stands as unsubsidized, a testament to the US's energy-producing prowess. Indeed, this advantage derives from the nation's low cost of input materials for electricity production, a luxury that our friends in Germany cannot relish. Even after Germany's efforts to subsidize its industry's electricity, the US maintains a remarkable 33% cost advantage over its European counterpart. For heavy industries, such as Chemicals and Metals, operating within a modest profit margin of 5% to 15%, this 33% cost advantage bears tremendous significance, nay, significance beyond measure!

Whispers abound that, should the US embark on further drilling endeavors, a tantalizing industry tariff of 2 US cents may be within reach. Ah, the wonders of imagination!

Yet, amid this unfolding saga, I cannot help but lament the lack of disciplined decision-making and adequate analysis in this case. Indeed, the situation is nothing short of appalling and revolting. To remedy this woeful state, I beseech the duo responsible in Germany to delve into the wisdom of Michael Porter, an eminent scholar. His work, "The Competitive Advantage of Nations" (1990), is an illuminating masterpiece, brimming with lessons that could have proved invaluable had our duo been inclined to elevate Germany's 20th-century industries and economy into a 21st-century beacon of competitiveness. Ah, what might have been, had they embraced the wisdom of a seasoned scholar!

Michael Porter's seminal work transcends the boundaries of time and space, inviting us to explore the factors contributing to the competitiveness of nations and industries alike. Here, allow me to share three key takeaways and their potential application for a nation endeavoring to reinvigorate its energy-consuming industries amid the challenge of high energy costs.

Factor Conditions, a notion of great elegance, encompass a nation's endowments, encompassing natural resources, labor, infrastructure, and capital. The key lies in the adept development and strategic leverage of these factors to forge a formidable competitive advantage. In the context of soaring energy costs, a nation should turn its focus to enhancing energy efficiency, investing wisely in renewable energy sources, and fostering an environment conducive to research and development, thereby reducing energy consumption across industries.

Demand Conditions, a vibrant force shaping innovation and competitiveness, emanate from a nation's domestic market. By nurturing a robust appetite for energy-efficient products and services, industries will be enticed to embrace sustainable practices. Governmental incentives, energy efficiency standards, and support for energy-saving initiatives can ignite the flame of demand, illuminating the path to a brighter and more sustainable future.

Related and Supporting Industries, steadfast pillars buttressing specific sectors, play a pivotal role in shaping competitiveness. In the realm of high energy costs, the cultivation of a thriving eco-system comprising energy-efficient technology providers, equipment manufacturers, and renewable energy suppliers becomes paramount. These partners in progress shall provide industries access to cost-effective solutions and transformative technologies, all driving a reduction in energy consumption.

Indeed, the key to resuscitating energy-intensive industries amidst the challenge of soaring energy costs lies in a harmonious symphony of energy efficiency promotion, sustainable investments in renewable energy, and a relentless pursuit of innovation. These orchestral notes resonate within the corridors of governmental policies and industry collaborations, culminating in a crescendo of research and development, all of which fortify the bastions of energy-intensive sectors while championing environmental sustainability.

Alas, my dear audience, I must urge caution against venturing down the treacherous path of silly subsidies. To halt the inexorable march of the future with such frivolity would only lead to a cost disadvantage of 33% to 66%, a bitter pill to swallow for any nation.

Furthermore, let us cast our gaze upon the curious phenomenon wherein the duo, despite possessing the power to enact substantial impact, have chosen to disregard certain areas. And the most bewildering aspect is that these areas bear no cost, not a single coin, nada, rien, zero! Allow me to illuminate this enigma.

In the realm of Germany's wind farm ambitions, the approval timeline dances to the tune of various factors – specific location, project size, potential environmental impacts, public consultations, and regulatory processes. As of our most recent analysis and inquiry in June 2023, the timeline for obtaining necessary approvals typically extends over several years, an intricate dance of intricacies indeed.

The unfolding process encompasses a delicate choreography of Project Development and Site Selection, an enthralling phase involving the identification of potential wind farm sites, meticulous feasibility studies, and active engagement with local stakeholders. Then, as the curtain rises on the Planning and Permitting phase, the ensemble of developers must seek the blessings of various authorities - local municipalities, state entities, and even potential federal agencies. Environmental impact assessments and public consultations add yet another layer of complexity to this grand performance.

Oh, the harmony of Grid Connection, a vital step ensuring seamless integration with the electricity grid. Coordination with grid operators to assess capacity and potential upgrades takes center stage, for a symphony is only as strong as the connections that bind its notes.

And finally, with a flourish of the conductor's baton, the Construction and Commissioning phase commences. Once all necessary permits and approvals are secured, the grand spectacle unfolds – the erection of wind turbines, the installation of infrastructure, and a rousing symphony of tests.

Finally, after this captivating performance of approvals and red tape, the Operational Phase begins, as the wind farm unleashes its majestic ballet of electricity production.

Yet, my esteemed audience, let us dare to dream of an alternative stage where efficiency reigns supreme! A world where bureaucratic entanglements dissolve, and approvals are swift, sure, and as graceful as a prima ballerina. Why not embrace the pinnacle of governance innovation and approve everything within a mere three months, within well-prepared zones? Ah, the wonders that await, a standing ovation for streamlined efficiency!

In conclusion, dear audience, the insights garnered from our recent analysis paint a vivid canvas of opportunity and transformation. Let us be guided by the timeless wisdom of Michael Porter, for within his scholarly tomes lie the keys to unlocking our Germany's potential. As we navigate the approval timelines of grand wind farm endeavors, may we seek the dawn of a new era, where bureaucratic hurdles dissipate, and progress surges with unfettered grace. Let us, too, shatter the veils concealing hidden distributions, ensuring a fair and equitable future for all.

V. Pollutant Pandemonium: State-Supported Favoritism For A Cleaner... Wallet?

Years ago, I embarked on a journey to the charming suburbs of Munich, near the historic city of Dachau. Now, when you think of Dachau, what comes to mind? A concentration camp, perhaps? Well, hold onto your "lederhosen", because Dachau has undergone a transformation that would make Cinderella jealous. It's now a delightful place, brimming with warmth and welcoming inhabitants. Picture yourself strolling through the streets, basking in the fusion of Bavarian charm and the tranquil ambiance of a quaint village. If that doesn't entice you to visit, I don't know what will.

During my visit, I had the pleasure of indulging in the company of my dear friends in their lovely small garden. We sipped on Bavarian beer, letting its golden nectar tickle our taste buds, while relishing the aromatic embrace of Cuban cigars. Ah, the pinnacle of sophistication and enjoyment. But just when we thought life couldn't get any better, an unexpected twist unfolded, like a plot twist in a gripping novel.

Suddenly, an unpleasant stench, a malodorous masterpiece, invaded the air around us. It was a scent reminiscent of a blend of feces and urine, a perfume fit for the most discerning noses. Perplexed, we turned to our hosts, seeking an explanation for this olfactory assault. And what did we discover? A nearby farmer, in all his aromatic glory, spreading liquid manure in his fields. Talk about a fragrant surprise!

Now, I can sense your curiosity building, dear reader. What exactly is liquid manure, you ask? Well, it's a delightful mixture of feces and urine from our fine farm-dwelling friends, primarily pigs and cattle. It's like a symphony of bodily excretions, blended with water, dissolved nutrients, organic matter, and minerals. Truly, a recipe for fertilization success. And oh, how widely it is used in agriculture. It's practically the Beast and the Beauty of fertilizers, commanding attention and admiration wherever it goes.

But here's the kicker: an alarming study has shed light on the excessive production of this liquid manure by the animal industry. It's like they're vying for the title of Liquid Manure Production Champions. To transport this precious elixir, they resort to using large liquid manure tankers, traversing significant distances, up to a staggering 220 kilometers. It's like a mobile spa for manure, on wheels. And you won't believe what's happening next.

Neighboring countries, those cheeky Netherlands, for example, have clamped down on the use of liquid manure in agriculture. They've imposed strict regulations, as if they're the fashion police of fertilizers. And what do our dear German farmers do in response? They import the liquid manure, like it's the hottest commodity on the market. The rumors, my friends, the rumors. But alas, this problem is not a mere figment of our imaginations. Oh no, it's as real as the stench that invaded our peaceful gathering.

Let's talk about the impact, shall we? Brace yourself, because it's about to get dirty. One of the main consequences of using this liquid manure is its effect on drinking water quality. You know that thing we humans need to survive. Groundwater, the primary source of Germany's drinking water, is under attack. Alarming levels of nitrate, the villain of this tale, have been detected in groundwater, rivers, and lakes, especially in regions where animal husbandry is a booming business. It's like a nitrate invasion, infiltrating our precious water sources.

But wait, there's more. When liquid manure is used as fertilizer, the nitrogen it contains can do some serious damage. It hangs around in the soil, playing a dangerous game of conversion, transforming into harmful nitrites. And you know where those nitrites end up? You guessed it, in our groundwater. It's like a toxic treasure hunt, with the grand prize being contaminated water. In fact, in North Rhine-Westphalia, 40 percent of groundwater is swimming in nitrate levels that would make your head spin.

Now, let's talk about the expansion of agricultural animal husbandry. To maximize their profits, farmers are going all-in, expanding their livestock facilities. And guess what that means? More manure, my friends. More liquid manure production, more pollution of our precious groundwater. To put it into perspective, the production of a mere 400 grams of pork yields 10 liters of liquid manure. And a single liter of milk from cattle? It generously gifts us with 3 liters of manure. It's like the animals are saying, "Here you go, a little present for your troubles." Thanks, guys.

But wait, there's more! These companies, the ones swimming in liquid manure, they receive subsidies. Yes, you heard me right. Subsidies. It's like they're being rewarded for their aromatic endeavors. And what does that mean for us, the consumers of drinking water? Well, our water treatment facilities must work overtime, retrofitting themselves to filter out those pesky nitrates. And guess who pays for that? That's right, we do. It's like a never-ending cycle of manure-induced expenses, a price we have to pay for clean drinking water.

So, dear reader, the impact of liquid manure on our drinking water is no joke. It's like a comedy of errors, with farmers spreading their aromatic elixir far and wide, while we, the consumers, bear the consequences. It's time to tackle this issue head-on, to regulate and incentivize sustainable agricultural practices. Because, let's face it, we all deserve clean and affordable drinking water. And we certainly don't need any more surprises in the garden while enjoying our Cohiba cigars.

Now, my dear audience, let's delve into another concerning consequence of the liquid manure extravaganza: the health hazards posed by antibiotics. Brace yourselves, for we're about to enter a world where pigs and cattle hold the keys to our antibiotic-laden future.

In the year 2020 alone, German livestock facilities collectively administered a mind-boggling 701 tons of antibiotics. That's right, my friends, tons of antibiotics flowing through the veins of our furry farm-dwelling friends. But what happens when these animals excrete a portion of the medication unchanged? Ah, that's where things get interesting.

You see, these residues of antibiotics find their way into the fields through the magical potion we call liquid manure. It's like a clandestine operation, covertly infiltrating our groundwater. And what did a study conducted on liquid manure samples reveal? Brace yourselves, for this revelation might send chills down your spine. In seven out of eleven cases, the samples contained pathogens resistant to the reserve antibiotic colistin. Yes, you heard me right, my dear audience. We have resistant germs lurking in the very liquid that nourishes our crops and seeps into our drinking water.

But wait, there's more! Groundwater, especially in areas densely populated with livestock farms, has been found to contain traces of antibiotics. It's like a chemical cocktail, straight from the farm to our glasses. And why is this development highly problematic, you ask? Well, let me enlighten you. These antibiotics, my friends, are not your run-of-the-mill variety.

Oh no, these are reserve antibiotics, the kind we save for dire situations in human medicine when conventional antibiotics fail due to resistance. We're talking about reserve army of antibiotics that are now under attack, slowly losing their effectiveness. It's like a battle between life-saving drugs and those pesky germs, with potentially fatal consequences.

The rise of antibiotic resistance has reached critical levels, my friends. It's like a global pandemic of its own, with thousands of fatalities from once easily treatable bacterial infections. In fact, more than 1.2 million people worldwide succumb to infections caused by antibiotic-resistant germs. It's a battle we cannot afford to lose. Recognizing the severity of the issue, the World Health Organization (WHO) labeled antibiotic-resistant germs as one of the top ten global health threats in 2019. It's like a ticking time bomb, waiting to explode and wreak havoc on our healthcare systems.

But let's take a step back, my dear audience, and look at the bigger picture. The unregulated production and utilization of liquid manure in the animal industry have far-reaching implications. We're not just talking about compromised drinking water quality and increased costs; we're talking about the emergence of antibiotic-resistant pathogens that could send us spiraling back to the dark ages of medicine.

It's time for action, my friends. We need to raise awareness, implement sustainable practices, and consider alternative fertilization methods. We must mitigate the adverse effects on both the environment and human health. It's a battle we can win, but it requires a united front and a commitment to change.

Access to clean and safe drinking water is a fundamental human right, my dear audience. It's like oxygen for the soul, a necessity for our well-being. But the use of liquid manure in agricultural practices poses significant challenges to this sacred right. We've explored the consequences of its abundant spreading on groundwater and the implications for water utility customers. We've drawn analogies to the concept of a poll tax, highlighting the unfair burden imposed on those who must bear the cost of additional water treatment required to mitigate the negative consequences of liquid manure.

But it doesn't end there, my friends. The impacts of antibiotic resistance go far beyond the realm of drinking water. The massive use of antibiotics in pig farming has serious consequences for us all. Environmentalists have made a concerning discovery, finding resistant germs in liquid manure samples from 19 German pigsties. Yes, you heard it right. Pigsties, my friends, the very dwellings of our curly-tailed companions. This poses an increasing danger to us humans, and even vegetarians and vegans are not spared from the potential perils.

A random analysis conducted by the environmental protection organization Greenpeace revealed that liquid manure from pigsties releases resistant germs and antibiotics into the environment on a significant scale. Laboratory analyses performed on 19 liquid manure samples extracted from the aforementioned pigsties have substantiated a disconcerting revelation. Among these samples, 13 exhibited the presence of bacteria showcasing resistance to antibiotics commonly prescribed for human use. It's like a microbial rebellion, with germs gaining the upper hand in the battle against our life-saving drugs.

But the plot thickens, my friends. Among these samples, six contained bacteria resistant to not just one, but three groups of antibiotics. It's like a germy game of resistance, evolving into a multi-player nightmare. And if that's not enough to make you shudder, one sample revealed the presence of a multi-resistant MRSA germ. MRSA, my friends, is no stranger to horror stories. It's a type of bacteria that can cause skin infections, wound infections, and inflammation of organs. It's like a menacing beast, lurking in the shadows, waiting to strike the vulnerable.

But it doesn't stop there, my friends. Residues of antibiotics were detected in 15 of the liquid manure samples. These antibiotics, typically administered to animals to enhance their immune systems, are finding their way into the very land where our food grows. It's like a silent invasion, with each bite potentially exposing us to the remnants of a chemical warfare waged in the name of livestock health.

It's time to take a stand, my dear audience. The use of antibiotics in pig farming poses a problem not just for meat consumers but for all of us. It's like a ticking time bomb, threatening to increase the incidence of deaths from previously treatable infections. We cannot let this scenario unfold. Greenpeace, the champion of environmental causes, advocates for a significant reduction in the use of antibiotics through the establishment of better husbandry conditions. It's like a call to arms, urging us to take responsibility and fight for a healthier future.

So, my friends, we've explored the impact of liquid manure on our drinking water and the insidious threat of antibiotic resistance. These challenges demand urgent action, for the stakes are high. Our health, our well-being, and the health of our environment hang in the balance. By raising awareness, implementing sustainable practices, and considering alternative fertilization methods, we can mitigate the adverse effects on both the environment and human health. Together, we can safeguard the provision of clean and affordable drinking water, protecting the health and welfare of present and future generations. Let's embark on this journey, my dear audience, and write a better future for ourselves and our planet.

Ah, my friends, we've reached the final act of our narrative on pollutant favoritism. In this grand finale, we shall fly over the potential reasons behind this intriguing phenomenon. Buckle up, for we're about to raise questions about economic considerations, lobbying and special interest influence, limited alternatives, and political considerations. It's a tale of power, complexity, and the delicate dance of governance.

First, let us turn our attention to economic considerations, my dear audience. Agriculture, a vital sector in many economies, relies heavily on cost-effective fertilizers to ensure bountiful harvests. Liquid manure, with its readily available and relatively inexpensive nature, becomes an attractive option for farmers. Governments, eager to maintain food security, preserve rural livelihoods, and promote economic growth, may prioritize supporting the agricultural industry. The potential economic benefits associated with liquid manure usage can overshadow the environmental and health concerns related to drinking water contamination. It's like a trade-off, where the scales tip in favor of economic prosperity, leaving the environmental repercussions in the shadows.

But let us not forget the power of lobbying and special interest influence, my friends. The agricultural sector is no stranger to powerful lobbies and interest groups advocating for their members' interests. These groups, armed with their persuasive tactics, may exert considerable pressure on governments to maintain the status quo. They resist stricter regulations or alternative practices that could disrupt established agricultural systems. In some cases, the influence of lobbying efforts and campaign contributions may sway government decisions, potentially favoring the interests of agricultural stakeholders over environmental considerations. It's like a dance of power, where the voices of the few shape the destiny of the many.

Now, let us ponder the limited alternatives and transition challenges that governments face, my dear audience. Transitioning away from liquid manure and implementing alternative fertilization methods is no easy feat. It requires substantial investments in research, infrastructure, and training. Governments, cautious about disrupting existing agricultural systems, may hesitate to regulate or restrict liquid manure usage. This hesitation may stem from the fear of potential economic repercussions and the difficulties associated with managing a transition. It's like a tightrope walk, balancing the need for change with the realities of feasibility and practicality. The winds of change may be blowing, but the journey is not without its hurdles.

Ah, but what about the ever-present force of political considerations, my friends? Elected officials, driven by short-term political gains, may tread cautiously when it comes to liquid manure regulation. The fear of alienating constituents, particularly those employed in the agricultural sector or those whose livelihoods depend on it, can be a powerful motivator. Restricting or regulating liquid manure usage may be seen as an unpopular decision, potentially leading to negative electoral consequences. Governments, mindful of their political survival, may opt to maintain the status quo, even if it means compromising on environmental and public health concerns. It's like a dance with the voters, where steps must be taken carefully to avoid a misstep.

Now, my friends, as we approach the end of our tale, let us remember that corruption and favoritism, while present in some cases, are not the sole driving forces behind this phenomenon. We must approach the issue with a nuanced perspective, for government decisions are shaped by a complex interplay of economic, political, and societal factors. Balancing the needs of various stakeholders, addressing economic concerns, and navigating the challenges associated with transitioning to sustainable practices can present significant hurdles for policymakers. It's like a puzzle, with many pieces to fit together to create a sustainable future.

It requires a comprehensive approach, my dear audience. A symphony of stakeholder engagement, scientific research, and the development of incentives and support systems. We must facilitate a sustainable shift away from liquid manure usage while ensuring the well-being of both the agricultural sector and the environment. It's a delicate dance, but one that must be choreographed if we are to harmonize the needs of the present with the demands of the future.

And so, my friends, we bid adieu to our tale of pollutant favoritism. We've explored the depths of drinking water contamination, danced with antibiotic resistance, and navigated the complexities of governance. But the journey does not end here. It is up to us, as informed citizens and conscientious beings, to carry the torch of change. Together, let us seek a future where the balance between economic prosperity, environmental stewardship, and public health is carefully maintained. Let us be the agents of change, for a cleaner, safer, and healthier world awaits.

VI. The Golden Goose Scheme: Government Mandates That Make The Few Rich, Paid By The Many

Once upon a time, in the enchanting land of Veridonia, where politicians seemed to possess more ambition than intellect, a comical situation unfolded. The country, known for its pristine landscapes and hardworking citizens, found itself under the rule of a second-class political elite who seemed to have stumbled into positions they were ill-prepared to handle. Their understanding of complex global challenges like globalization and climate change was as flimsy as a paper umbrella in a storm.

Nonetheless, fueled by their good intentions, and perhaps a dash of delusion, these leaders launched a series of progressive programs in Veridonia. They mandated the use of electric vehicles, regulated the presence of bicycles on streets, and plastered photovoltaic panels on rooftops with the enthusiasm of children playing with Lego bricks. They even imposed strict regulations on wastewater and plastic usage, as if trying to build a utopia from recycled bottles and shower water.

The citizens of Veridonia, ever hopeful, embraced these initiatives with open arms. They believed their country was at the forefront of the battle against climate change. Quoting Voltaire's famed work "Candide," they sang in unison, "All seems to be for the best." But alas, trouble loomed on the horizon. One fateful day, news spread like wildfire through Veridonia: the government had decided to mandate the installation of zero-emission heating technology in every household. Their reasoning appeared logical at first glance – achieving net-zero objectives required carbon neutrality in heating, and the escalating prices of natural gas and oil due to the Ukraine conflict seemed reason enough to switch.

However, amidst the buzz, an old sage named Mister Phantasmagoricus emerged. He was a knowledgeable and financially astute individual, wise beyond his years, and possessing a knack for seeing through the façades of political charades. He warned the government about three grave consequences lurking within their policy decision.

"Firstly," he proclaimed with a twinkle in his eye, "mandating the new technology would burden homeowners with exorbitant refurbishing costs. It's as if the government expects every citizen to possess pockets as deep as the Mariana Trench!"

"Secondly," he continued, his voice dripping with sarcasm, "the narrow timeframe for replacements will undoubtedly create a delightful supply-demand imbalance. Ah, the joys of seeing prices skyrocket while the product availability dwindles. It's as if Veridonia were competing in the Olympics of economic absurdity!"

"And lastly," he chuckled, "one might suspect that the government, by mandating a specific system and technology, has been swayed by the sweet whispers of lobbyists. Oh, but we wouldn't dare utter the word 'corruption' here in Veridonia, would we? That's reserved for other, less refined countries."

Little did Mister Phantasmagoricus know that his warnings would soon be validated. A captivating headline materialized in the financial section of newspapers and stock market updates. It revealed that the leading heating technology company in Veridonia had been acquired by a wealthy foreign group for a record-breaking sum. As Mister Phantasmagoricus delved into the numbers, he realized the acquisition price was but direct result of the government's mandate.

The revenue forecasts of the acquired company ascended to unprecedented heights, propelled by exaggerated prices and an unprecedented surge in market share. It was as if Veridonia's citizens had become unwitting patrons of this grand theatrical production. Every household was obliged to purchase the mandated system, and the cash flowed like a river in flood, sweeping away any doubts or suspicions.

For those who believed in coincidences, laughter erupted like fireworks. But for Mister Phantasmagoricus, events like these were no mere happenstance. He suspected that a clandestine alliance had formed between the Veridonian government and the puppeteers behind the target company. An orchestrated dance to perfection, leaving truth concealed in the shadows, forever dancing out of reach.

Let it be known that this story is a work of pure fiction. Any resemblance to real-life events, persons, or organizations is purely coincidental. The characters and their fantastical names are, of course, figments of imagination, created solely for the purpose of weaving an enthralling tale.

VII. Artistic Echoes Of Climate Change: Toxic Tales

CHAPTER *Nine*

Green Neocolonialism 4.0: The Good, The Bad And The Ugly?

"Everything is connected. Everything is a part of us, and we are a part of everything. What we do to the land, we do to ourselves."
– Oren Lyons (Onondaga)

I. Green Neocolonialism 4.0: The Irony Of Saving The Global South From Itself

Amidst the complex dance of financial arrangements within climate investments, a distinct pattern emerges – one that underscores the intersection of neocolonialism, international finance, and environmental initiatives. Nowhere is this confluence more evident than in the phenomenon of land grabbing in the Global South, a telling example of the intricate dynamics at play.

In the pursuit of climate investments and the quest for renewable energy, powerful entities from the developed world descend upon the Global South, fixating their gaze upon vast tracts of land. Cloaked in the guise of green progress and sustainable development, they offer promises of economic growth and job creation. Yet, the reality that unfolds beneath this captivating façade paints a starkly different picture.

Time and again, local communities, who have long sustainably lived off and managed these lands, find themselves displaced and marginalized to make way for these grandiose projects. The decisions about land use and resource allocation are wrested away from their hands and placed firmly in the control of external actors, driven by profit motives and a narrow understanding of what qualifies as "green" or "sustainable."

The consequences of these land-grabbing ventures are far-reaching. Not only do they disrupt the livelihoods and social fabric of local communities, but they also perpetuate a form of neocolonial control. Financial institutions and investors behind these projects amass substantial power and influence, dictating the terms of engagement and reaping the lion's share of benefits, while the communities affected bear the brunt of the social, economic, and environmental costs.

Yet, the entanglements go deeper still. The financial arrangements that underpin these climate investments serve to reinforce the dependence and vulnerability of the Global South. Often, these projects are financed through loans or investment deals laden with onerous conditions and terms. Consequently, a cycle of indebtedness takes hold, perpetuating the neocolonial narrative, where the Global South becomes ensnared in repaying its so-called "climate debt" through the extraction of natural resources or the exploitation of its labor.

Consider the case of large-scale hydroelectric dam projects, often lauded as clean energy solutions to combat climate change. While they may appear virtuous on the surface, these projects frequently entail the displacement of local communities, the loss of biodiversity, and significant environmental impacts. Moreover, the financial arrangements behind such endeavors typically involve loans from international financial institutions or foreign governments, shackling the Global South with astronomical debts that prove formidable to repay.

Thus, we witness an intricate dance of financial arrangements within the realm of climate investments – a dance that reinforces neocolonial dynamics, relegating the Global South to the role of resource provider and recipient of external control and influence. To ensure a more equitable and just transition to a sustainable future, it is imperative to critically examine and challenge the power imbalances deeply embedded within climate finance.

Ah, the delicate choreography of international finance continues. While the melodies of green development and sustainability reverberate throughout the corridors of global finance, the Global South finds itself entangled in a web of dependence and vulnerability, twirling to a tune that seems strangely familiar.

Take, for instance, the renewable energy projects that beckon developing countries. Multinational corporations and financial institutions from the developed world descend upon these nations, bearing promises of investment and loans to erect solar or wind farms. On the surface, it appears as a harmonious partnership – an embrace of cleaner energy sources intertwined with an infusion of capital into the local economy. Yet, beneath this seemingly benevolent façade lies a different story altogether.

The financial arrangements accompanying these projects often tip the scales, favoring the interests of investors and perpetuating the neocolonial narrative. Loans and investment deals come laden with stringent conditions and exorbitant interest rates, burdening the host country with mounting debts. These financial obligations swiftly consume a significant portion of the country's budget, diverting resources away from crucial social sectors such as education, healthcare, and infrastructure.

To exacerbate matters further, the terms of these loans may require the Global South to prioritize repayment over the well-being of its citizens. In some instances, countries are coerced into allocating a substantial portion of their export earnings or natural resources as collateral, effectively locking them into a cycle of extraction and exploitation. Thus, the very systems of resource dependency and labor mistreatment that the world claims to be distancing itself from are perpetuated under the guise of progress.

The case of palm oil production stands as a poignant example. The global demand for this versatile commodity, fueled by the food and biofuel industries, has triggered the expansion of palm oil plantations in countries like Indonesia and Malaysia. Financing for these plantations often comes from international banks and investors seeking high returns. However, the exponential growth of palm oil production has resulted in deforestation, habitat destruction, and egregious human rights abuses, with local communities paying the steep price for this supposed "green" endeavor.

International financial institutions cannot evade their role in perpetuating this neocolonial narrative. Entities such as the World Bank and regional development banks extend loans and grants for climate-related projects in the Global South. Though their intentions may ostensibly revolve around addressing climate change, the conditions and terms attached to their financial support inadvertently reinforce the existing power dynamics.

These institutions frequently prioritize the interests of their member countries and adhere to a specific economic model that may not align with the needs and aspirations of the Global South. The conditionalities tied to loans often necessitate policy reforms that favor deregulation, privatization, and market-oriented approaches, further entrenching the neocolonial framework.

In essence, the financial arrangements interwoven with climate investments serve to deepen the dependence and vulnerability of the Global South, perpetuating a cycle of indebtedness and extraction. To counter these deeply embedded power imbalances, it is crucial to challenge and reshape these dynamics, cultivating more equitable and sustainable financial models that empower countries to chart their own paths towards climate resilience and development.

In their seminal work, Fairhead, Leach, and Scoones (2012) delve into the concept of "green grabbing" and its reverberations in the appropriation of nature. Their enlightening analysis dissects the phenomenon of large-scale land acquisitions and the expansion of conservation and environmental initiatives, illuminating the displacement and marginalization experienced by local communities (Sovacool et al., 2021).

Their exploration challenges the notion that environmental conservation and sustainability inherently align with the interests of local communities or promote social justice. Instead, they argue that green initiatives and policies can be easily influenced by powerful actors, including governments, corporations, and international organizations, who exploit the rhetoric of environmental protection to advance their own agendas.

The authors offer a compelling examination of green grabbing, where land and resources are appropriated by external actors in the name of conservation, biofuel production, or carbon offset schemes. Through their research, they reveal that such appropriations often result in the loss of livelihoods, displacement of communities, and a decline in local control over resources.

As we scrutinize the burden placed upon the Global South in the pursuit of saving the planet, it is vital to dive into additional examples and references that shed further light on this pressing issue.

Criticism often targets the Global South for its carbon emissions, despite the fact that developing countries contribute comparatively less to global emissions than their developed counterparts. According to a report by the World Resources Institute (WRSI,2022; WRI, 2022), developing countries collectively accounted for only 60% of global carbon dioxide emissions in 2023, with China as the largest emitter. Developed nations, on the other hand, have historically exhibited significantly higher emissions per capita.

The burden of implementing climate change mitigation measures predominantly falls upon the Global South. The Paris Agreement places the responsibility on developing countries to make concerted efforts toward reducing emissions and transitioning to renewable energy sources. In contrast, developed countries have been sluggish in fulfilling their commitments, both in terms of emission reductions and providing financial support for adaptation and mitigation in developing nations.

The Clean Development Mechanism (CDM) under the Kyoto Protocol aimed to foster sustainable development in the Global South while reducing emissions. However, certain CDM projects in developing countries have drawn criticism for perpetuating inequalities and primarily benefiting multinational corporations. Research by Sovacool and Linnér (2016) sheds light on the power imbalances and environmental injustices associated with CDM projects. They emphasize the tendency of these projects to provoke land grabbing, displace communities, and provide limited local benefits.

The Global South disproportionately bears the brunt of climate change impacts, leading to loss and damage in vulnerable regions. Despite this, developed countries have been reluctant to provide adequate financial assistance for loss and damage, including compensation for climate-induced losses. The discrepancy in support became a contentious issue during the negotiations of the Warsaw International Mechanism for Loss and Damage (WIM) under the United Nations Framework Convention on Climate Change (UNFCCC, 2023).

These examples vividly illustrate the unequal distribution of responsibilities and burdens in addressing climate change. While the Global South is expected to prioritize environmental conservation and adopt sustainable practices, the developed world often falls short in taking significant actions to reduce their own carbon footprints and provide necessary support to developing nations (Sovacool et al., 2017).

It is imperative to acknowledge that burden and responsibility should be equitably shared among all nations, considering historical emissions, development needs, and capacity building efforts. Achieving a truly sustainable and just global approach to combating climate change necessitates collective action and a concerted effort to address the underlying power imbalances that perpetuate this burden on the Global South.

In the pages that follow, we embark on a journey through the intricate web of global finance, risk management, and the enduring legacy of colonialism. We seek to shed light on the often-veiled realms of neocolonialism and capitalism in the context of global finance and risk management.

Our expedition began by acknowledging the profound impact of colonialism and the transatlantic slave trade on Africa's present realities. These historical forces, discussed in previous chapters, have cast long shadows over the continent, shaping its trajectory towards the present status-quo and leaving behind lasting consequences. Now, as we turn our attention to the contemporary landscape, we explore the manifestations of neocolonialism and capitalism, which continue to influence Africa's position in the global economic order including Africa's ability to migrate towards a net-zero economy.

Within the realm of international finance, risk management assumes a pivotal role, serving as the custodian of stability and regulation. It is in this arena that we shift our focus, for it is here that the dynamics of power and control are often obscured. Risk management, executed through regulations such as the Basel Regulations, carries the weight of responsibility, safeguarding the financial system against volatility and collapse. Yet, it is vital to acknowledge that these regulations can inadvertently perpetuate protectionism and colonialism in the lending practices of European banks towards Africa.

The thesis we venture to postulate revolves around the implications of Basel Regulations and insurance regulation on the lending landscape in Africa. Through a careful analysis, we aim to draw inferences on two critical dimensions: protectionism and neocolonialism. By scrutinizing the influence of these regulations, we uncover whether Africa is relegated to the role of a mere supplier of raw materials, hindering its economic diversification and self-sufficiency.

Furthermore, we must confront the pressing issue of Africa's decarbonization efforts. As we have previously agreed, investment plays a crucial role in combating climate change. However, in a capitalist society, significant disruptions seldom occur without equity investment or loans. Here lies a concern: Could Basel Regulations inadvertently slow down or even halt Africa's decarbonization endeavors? To answer this question, we navigate the intricacies of compliance costs, risk perception, and the limited financial resources that pose significant challenges to the continent's sustainable development.

I believe it might be prudent, even at this advanced stage of our conversation. to present my interpretation of the term neocolonialism. This way, we can establish a shared understanding for our upcoming discussion. Neocolonialism refers to the continued economic dominance and control exerted by former colonial powers or other developed nations over less developed regions, often employing indirect means. It is a phenomenon deeply intertwined with the concept of capitalism, which is an economic system characterized by private ownership, profit-driven production, and free market competition.

In the present context, the legacy of colonialism has left an indelible mark on global power dynamics and economic relationships, particularly between the developed world and Africa. While the era of formal colonization may have come to an end, the enduring effects of colonial rule continue to persist, albeit in more subtle and obscured forms. Neocolonialism is an extension of these historical power imbalances and manifests itself through various means, including economic control, resource exploitation, and perpetuation of unequal power dynamics.

With these definition in mind, let's go after it.

II. Basel Regulations And Insurance Regulation: Implications For European Banks Lending To Africa

The complexity and interconnectivity of the global financial system necessitate a comprehensive regulatory approach, particularly in the realm of international finance. Among the prominent frameworks in this regard are the Basel Regulations, which have emerged as a cornerstone for regulating banks and financial institutions on an international scale. Spearheaded by the Basel Committee on Banking Supervision, a collective composed of central banks and regulatory authorities from various countries, the Basel Accords have laid the foundation for these regulations.

The Basel Accords encompass a series of guidelines and standards that pertain to risk management, capital adequacy, and liquidity requirements for banks. Their primary objectives are to bolster the stability of the banking sector, fortify risk management practices, and ensure that financial institutions possess adequate capital buffers to weather potential financial shocks.

Nevertheless, the impact of Basel Regulations extends beyond the individual banks themselves. These regulations possess far-reaching implications for cross-border lending and investment activities, encompassing the relationships between European banks and African nations. The regulatory requirements imposed by Basel can significantly influence banks' risk appetite and lending behavior, potentially affecting their willingness to engage in lending activities in specific regions or industries.

In the case of African countries, the impact of Basel Regulations on lending practices can be particularly profound. The regulatory framework may impose higher capital requirements or demand increased risk-weighted assets for loans extended to African nations, particularly those perceived to carry elevated risk levels. This circumstance presents a host of challenges for African countries, hindering their ability to access affordable credit and impeding their capacity to invest in critical infrastructure, stimulate economic growth, and address pressing developmental priorities.

Let us delve for show-casing purposes into the experiences of this seasoned banker as we explore the ramifications of Basel Regulations on loans extended to Africa. The first hurdle that emerges is the imposition of higher capital requirements. To shield against potential losses, banks must maintain a robust capital buffer. However, when it comes to lending to regions deemed as higher risk, such as Africa, the stakes are raised. The banker realizes that even more capital is mandated as a precautionary measure. This amplified capital requirement renders lending to African countries a more costly endeavor for banks. The consequences are palpable – higher interest rates for borrowers or, in some cases, an unsettling reluctance to extend credit altogether.

The banker's journey through the labyrinth of Basel Regulations continues, uncovering another obstacle: increased risk-weighted assets. These regulations determine the risk weights assigned to different types of assets, including loans made to specific regions or countries. Should African nations find themselves classified as high-risk, European banks must assign higher risk weights to loans directed towards these territories. Consequently, banks are compelled to hold more capital against such loans, augmenting the cost of lending. The realization dawns upon the banker that European banks may grow more cautious about lending to African countries due to the heightened perceived risk and the accompanying capital requirements.

As the banker peers deeper into the abyss, he unearths the profound impact of these regulations – limited access to affordable credit. The combination of higher capital requirements and increased risk weights conspires to restrict African countries' entry to affordable credit. European banks, now donning an air of hesitancy, find themselves less inclined to provide loans to African nations, particularly for ambitious infrastructure projects and crucial developmental initiatives. The harsh reality becomes apparent – The increased cost of capital and the amplified risk associated with lending to African countries engender higher interest rates, diminished loan amounts, or even a complete cessation of lending services. These adverse circumstances make it arduous for African countries to secure the necessary funding for their critical projects.

The banker, now acutely aware of the consequences, sets his sights on the implications for infrastructure development. The limited availability of affordable credit, a by-product of Basel Regulations, poses a formidable barrier to Africa's pursuit of essential infrastructure projects. The realization strikes – infrastructure development demands substantial investment. However, with access to affordable loans restricted, African countries are left grappling with a daunting challenge. Building climate saving assets, roads, bridges, power plants, and improving transportation networks become an uphill battle. The consequences reverberate throughout the region – economic growth is hindered, job creation is stifled, migration towards net-zero is delayed or cancelled and overall development is delayed.

Now let us turn to another aspect of Basel Regulations. The risk perception associated with lending to Africa is influenced further by a confluence of historical factors and prevailing economic conditions. These factors further exacerbate the cautious approach adopted by European banks. Concerns surrounding political stability, governance issues, and the perceived risk of default loom large, dissuading banks from extending loans to African countries and perpetuating a cycle of limited investment and economic dependency.

Our banker's journey will probably continue by inferring that political stability and governance issues in African countries are significant concerns that discourage him and his peers from extending loans to these nations. The landscape of African politics he will infer is often tumultuous, with changes in government, social unrest, and conflicts that create an atmosphere of unpredictability. This volatility can disrupt economic activities and put the repayment capacity of borrowers at risk. Our banker and his peers, cautious and risk-averse, find themselves hesitant to venture into lending to African countries plagued by unstable political climates.

However, it is important to consider how Africa's political and governance risks compare to those of other nations like Russia, Ukraine, Vietnam, Argentina, and Brazil. Are the risks truly larger in Africa? Well, let's take a moment to ponder and evaluate.

Our banker's analysis may continue with review of governance issues and probably will conclude with his peers that governance play a significant role in European banks' reluctance to lend to African countries. Corruption, lack of transparency, and ineffective regulatory frameworks raise valid concerns about the accountability and credibility of borrowers. Banks become skeptical about the responsible utilization of loan funds and the ability of African governments to implement sound economic policies and manage their debt obligations. These governance challenges present obstacles for African countries in attracting investment and securing favorable lending terms.

Now, when we talk about the perceived risk of default, it's crucial to understand that historical factors and prevailing economic conditions have painted African nations as high-risk borrowers. Public debt levels, limited fiscal discipline, and vulnerability to external shocks all contribute to this perception. European banks, mindful of potential defaults, tighten their lending criteria, demand higher interest rates, and may even require additional collateral from African countries. These measures, while aimed at mitigating risk, further restrict access to affordable credit, perpetuating a cycle of limited investment and economic dependency.

Delving into a concrete example helps illustrate how regulations on capital adequacy can profoundly affect insurance companies' investment or lending activities in Africa. Imagine an insurance company contemplating providing a loan for a significant infrastructure project in an African nation, like the construction of a Wind park. In such a scenario, the insurance company must ensure it maintains the necessary level of capital adequacy to safeguard against potential investment-related losses.

Regulations typically outline the minimum capital adequacy ratio that insurance companies must uphold. This ratio represents the relationship between their available capital and risk-weighted assets. Risk-weighted assets take into account the risk profiles associated with the investments or loans held by the insurance company. These regulations are in place to guarantee that insurance companies possess adequate financial resources to absorb potential losses and fulfill their obligations to policyholders.

Within the context of investing or lending in Africa, the insurance company must assess the risk profile of the Wind Park project. Factors such as the country's political stability, regulatory environment, economic indicators, and project-specific risks are thoroughly evaluated. Based on this assessment, the insurance company assigns a risk weight to the loan or investment.

For instance, if the African country is deemed to carry a higher level of risk due to political instability or weak regulatory frameworks, the insurance company may assign a greater risk weight to the loan or investment. Consequently, this would necessitate a higher capital requirement for that particular exposure, as the insurance company must hold more capital against the perceived elevated risk.

The impact of capital adequacy regulations manifests in two significant ways. Firstly, it affects the availability of capital for investment or lending purposes. Insurance companies must ensure they possess enough capital to comply with regulatory requirements. However, this can potentially limit the amount of capital accessible for investment in Africa, consequently impacting the scale or number of projects the insurance company can financially support.

Secondly, capital adequacy regulations exert influence over the risk appetite of insurance companies. When faced with higher capital requirements for riskier investments, insurance companies tend to adopt a more cautious approach. They may become more selective in extending loans or investing in projects in Africa, implementing measures such as higher interest rates or stricter collateral requirements to compensate for the perceived risks involved.

III. Impact Of These Regulations On Protectionism And Perpetuating Colonial Dynamics

Come, my friends, and embark on a journey into the enigmatic realm of international finance, where the delicate dance of power and influence shapes the destiny of nations. Here, we delve into the intricate dynamics surrounding sovereign guarantees and bank guarantees, those elusive instruments that hold the key to securing financing for ambitious projects and ventures. These instruments, often touted as essential tools for mitigating risks and enabling African projects to pass rigorous risk assessments, warrant closer examination. As someone who has personally encountered requests for these guarantees, which I must admit I find rather amusing, I will shed light on their nature and implications.

First, let us delve into the concept of a sovereign guarantee. As the name suggests, this instrument involves a sovereign entity providing a guarantee, typically backed by its full faith and credit, to secure a financial transaction or project. In essence, the sovereign entity acts as a guarantor, reassuring lenders or investors that they will be compensated in the event of default or non-performance.

But beware, for within this labyrinth of financial institutions, the Bretton Woods duo, the International Monetary Fund (IMF) and the World Bank, reign supreme, wielding immense authority over global economic policies and the fate of nations.

Picture the scene: the aftermath of World War II, a world yearning for stability and development. It is here that the Bretton Woods institutions take their first steps, armed with a noble mission to foster economic growth and safeguard financial systems. A grand objective indeed, but one that comes with its share of consequences.

Sovereign guarantees, throughout history, have been a lifeline for countries seeking financial backing for their ambitious endeavors, be it infrastructure projects, investments, or the daunting task of servicing debt. These guarantees, backed by the full faith and credit of nations, provided a sense of security to lenders and investors, encouraging them to open their coffers.

In their quest for financial stability and responsible management, the IMF and the World Bank prescribe a strict regimen of fiscal discipline, structural reforms, and debt reduction measures. They demand adherence to their doctrines, known as structural adjustment programs, where countries are forced to swallow bitter pills of austerity, privatization, and economic liberalization.

African countries, hungry for financial assistance from these behemoths, find themselves entangled in a web of conditions. The pursuit of fiscal prudence and debt reduction has unintended consequences, my friends. The ability of African nations to issue sovereign guarantees, once a lifeline for their ambitious projects, dwindles under the disapproving gaze of the Bretton Woods institutions. These guarantees, seen as potential catalysts for debt accumulation and hindrances to the progress made under the structural adjustment programs, are discouraged and marginalized.

But the tale does not end there, for the IMF and the World Bank, in their wisdom, preach the gospel of private sector participation, beckoning countries to build domestic capacity, attract private investments, and embrace transparency and accountability in public finances. The intentions are noble, my friends, for they seek sustainable economic growth and resilience. However, the unintended consequence is that the African nations, with their fledgling financial markets and weaker private sectors, struggle to court the necessary investments and alternative financing options. The scales of access to capital tip further out of their grasp.

Thus, the consequences are twofold, my friends. First, African countries find themselves grappling with mounting difficulties in securing financing for their ambitious projects, be it the construction of vital infrastructure or the pursuit of renewable energy initiatives. The scarcity of sovereign guarantees becomes a suffocating shroud that stifles investment and hampers progress in the region.

Secondly, the very shift towards private sector participation, advocated by the Bretton Woods institutions, creates a stark disparity in accessing capital. African countries, lacking the robust financial mechanisms of their developed counterparts, find themselves struggling to attract investments and alternative financing options. The doors that once swung open with the allure of sovereign guarantees now remain closed, leaving African nations yearning for financial backing, but often finding it just out of reach.

And so, my friends, we come to a poignant realization. The policies and conditions imposed by the Bretton Woods institutions, born from their pursuit of financial stability and responsible economic management, inadvertently cast a shadow upon the ability of African countries to issue sovereign guarantees. While their intentions were noble, the scale tipped in favor of private sector participation and debt reduction, at the expense of Africa's capacity to leverage sovereign guarantees for vital development projects. Consequently, African nations are left grappling with the arduous task of seeking alternative approaches to financing, with limited success in addressing their pressing developmental needs.

The stage is set, my friends, and the consequences of this intricate dance between international financial institutions and the aspirations of African nations continue to unfold. As we bid farewell to this chapter, we part ways with a deeper understanding of the forces at play and the challenges faced by those who seek to harness the power of sovereign guarantees in pursuit of a brighter future.

Come, my curious companions, and let us delve into the fascinating realm of bank guarantees, where trust and financial security intertwine. Here, we unravel the intricate mechanism behind these instruments, exploring the guarantors, the means of guarantee, and the elusive origins of the assets that underpin them.

At its core, a bank guarantee is a promise made by a financial institution, typically a bank, to stand as a guarantor for the obligations of a third party. In simpler terms, it is a commitment that assures the recipient of the guarantee that they will be compensated if the party being guaranteed fails to fulfill their contractual obligations.

Now, let us turn our attention to the guarantor, the institution that assumes the responsibility of providing the guarantee. Typically, this role falls upon established and reputable banks that have the financial capacity and credibility to back such commitments. These banks, armed with their formidable resources and extensive networks, step forward as the guardians of financial promises.

But how does a bank guarantee actually work, my inquisitive companions? Ah, fear not, for I shall guide you through the labyrinthine path of this process. When a bank issues a guarantee, it effectively assumes the risk on behalf of the party being guaranteed. In essence, it is saying, "If our client fails to fulfill their obligations, we will step in and honor the commitment on their behalf."

Now, let us uncover the origins of the assets that support these bank guarantees, for they are the bedrock upon which this mechanism rests. Banks, with their intricate knowledge of finance and their ability to tap into various sources of funds, allocate a portion of their assets to serve as collateral for the guarantees they issue. These assets can take many forms, my friends, ranging from cash reserves to government securities, bank deposits, or even tangible assets such as real estate or valuable commodities.

These assets, carefully selected and managed by the issuing bank, serve as a form of security and ensure that the guarantee has substance and value. They provide re-assurance to the recipient of the guarantee that there are tangible resources backing the bank's commitment, bolstering confidence and instilling trust in the process.

But where do these assets come from; you ask? Ah, therein lies the mystery, my companions. Banks, as multifaceted financial institutions, have a myriad of sources from which they gather assets. These sources can include customer deposits, profits generated from their various business activities, access to interbank lending markets, and, in some cases, even support from central banks or government entities.

When a bank issues a guarantee, my curious companions, it embarks on a complex process that involves assessing risks, evaluating the creditworthiness of the party being guaranteed, and meticulously crafting the terms and conditions of the guarantee. But, alas, these endeavors are not without cost.

Indeed, issuing a bank guarantee incurs certain expenses that must be borne by someone within this intricate dance. But who, you may ask, is responsible for paying these costs? Let us peer deeper into the heart of this matter and uncover the truth.

The allocation of costs for issuing a bank guarantee can vary depending on the specific arrangements agreed upon by the parties involved. While it's common for the applicant – the party requesting the bank guarantee – to bear the costs, it is not a universal rule. The actual agreement is subject to negotiation and can be influenced by factors such as industry practices, local regulations, and the terms specified in the contractual agreement between the applicant and the beneficiary. In some cases, the beneficiary may agree to bear the costs, especially if it is a standard practice in a industry or region. Therefore, it is crucial for the parties to clearly define and agree upon the cost-sharing arrangements during the negotiation and drafting of the relevant agreements.

This fee, my companions, is known as the guarantee fee or commission. It serves as compensation to the issuing bank for assuming the risk and providing the financial security that the beneficiary seeks. The guarantee fee varies depending on the nature of the guarantee, the amount involved, the perceived risk, and the prevailing market conditions. It is a reflection of the resources, expertise, and reputation of the bank providing the guarantee.

However, let us not forget that the issuing bank also incurs certain costs in the process of issuing the guarantee. These costs include administrative expenses, due diligence activities, legal and documentation fees, and the ongoing monitoring and management of the guarantee. While the issuing bank may bear these costs upfront, it is customary for them to be recouped through the guarantee fee charged.

In this intricate waltz, my daring companion, the interplay of trust, transparency, and the promise of shared benefits assumes paramount significance. Should you traverse these realms with grace, the elusive bank guarantee may indeed become attainable, even amid adversities. However, one must candidly ponder the likelihood of an African project developer engaging in a successful dance with European banks in this scenario.

Ah, my daring friend, if you find yourself branded as a high-risk customer like Africa in the eyes of the banking realm, securing a bank guarantee may prove to be a daunting task. However, fear not, for there are circumstances under which a bank might still be willing to issue a guarantee on your behalf, even amidst the perceived risks.

Firstly, let us delve into the considerations that a bank takes into account when evaluating the feasibility of issuing a bank guarantee for a high-risk customer. These considerations revolve around assessing the nature of the risk, the mitigating factors, and the potential benefits that may outweigh the perceived dangers.

One crucial factor is the purpose and scope of the guarantee. If the guarantee is required for a transaction or project that presents significant potential for profitability or economic growth, the bank may be more inclined to assess the risk and evaluate the potential rewards.

Another vital aspect is the financial standing and credibility of the high-risk customer. If the customer can demonstrate a track record of fulfilling financial obligations, providing collateral or assets as security, or showcasing a solid business plan, the bank may be more willing to consider issuing a guarantee despite the perceived risks.

Additionally, the presence of mitigating factors can sway the decision in favor of issuing a guarantee. These factors may include the involvement of reputable third parties, such as insurance companies or government entities, who are willing to share the risk or provide additional security measures.

Furthermore, the bank may employ risk management techniques such as limiting the amount or duration of the guarantee, imposing stricter conditions or collateral requirements, or charging higher guarantee fees to compensate for the perceived risks.

It is important to note, my audacious companion, that each bank has its own risk appetite and internal policies governing the issuance of guarantees. What may be acceptable to one bank may not be to another. Hence, it is essential for the high-risk customer to engage in open and transparent discussions with the bank, providing all necessary information and addressing any concerns proactively.

In summary, my intrepid friend, while being deemed a high-risk customer may present challenges in obtaining a bank guarantee, there are circumstances under which a bank may still be willing to issue one. These include assessing the purpose and potential benefits of the guarantee, evaluating the customer's financial standing and credibility, considering mitigating factors, and employing risk management techniques to address the perceived risks.

It is a delicate dance, my adventurous companion, where trust, transparency, and the potential for mutual gain play crucial roles. If you can navigate these realms with finesse, you may yet secure that coveted bank guarantee, even in the face of adversity.

Ah, discerning observer, your incisive insights pierce through the layers of a profound spectacle, particularly when viewed from the African perspective. The stark contrast between lofty expectations and the unforgiving truths unfolds a tapestry both bewildering and exasperating. It paints a poignant portrait of climate projects in Africa consigned to oblivion, entangled in the restrictive webs of financial access, where the prerequisite of bank guarantees prior to funding stands, as an impervious barrier. Let us embark on unraveling this enigma, acknowledging the bitter truth that demands our contemplation.

In the realm of African projects and ventures seeking funds, the requests for sovereign guarantees or bank guarantees have become a recurring absurdity. A wry smile may grace your lips as you realize the sheer audacity of such demands. How can a sovereign nation present a guarantee when bound by the shackles of debt covenants imposed by the IMF?

And how can private ventures conjure bank guarantees when their coffers bear witness to emptiness?

The tale begins with the IMF, that mighty institution wielding power over sovereign nations. With their gaze fixed on debt covenants and fiscal responsibility, they dictate terms that leave little room for maneuvering. The weight of these requirements bears down heavily on the shoulders of African nations, rendering them incapable of issuing sovereign guarantees. A paradoxical twist, indeed, for those who demand guarantees are the very ones preventing their issuance.

As for the private ventures, their empty pockets and lack of assets form an impenetrable fortress, shielding them from the fantastical world of bank guarantees. How dare they ask an empty pocket man for money? It is a charade of the highest order, a dance of illusion that mocks the very essence of financial reality.

Oh, the absurdity of it all! The chasm between the demands and the stark truth is as vast as the Sahara itself. It is a theater of the absurd, where the stage is set with hollow promises and the players are left grasping at mirages. The farce is laid bare for all to see, and yet, the cycle continues, perpetuated by those who remain blissfully ignorant of the impossibility they demand.

So, my discerning companion, let us cast aside the mask of illusion and confront this farce with unwavering clarity. The demands for sovereign guarantees and bank guarantees in the face of IMF-imposed debt covenants and empty pockets are nothing short of a tragicomic spectacle. A testament to the disconnect between expectations and reality, it serves as a stark reminder of the challenges faced by African projects and ventures in their quest for financial support.

IV. Will The Looming Sovereign Debt Crisis In Sub-Saharan Africa Be The Nail On The Climate Coffin?

Amidst the looming sovereign debt crisis in Sub-Saharan Africa, a pertinent question arises: Could this financial burden become the final blow to climate action in the region? The intricate web of financing models for renewable projects reveals a potential hurdle that could severely impede the timely implementation of green infrastructure and investments in renewables in Sub-Saharan Africa.

Sub-Saharan African nations are grappling with a significant debt burden, with countries like Ghana, Zambia, Angola, Nigeria, and Kenya spending a substantial portion of their budget on debt interest payments alone. This precarious situation is exacerbated by the growing disparity between the US dollar and local currencies, increasing the risk of sovereign debt defaults. The devaluation of local currencies has resulted in an additional $10.8 billion in currency costs, directly impacting the region's GDP.

While discussions around climate finance, debt relief, grants, and aids are underway, it is imperative to recognize the limitations of these instruments. They often serve as mere placebos, offering temporary respite without addressing the fundamental challenges at hand. The time has come to approach the issue of financing renewable projects in Sub-Saharan Africa from a fresh perspective.

One suggestion is for Sub-Saharan African countries to explore funding models that do not further burden their sovereign debt. Rather than relying on traditional financial instruments, an alternative approach could involve considering "Equity in Kind." This innovative concept entails leveraging tangible and intangible assets, such as land, access to the grid, concession rights, and security, as cash equivalents. By recognizing the inherent value of these assets, African nations can actively contribute to the development of green infrastructure without incurring additional debt.

Another key consideration is the granting of power purchase agreements (PPAs) in local currencies instead of foreign denominations. The financial strain of PPAs issued in US dollars, euros, or pounds can be mitigated by developing adequate hedging mechanisms in collaboration with investors, lenders, and developers. This approach helps protect against currency risks associated with local currency PPAs.

To reduce default risks and enhance investor confidence, Sub-Saharan African countries can explore the development of syndicated PPA instruments. By bundling energy demand from the private sector, these syndicated PPAs provide a compelling alternative to sovereign PPAs in public-private partnership projects. This shift in focus helps alleviate the burden on governments while maintaining the momentum of renewable energy initiatives.

Furthermore, it is crucial for Sub-Saharan African countries to prioritize the development of local expertise and job creation within the renewable energy sector. Rather than relying solely on imported renewable parts and know-how, these nations should insist on a minimum of 40% local content in renewable projects. This approach not only fosters economic growth but also prevents the export of wealth and debt importation.

By striving to become major players in the export of green power, particularly stored or liquefied green hydrogen, Sub-Saharan African countries can assert their position in the global energy landscape. This untapped opportunity not only contributes to economic growth but also strengthens the region's voice in the ongoing dialogue surrounding climate change.

Taking inspiration from successful funding models like Ethiopia's Renaissance Dam, Sub-Saharan African countries can explore innovative financing options. This includes bond issuances to local companies and citizens, embracing crypto-currencies and tokenization, and tapping into the vast resources of the global African diaspora. These approaches unlock new avenues for financing and empower local communities to actively participate in sustainable development.

Ultimately, investments in green infrastructures should be viewed to create local wealth and foster economic prosperity. By harnessing the returns from such projects, Sub-Saharan African countries can establish a new kind of social safety net, breaking free from traditional dependency models.

As Sub-Saharan Africa navigates the challenges of sovereign debt and climate action, a collective commitment to collaboration, innovation, and local empowerment is paramount. Through these concerted efforts, the region can overcome barriers, unlock its immense potential, and forge a sustainable, green future.

The quest to shift towards green technologies and confront the climate crisis hangs on a precipice, demanding immense investments to combat the pressing environmental perils that loom large. Yet, the tried and tested methods of financing may prove inadequate in the face of these daunting financial requirements. In this complex scenario, the emergence of innovative and sustainable sources of finance, such as debt buybacks, assumes a pivotal role in securing the necessary capital. Debt buybacks entail a government reclaiming its outstanding debt securities from investors, oftentimes at a discounted rate, and the implications are far-reaching.

Through the strategic utilization of debt buybacks, governments can effectively alleviate the historically exorbitant interest payments associated with servicing their debt burdens. This shrewd move liberates substantial financial resources that can be channeled towards investments in green technologies, renewable energy ventures, sustainable infrastructure projects, and an array of climate-centric initiatives. By diminishing interest payments through debt buybacks, governments can deftly redirect the funds that would have been otherwise funneled into debt service, fostering a conducive climate for investing in sustainable solutions.

To illustrate the practicality of this concept, let us delve into a hypothetical scenario where a government groans under the weight of towering debt levels and grapples with the onerous interest payments they entail. Enter the debt buyback mechanism – a strategic maneuver that empowers the government to reacquire a portion of its outstanding debt at a discounted valuation, consequently mitigating interest obligations. The consequential savings serve as a wellspring of funds, poised to finance ground breaking projects in the realm of renewable energy, be it solar farms basking under the sun's radiance or wind turbines gracefully harnessing the ever-present gusts. Furthermore, the capital gleaned through debt buybacks can be judiciously allocated to fuel sustainable transportation infrastructure initiatives, propel energy-efficient building ventures, or embolden a diverse spectrum of investments that embrace the climate cause.

The allure of debt buybacks lies in their dual capacity to simultaneously alleviate the financial burdens weighing on governments while accelerating the transition towards a low carbon economy. By curtailing interest payments, governments unlock fiscal breathing room, nurturing an environment ripe for channeling resources into sustainable initiatives and revolutionary technologies. This metamorphosis marks an indispensable stride in our collective endeavor to effectively combat the climate crisis.

However, it is imperative to tread this path with utmost caution, recognizing the need for a comprehensive and responsible fiscal framework to underpin the implementation of debt buybacks. Governments must meticulously evaluate their debt sustainability, accounting for crucial factors like economic growth, fiscal stability, and the formulation of long-term debt management strategies. Moreover, transparency and adherence to international financial regulations must be steadfastly observed during the execution of debt buybacks.

While debt buybacks present a promising avenue for financing green technologies and addressing the climate crisis, it is crucial to perceive them as one piece of a broader financial strategy. Exploring and integrating other sustainable finance mechanisms becomes paramount in this holistic pursuit. Green bonds, climate funds, carbon pricing mechanisms, and public-private partnerships all beckon as essential components of the overarching financing framework.

Thus, it becomes abundantly clear that the colossal sums demanded to invest in green technologies and combat the climate crisis necessitate a tapestry of innovative and sustainable sources of finance. Debt buybacks emerge as a potent tool, wielding the power to slash historically astronomical interest payments, unfettering funds for investments in renewable energy, sustainable infrastructure, and a myriad of climate-related initiatives. However, this potent weapon must be wielded judiciously, within the confines of a comprehensive financial strategy that embraces the multitude of sustainable finance mechanisms at our disposal.

In the realm of climate change and the mounting debt burdens of developing countries, a complex dance unfolds, revealing the precarious balance between financial constraints and the urgent need for environmental action. As these nations grapple with the devastating impacts of climate change, their ability to bear the financial weight of necessary adaptations hangs in the balance. It is here, in this delicate equilibrium, that we recognize the pressing demand to bolster support for climate response and adaptation in the developing world.

The burgeoning debt loads straining these countries' economies paint a bleak picture of their fiscal health. Hindered by limited resources, they find themselves ill-equipped to absorb the immense costs associated with addressing climate change's far-reaching consequences. From resilient infrastructure to agricultural innovations, the financial burdens of these essential adaptations prove overwhelmingly burdensome for nations already burdened with debt.

Yet, despite the seemingly insurmountable obstacles, there emerges a glimmer of hope – a recognition that the international community must rally together, extending a helping hand to uplift the developing nations in their time of need. This support must be multifaceted, offering financial backing, technological transfers, capacity building, and knowledge sharing, empowering these countries to mount a formidable response to climate impacts.

Enter the Green Climate Fund (GCF), a beacon of promise amidst the storm. Conceived under the auspices of the United Nations Framework Convention on Climate Change (UNFCCC), the GCF seeks to mobilize financial resources destined for climate initiatives in developing countries. Its mission is clear: to galvanize these nations in their quest to mitigate greenhouse gas emissions and fortify their defenses against the wrath of climate change. Armed with grants, concessional loans, and innovative financial instruments, the GCF endeavors to transform aspirations into tangible actions on the ground. Even as a beacon of hope, one wonders where the real impacts of GCF truly lie. It seems to be merely a drop on a hot stone so far. Let's strive for greater effectiveness.

But financial aid alone is not enough. To ensure an inclusive and effective response, the developing countries themselves must be active participants, actively engaged in the reform efforts that shape their destinies. Their voices, experiences, and unique perspectives must resound within the hallowed halls of decision-making, policy formulation, and implementation. Technical assistance and capacity-building initiatives must accompany financial aid, empowering these nations to navigate the treacherous waters of climate change with skill and determination.

As we venture forth, let us heed the call to intertwine climate and development goals inextricably. The two are not disparate entities, but rather two sides of the same coin, interdependent and inseparable. For it is through their harmonious convergence that true progress can be achieved – a symphony of reduced emissions, enhanced resilience, poverty alleviation, and sustainable economic growth.

In the grand narrative of climate change and debt burdens, the plot thickens. The fate of developing countries hangs precariously in the balance, their economic future intertwined with their ability to combat climate change. With the right combination of financial support, direct engagement, and synchronized objectives, we may yet find a path forward – a path that leads us toward a more sustainable, equitable, and prosperous future.

V. Artistic Echoes Of Climate Change: Raids Concealed In Silence

CHAPTER

Ten

Climate Justice: Who Orders, Pays

"My humanity is bound up in yours, for we can only be human together."
-Archbishop Desmond Tutu (Ubuntu philosophy)

I. Who Orders, Pays

In the riveting theater of understanding climate justice, one finds a compelling analogy nestled within the folds of a restaurant's menu. Imagine, if you will, stepping into an establishment of gastronomic delights, where the options span a sumptuous three-course affair. A culinary indulgence beckons, and a choice is made. Yet, picture this: as the plates are cleared and the dessert forks retire, a peculiar notion takes hold. What if, perchance, it were not you, the connoisseur of flavors, who footed the bill, but another unsuspecting diner?

This gastronomic metaphor unfurls a tapestry woven with the threads of climate justice, an intricate concept that grapples with the stark disparities in the fallout of climate change upon diverse communities and individuals. The narrative's crescendo pivots upon those who bear the lightest culpability for the climatic conundrum being thrust into the vortex of its most dire consequences.

Just as the very fabric of dining etiquette shuns the notion of one relishing an extravagant repast while cunningly passing the financial baton, so too does the discourse of climate justice decry the audacity of those who, historically, have exhaled the lion's share of greenhouse gases (GHGs). The charge levied against this cohort resonates: Is it not a glaring inequity to cast upon the shoulders of marginalized and nascent societies, who have cast but a modest pebble into the tempestuous sea of emissions, the Herculean task of mitigating the tempest's toll?

Within this climatic tableau, the restaurant metaphor unveils a profound parable of responsibility, justice, and the symphony of shared burdens. The heart of the matter beats in harmony with a fundamental precept – those who have supped lavishly upon the banquet of activities that kindle climate change should not evade the onus of ushering in its redress. This harmonizes splendidly with the strain of wisdom known as "common but differentiated responsibilities," a sonnet that acknowledges the diverse capacities and historical footprints of nations and individuals in the grand theater of climate change.

In sum, this culinary analogy stands not only as a palatable illustration but also a clarion call, reverberating through the corridors of conscience. It deftly distills the essence of climate justice, reverently inviting us to ponder the imperative of an equitable apportionment of responsibilities and endeavors. Just as a feast finds its truest fulfillment in a shared banquet, so too does the journey toward climate justice beckon us to set a table where all partake in the labor of stewardship.

My dear companions, allow us to depart from the realm of eateries and instead immerse ourselves in the realm of data. Data, akin to models, comes accompanied by assumptions, for models, one must remember, are built upon such foundations. Here, however, the departure lies in the origins of these data and assumptions, stemming from my own haven of analysis. These assertions I present hold not only the ring of feasibility but, to a certain extent, conservatism – particularly when examining the prolific polluters, be they societies or individuals.

With your kind permission, I shall lay before you two case studies, or rather, models, as a prelude to the ensuing discourse. Indeed, I await with anticipation the forthcoming dialogue, as together we shall construct and mold the very framework of what the developed world, birthed in the industrial revolution, has woven into the fabric of our climate and world. A creation of devastation, one must admit, which has hitherto gone unpaid, for the simple reason that the ledger of nature's cost or the cost of depleting Earth's resources remains starkly bereft of parameters.

Indulge me, if you would, in including within this developed world not just the ancestral progenitors of industrialization but also nations such as China and India. A group whose legacy as net contributors to pollution may not stretch back to the days of the mechanized age, yet they too have joined this unfortunate chorus.

Furthermore, I dare to tread upon the ground of modeling and contraposing these climate costs against those of, let's say, Sub-Saharan Africa. Here, the stage is set not only on a societal canvas but also on the individual scale – a scope that will see the climate-shattering expenses of the wealthy, the well-heeled, and the well-connected juxtaposed against the ordinary souls that make up the masses.

This meticulous analysis, fellow thinkers, shall serve as the cornerstone upon which I intend to erect my proposition. A proposal that harbors the essence of Climate justice, destined to be not only an intranational endeavor but one that extends its tendrils across the globe. Let us now embark on this journey, as the numerical plans beckon, inviting us to unravel their secrets, to glean from them the insights that shall guide us towards a world where the balance sheet of climate's toll finds reconciliation.

In the sprawling tapestry of human progress, there are threads weaved from innovation and threads spun from necessity. Yet, interwoven within this grand narrative, there's an undeniable truth that we can no longer ignore – the consequences of our actions on the very planet that cradles our existence. Amidst the hum of factories and the rush of vehicles, the carbon footprints of nations rise and fall like tides, leaving marks on the atmosphere, imprints of our choices etched into the very air we breathe.

In the land where stars and stripes unfurl in the wind, the United States, a powerhouse of industry, casts a formidable shadow on the carbon ledger. In the year 2021, it's said to have emitted around 5 billion metric tons of CO_2-equivalent greenhouse gases (GHGs), a haunting reminder of the price of progress. From the towering chimneys to the sprawling highways, the carbon footprints stretch across the expanse, tracing the contours of energy production, transportation, and industrial endeavors.

Across the Atlantic, where history breathes through cobblestone streets and modernity echoes in the skyscrapers, the European Union stands as a confluence of nations with a shared future. In the same year, whispers of approximately 2.73 billion metric tons of carbon dioxide filled the air, a sign that progress knows no borders. From the buzz of renewable energy to the rumble of engines, the carbon footprints mingle with the stories of energy, transportation, and the dance of industries striving for prosperity.

And in the East, where the Great Wall stood for centuries, a different kind of wall emerges – an invisible one made of carbon emissions. China, with its vibrant cities and teeming population, casts a long shadow in the carbon chronicles. A staggering 11.47 billion metric tons of carbon dioxide emissions in 2021 paint a portrait of industrial might and energy hunger, a narrative etched in coal and powered by ambition.

On the subcontinent of colors and contrasts, India's journey is a canvas of complexity. With a carbon footprint of around 3.9 billion metric tons of CO_2-equivalent in 2021, the story is one of striving to meet the needs of millions while navigating the path to sustainability. From ancient practices to modern innovation, carbon footprints merge with the tales of energy, industry, and the aspirations of a nation.

In the land where vast landscapes stretch beyond the horizon, Russia's carbon story is one of enigma and energy riches. About 1.711 billion metric tons of CO_2-equivalent greenhouse gases (GHGs) are said to have been released in 2021, a symphony of carbon woven into the very fabric of oil and gas extraction, alongside the pulse of industrial endeavors.

And across the vast expanse of northern beauty, Canada's carbon ledger tells a story of landscapes both wild and urban. Approximately 540 million metric tons of CO_2-equivalent greenhouse gases (GHGs) wafted into the atmosphere in 2021, woven into the intricacies of energy production, transportation arteries, and the heartbeat of industries.

These carbon footprints, these whispers of numbers, represent the intricate dance of progress, the rhythm of innovation, and the pulse of economies. But beneath these figures lies a weightier question: What is the cost of these footprints? As our planet's balance tips, as the consequences of these emissions become more evident, a reckoning is on the horizon.

Now, as we stand on the precipice, a calculated proposition emerges. Imagine a world where each ton of carbon carries a price – a price paid not only to Mother Earth but to all those who tread lightly, who sow seeds of sustainable growth. At $30 per ton of CO_2, a number that dances between realism and aspiration, the ledger takes a new turn.

In the land of stars and stripes, that colossal 5 billion metric tons of carbon dioxide emissions translates to a significant weight on the conscience, resulting in an approximate bill of $150 billion. Across the Atlantic, the European Union, with its 2.73 billion metric tons, would be contemplating a bill of around $81.9 billion. In the sprawling landscapes of China, the colossal carbon footprint of 11.47 billion metric tons would lead to an astonishing bill of $344.1 billion.

Meanwhile, India, with its striving dreams and its 3.9 billion metric tons of carbon emissions, faces a cost of about $117 billion. Russia, entwined in its own intricate dance of energy and industry, would see a bill of $51.3 billion for its 1.711 billion metric tons. And Canada, balancing urban aspirations with vast natural beauty, would reckon with a cost of around $16.2 billion for its 540 million metric tons.

Collectively, the numbers add up, a sum that echoes the collective responsibility of nations towards the very planet that sustains them. As each country contemplates its contribution to the carbon ledger, the echoes of a world in flux become louder, a reminder that the future is not just a chapter but a story that each one of us has a role in shaping.

Ladies and gentlemen, gather 'round as we explore the saga of numbers, those tantalizing bits of information that tell tales of our progress – or perhaps, regress? Look upon the figures above, my dear audience, and marvel at the carbon footprints that dance through the years like forgotten promises.

Ah, but what's this? Only one year's worth of emissions, you say? How quaint! Allow me to take you on a journey through time, a journey where countries have been emitting greenhouse gases (GHGs) like there's no tomorrow, all while their supplier, good ol' Mother Earth, looks on with an arched eyebrow and a ledger that grows heavier with each passing year.

Remember our chapter on "unsustainable growth"? Well, here we stand, with these nations happily puffing away, treating the atmosphere as their personal ashtray, all the while dodging the bill that's steadily accruing. It's like a modern-day version of the prodigal son, except instead of returning to the homestead, these countries are racking up a debt that stretches back to the industrial revolution.

Oh, but wait, you might ask, why don't these countries just pay up? Well, it seems they've been quite the recalcitrant customers when it comes to settling their account with Mother Earth. Perhaps they think they can just swipe their credit card, rack up a balance, and then ignore the monthly statements. And who's left footing the bill? That's right, our dear planet, who's been providing resources and a haven for all these shenanigans.

Now, if I were inclined to dive deep into the abyss of numbers and financial intricacies, I'd map out a journey from 2021 all the way back to the industrial revolution. I'd construct a maze of discounted cash flows, adorned with the finest weighted costs of capital, and present a financial analysis that would make your head spin faster than a roulette wheel. But alas, my friends, I am but a humble narrator, and such an endeavor requires the spirit of a young scientist with a penchant for complex models and a craving for truth.

But fear not, for I have a simpler proposal. Let's embark on a primitive experiment, an exercise that hammers home a point with all the subtlety of a sledgehammer. Imagine, just for a moment, that the debts these recalcitrant countries owe were tallied up, year after year, until the dawn of the industrial revolution. What would we find? A staggering sum, a financial reckoning so profound that it would make even the most steadfast economist break into a cold sweat.

So, my esteemed audience, as you go about your day, ponder this brutal reality. The numbers may dance, the emissions may rise, and the Earth may continue to bear the burden. But one day, oh, one day, the debts will be called in, and then we'll see who truly has been playing a high-stakes game of carbon roulette.

Ah, the realm of numbers and finance, where debts and correlations weave their intricate dance. Let us embark on this numerical journey, a tale of recalcitrant customers, investment banking rules, and haircuts that could rival the most stylish salons.

Imagine, if you will, a world where countries owe Mother Earth for their carbon emissions, and the ledger stretches back to 2012. It's like a grand financial reckoning, except this time the currency is carbon, and the stakes are higher than Wall Street's skyscrapers. Our starting point is a decade before 2021, back to the year 2012 – where the figures tell a story that's both cunning and insightful.

Now, let's introduce a rule from the world of investment banking, a rule that's as classic as a pinstriped suit: the haircut model. Just as a barber trims excess locks, our model will trim away a portion of the debt, acknowledging the reality that some recalcitrant customers might not be able to foot the entire bill. But what's the twist? Our haircut model is a beast of its own kind, fueled by correlations that bend time and test the limits of financial logic.

We'll start with the assumption that there's a correlation between the financial debts owed to Mother Earth in 2021 and those of each preceding year. This correlation, my friends, is like a mathematical dance partner – one-tenth multiplied by the difference between the year under scrutiny and the golden year of 2012. It's a concoction that's bold, audacious, and quintessentially financial.

So, armed with these assumptions, let's dive into the calculations. We'll use the carbon emissions data we've discussed earlier and slap on that hefty $30 per ton price tag. The result? A symphony of numbers that will either delight or dismay, depending on whether you're a carbon debtor or a planet-saving hero.

Country by country, we'll tot up the financial debts as they ripple through time, from 2012 to the present. And what will we find in this numerical tapestry? A story of debts amassed, a tale of haircuts taken, and a lesson in how correlations and haircuts can bend even the most unwavering of financial rules.

And finally, my fellow truth-seekers, we shall gaze upon the grand totals of all these countries combined, from 2012 to 2021. It's a sum that's more than just numbers – it's a mirror reflecting our actions, our choices, and the very cost of progress. So, hold on to your calculators, my dear readers, for this financial roller coaster promises to be a ride unlike any other.

Ladies and gentlemen, prepare yourselves for a cosmic voyage through the labyrinthine galaxies of finance, where numbers twirl like celestial bodies in a cosmic ballet and equations waltz with mathematical elegance. But fret not, dear audience, for I shall spare you the intricacies of mathematical choreography and dive straight into the heart of revelation. Yes, gather 'round as we unveil the quintessential number, the culmination of our numerical odyssey – the one we've all been yearning for.

With meticulous care, we've unraveled the threads of carbon emissions, danced with the enigmatic correlation factors, and even dared to venture into the realm of a "haircut." And now, my friends, brace yourselves, for the moment of truth is upon us – drum roll, please – the grand summation of all fears, the financial embodiment of our carbon transgressions:

Total Financial Debt accrued from 2012 to 2021 for All Countries discussed above: $5,918.57 billion, residing comfortably within the enchanting realm of the haircut case. But hold your applause, dear audience, for our journey has only just begun.

Now, let's shift our gaze to the microscope of perspective, where numbers reveal their true enormity. Imagine, if you will, that in that very year, this debt of $5,918.57 billion would amount to a staggering portion of... well, let your minds wander, for this numerical marvel is awe-inspiring – an entire economy's worth of obligations, pressing down like an elephant on a trampoline.

Ah, but that's not all! Let us delve deeper into the financial cosmos. Transport yourself to the crossroads of history, where at the monumental COP15 summit in 2009, pact was forged – a pact echoing with the urgency of whispering winds and the weight of shared global responsibilities. Developed nations stood together, raising their collective voice in a symphony of solidarity, pledging an astonishing goal: to mobilize a staggering $100 billion annually by the year 2020. Not for their own treasure chests, mind you, but as a beacon of hope for the developing nations, a lifeline to battle the tempestuous tides of the climate crisis.

What does that mean? Allow me to decode: These recalcitrant nations, in an act that rivals the noblest of endeavors, have decided to pay down their carbon debt, year by year, by contributing $100 billion. What a noble gesture, don't you think?

Yet, let's delve further into this financial labyrinth. Let's navigate through the intricacies where $5,918.57 billion in debt is coupled with earnest annual repayments of $100 billion – a dance of interest and principal. With an agreed interest rate of 5% per annum, we shall unravel the temporal tapestry to determine just how many years it will take to expunge this debt from the annals of financial history.

A staggering revelation unfolds before us – more than 73 years would be needed to clear this financial ledger, to set the scales of debt back to equilibrium. But, dear audience, there's a twist in this tale. An unsettling reality awaits – this isn't just about paying back an existing debt, for the recalcitrant countries continue to accrue more than $1,000 billion in carbon debt annually.

Now, picture a world where each passing year adds another $1,000 billion to an already towering debt. Meanwhile, the repayment efforts struggle valiantly, allocating a mere $100 billion towards stemming the tide of borrowing. As the wheels of compounding interest turn, the interest on this growing debt soars, fed by the ever-expanding principal. A chasm emerges between borrowing and repayment, and with every tick of the clock, the idea of repaying the entirety of the debt becomes more of a quixotic dream than a practical reality.

It's a lesson in financial folly, a stark reminder that unchecked borrowing and minimal repayment lead to a mountain of debt that can never truly be conquered. This is the tale of fiscal imbalance, a perilous dance on the precipice of fiscal ruin.

The essence is clear: A debt structure like this is a house of cards, destined to crumble under its own weight. Attempting to fill an ever-expanding chasm is a futile endeavor, much like trying to quench a bottomless thirst. The debt inflates beyond control, the borrowed funds spiraling into oblivion while the repayment efforts falter.

This, my friends, is a lesson etched in financial reality – a proverbial perpetual debt structure is the antithesis of sustainability. Just as nature yearns for balance, so must the financial world strive for equilibrium, lest it plummet into the abyss of insurmountable debt.

As you absorb this cosmic revelation, let it marinate – the climate debt, the narrative entwined with carbon footprints, and the price tag of our choices. A badge of honor, indeed, one that the ledger of Mother Earth demands we acknowledge – a testament to the dance between our actions and their repercussions. So, as we navigate this financial cosmos, remember the whisper of numbers, the echoes of the universe – heed the call of the ledger, for it speaks a language we must all understand.

And now, my fellow cosmic travelers, we arrive at a crossroads. The tale of unsustainable climate debt looms large, a cautionary tale that calls for audacious action. The words of Bob Marley echo in the wind: "Get up, stand up, stand up for your right." The time has come to rise, to take action, and to create a world where balance reigns and the ledger of climate finds harmony once again.

II. Who Orders, Pays: Shaping Behavior Through Accountability

In the grand theater of human progress, where the players don the masks of politicians, bureaucrats, and activists, I find myself amidst a quiet battle of epic proportions – a battle that transcends borders and generations: the battle against climate change. As the curtains rise on this global drama, a cacophony of voices clamors for attention, each advocating for its own solution, its own pathway to salvation. Amidst this chorus, a lone voice emerges, audacious in its candor, daring to suggest that perhaps audacity itself might hold the key.

"Let us be audacious," I proclaim, "and believe that we can contribute with these lines by inspiring those in power to craft equitable climate policies at both national and global levels." With these words, I sow a seed of hope – a seed that seeks not only to inspire change but to reshape the very mechanisms of change themselves.

In the dimly lit corridors of power, the change of behavior is heralded as a sacred chalice, an elixir capable of healing the planet's wounds. Yet, caution is my counsel, for history has taught us that paths paved with prohibition, bans, and taxes often lead to unforeseen consequences. Herein lies a lesson echoing through time: bans, like whispers of forbidden knowledge, breed dissent, fanning the flames of conspiracy and mistrust. Those who peddle in such theories weave their narratives, exploiting the fractures in the governance façade.

With a steady hand and a humble gaze, I share the fruits of my observations, recounting how bans and prohibitions, once wielded as instruments of justice, have frequently struck back with a vengeance. Democracy trembles under the weight of such actions, as shadows of disenfranchisement loom ominously. The unadulterated truth reveals that such measures, however well-intentioned, can become tools for exploitation, rendering the uninformed and uneducated as mere collateral damage.

Then, the spotlight falls on taxes – a paradoxical proposition. A mechanism, it seems, intended to regulate behavior in the noble fight for climate preservation. Yet, make a daring proposition: "Taxes aimed at regulating behavior in the fight to preserve the climate? I venture to assert that such a notion is highly unwise! skepticism takes center stage, a waltz of disbelief in the efficacy of taxation. In my eyes, governments embody insatiability, forever hungry for revenue to fuel their pet projects, while climate concerns remain obscured in the shadows.

In the land where beer and sausages reign, a vivid case study unfolds. Germany, I reveal, has woven a tapestry of taxes designed to guide its citizens towards the sunlit shores of climate consciousness. Petrol and gas taxes, energy taxes – the arsenal is impressive. Yet, beneath the shimmering façade, a stark reality surfaces: the lion's share of these revenues is consumed by the voracious maw of the national budget, leaving the aspirations of the climate in tatters. Roads are constructed, pensions secured, and political caprices indulged, while the environment stands as an unwitting pawn in the game of governance.

The script takes an unexpected turn, unveiling a tableau of power dynamics. "If you are not at the table, you are on the table," the age-old adage intones, as climate finds itself teetering on the precipice of obscurity. In a bitter twist of fate, it becomes a pawn – a mere excuse for raising taxes in the eyes of the masses. They say it's trendy to tax in the name of the environment.

Yet, within the contours of my skeptical exploration, a radical proposal emerges – an idea akin to a phoenix rising from the ashes. A proposal that envisions climate justice fought on two fronts: within the borders of every nation and across the expanse of global collaboration. My voice adopts a contemplative tone, an intellectual whisper that commands attention. "Every politics is local," I remind, urging the battle to be waged on home turf, where the flames of passion burn most intensely.

But the global stage beckons – a call for unity that transcends political boundaries. With a deft switch of my narrative brush, I outline two models: the "intra-country approach" and the "international approach." Through a symphony of thought, I elucidate the reasons for a global struggle, intricately woven with the necessity of localized action.

As the curtain descends on this chapter of audacious contemplations, I leave the reader with a sense of anticipation, a flavor of ideas ripe for exploration. My voice, now firmly established, whispers final thoughts, invoking the very essence of audacity. With each word, I paint a canvas of potential, daring all to ponder the uncharted territories of climate justice – both within our national borders and across the expanses of our shared planet.

Pinpointing the precise carbon footprint of billionaires is akin to charting a tempestuous sea; an intricate task that navigates the vast expanse of their lifestyles, consumption habits, and endeavors. To traverse this tumultuous terrain, I proffer a handful of broad approximations, drawing upon the reservoir of data available until September 2021.

In the realm of affluence, billionaires bear witness to access unburdened by constraints, and this access is most pronounced in their carbon footprints – a complex web woven by threads of private jets, opulent vehicles, sprawling estates, and lavish living. It's a tapestry that tells of the grandeur they indulge in and the extravagance they foster.

Consider, if you will, the private jets that ferry these elite denizens across the globe. These behemoths of the skies, while swathed in luxury, exude emissions on a scale commensurate with their opulence. A solitary sojourn aboard one of these marvels can engender the release of several metric tons of CO_2 equivalent, an environmental price that seemingly pales before the allure of exclusivity.

Yet, the luxury doesn't terminate with the journey. Luxury vehicles, a status symbol par excellence, traverse the roads with an air of power, but also with emissions that whisper of their energy-hungry hearts. Such grandeur, often, is not frugal in its carbon currency.

And then there are the estates that sprawl across the landscape, substantial home steads emanating prestige and wealth. But these architectural triumphs do not rise in isolation; they cast long shadows upon the Earth's resources, their energy-intensive interiors, their climatized ambiance, all resulting in a substantial carbon footprint.

Intricately interwoven are the threads of consumption – luxury goods coveted, international escapades frequented, events orchestrated in ostentation. All these whispers collectively, in the language of carbon, weave a tale of immense indulgence.

Billionaires, with their surfeit of resources, command carbon footprints that can be vast, perhaps even colossal ranging into hundreds, even thousands, of metric tons of CO_2 equivalent annually. Yet, as with all enigmas, the precision of these estimates is a dance with ambiguity, for it hinges upon the choices and circumstances that carve the individual's path through this world of opulence.

On a parallel note, in the tapestry of carbon footprints that the spectrum of society weaves, the 'average Joe' emerges. An entity whose footprint, while less conspicuous, holds resonance. The tapestry, here, is more modest, yet by no means negligible. From housing choices to the rhythm of transportation and the hum of energy consumption, it evokes a carbon footprint ranging from 10 to 20 metric tons of CO_2 equivalent per annum, in the context of Western Europe and North America.

Amid the labyrinthine expanse of Africa, where cultures meld and landscapes stretch into eternity, lies the embodiment of a carbon footprint that has its own story to tell. Here, the narrative is one that encompasses diversity, speaks of daily existence, and charts a path less carbon intensive.

In this theatre of existence, a dichotomy emerges between urban enclaves and rural retreats. Urbanity, though in its nascent stage for some, is marked by a blend of ambition and aspiration. As the urban dweller navigates the rhythms of life, transportation options may span the gamut, from shank's mare to the rickety charm of shared minivans. Public transit, a cornerstone of communal mobility, casts a modest yet impactful shadow on the carbon ledger.

However, the heartbeats of rural life echo with different footsteps. In these landscapes, the dance of traditional biomass – wood, crop residues – provides sustenance. Cooking, a ritual woven into daily life, is often accompanied by the gentle sighs of wood smoke, releasing whispers of carbon. Yet, these numbers often pale before the larger backdrop of rural lives, their carbon footprints colored by the embrace of simplicity and the constraints of circumstance.

Energy, an enabler of modernity, casts a feeble light across this canvas. Access to reliable energy sources remains uneven, with the echo of kerosene lamps and the hum of small generators intermingling with the growing presence of renewable solutions. The carbon footprint here, while modest, is not without its charm – a dance of pragmatism, one might say.

As the African narrative unfolds, the carbon footprint of an average individual, draped in the hues of diversity, weaves a story of subtle contrasts. On this continent, where aspirations flourish alongside traditions, where the urban pulse echoes in the shadow of rural landscapes, the carbon footprint stands as an intricate tapestry, woven with threads of sustainability, simplicity, and the burgeoning ambitions of a dynamic populace.

The estimate, nestled within these African narratives, casts a gentle breeze over the carbon ledger, with a footprint spanning around 1 metric tons of CO_2 equivalent per year for rural enclaves, and potentially reaching 3 metric tons for urban epicenters.

So, my dear friends, to delve back into the analogy that seems as timeless as human history itself – debt – it's worth pondering upon this debt not amassed in gold or silver, but in the ethereal currency of carbon emissions. The culprits, diverse as they are – societies, those conglomerations of lives; individuals, both those adorned in wealth and those walking the middle path; and even those quietly treading upon the soils of rural Africa – all contribute to this carbon debt.

And so, the question arises: why, in the name of reason, should the burden of this debt fall upon your shoulders? Why should you bear the yoke of carbon recklessness, a burden borne by those whose delight seemingly thrives in polluting the very air we share? This isn't your bill to foot; it's theirs. They ought to be held accountable for their own emissions.

But, you see, there's a way to tip the scales of justice. A way to make them reckon with the balance they've disrupted. A way to untangle the threads of accountability and let the winds of fairness blow freely.

It begins with the notion that the polluters themselves – the carbon culprits – must shoulder the responsibility for their actions. Their carbon debt, like any debt, requires payment. And they're not immune to this obligation.

So, how does this vision unfold? How can we, the bystanders of this ecological drama, spur the carbon footprints to face their dues? The mechanism is no feat of sorcery.

Allow me now to introduce a tool from my investment practice that serves as a fitting conclusion to this chapter. You may already be acquainted with the concept – an instrument known as the "elevator pitch." But what exactly is an elevator pitch? In its essence, it's a masterfully crafted and concise portrayal of a product, idea, project, or oneself. Delivered within the fleeting time frame of an elevator ride, which spans approximately 30 seconds to 2 minutes, this communication gem is designed to captivate the listener's attention, convey paramount points, and leave an indelible mark. Its brevity, engagement, and alignment with the context and audience are its hallmarks. Elevator pitches gracefully navigate networking events, business gatherings, job interviews, and other swift-exchange scenarios, where the art of rapid message delivery takes center stage.

Armed with this understanding, I now raise the curtain twice to unveil my two policies. As we draw back the first curtain, let's delve into my inaugural policy, delivered in the form of an elevator pitch.

Title: A Symphony for the Planet: The Global Climate Equity and Sustainability Framework (GCESF)

Ladies and gentlemen, esteemed delegates, and fellow stewards of our planet, allow me to paint a portrait of a world in transition, a world where climate justice takes center stage, where emissions reduction is a collective endeavor, and where sustainable development thrives. Imagine, if you will, a symphony of cooperation, innovation, and accountability – this is the heart of the Global Climate Equity and Sustainability Framework, or GCESF.

The Prelude: Vision and Purpose

> In the grand theater of global climate action, the GCESF takes the spotlight as a visionary strategy. It beckons us to advance climate justice, to spur emissions reduction, and to drive sustainable development in harmony. GCESF orchestrates an integrated approach to tackling climate challenges, threading together diverse elements to create a harmonious whole.

Movement 1: Unified Carbon Pricing Mechanism

> Our first movement opens with the Unified Carbon Pricing Mechanism, a cadence that resounds across borders. By establishing a uniform carbon price, say $30 per ton of emissions, GCESF sets a baseline for internalizing the costs of environmental harm. It's a unifying rhythm that resonates through both historical emissions and future projections, creating a global framework for addressing carbon debt.

Movement 2: Enforcement and Collection Authority

> As the piece progresses, the second movement enters with the majestic Enforcement and Collection Authority. Empowered by the United Nations, this body wields the baton of compliance. It conducts symphonic emissions audits, calculates excess emissions, and even imposes penalties on countries dancing out of step. These penalties, like a crescendo of automatic tariffs, build proportionally with carbon pricing, sending a clear message that the global community takes non-compliance seriously.

Movement 3: Ethical Private Climate Fund Managers

> A delicate interlude introduces the Ethical Private Climate Fund Managers. These virtuoso managers take the stage, entrusted with the orchestration of collected funds. Their commitment to transparency and investments aligned with climate goals brings a renewed sense of trust

to the global stage, a trust that's been challenged by traditional channels rife with inefficiencies and corruption risks.

Movement 4: Equitable Allocation Formula

As our symphony swells, the Equitable Allocation Formula sweeps in, a melody that rewards virtuous conduct. The limelight falls on countries within their allocated carbon budgets. These countries, like skilled musicians, are poised to receive funds allocated to their dedicated country climate funds. The allocation formula itself is a sophisticated tune, one that harmonizes with a country's percentage under their carbon budget. It's an ode to responsibility, a tribute to nations taking strides to balance the scales.

Movement 5: Non-Compliance Measures

In our penultimate movement, we confront the Non-Compliance Measures, an allegro of incentives and repercussions. Automatic tariffs ring out, a powerful refrain applied to goods from non-compliant nations. With each encore of non-compliance, these tariffs crescendo, urging corrective action to avoid a discordant trade disruption. Behind these measures stands a Global Review Panel, a vigilant guardian ensuring fairness and transparency, a beacon of integrity amid the crescendo of penalties.

The Finale: A Transformative Symphony

And now, dear audience, we arrive at our finale, a triumphant crescendo that encapsulates the essence of the Global Climate Equity and Sustainability Framework. It's a transformative symphony that links funding to compliance, an anthem of accountability. It's a composition that wields impactful sanctions as a force for change, a tribute to global unity against climate challenges.

In closing, as we contemplate this symphony for the planet, let us remember that while each movement has its strengths and weaknesses, it's the harmony of the ensemble that guides our future. It's the dedication of each player that will shape the destiny of generations to come. GCESF is our collective overture to a world where the notes of sustainability resonate in every corner. Thank you.

(Curtains close)

And now, I invite you to join me as we embark on the elevator pitch journey for the second policy.

Title: The Green Pioneers: Unleashing Climate Responsibility Through the National Carbon Dividend Equity Program (NCDEP)

Ladies and gentlemen, distinguished guests, and fellow stewards of our planet, I stand before you today to unveil a groundbreaking symphony of change, a new melody that resonates at the heart of our communities and echoes through the corridors of personal responsibility. Allow me to introduce you to the National Carbon Dividend Equity Program, or NCDEP, a visionary initiative that promises to redefine our relationship with the environment and inspire a grassroots revolution in combating climate change.

Setting the Stage: The NCDEP Vision

Imagine a world where each citizen is not just a spectator but a protagonist, where climate action transcends rhetoric and becomes a tangible endeavor. NCDEP is this very vision, a roadmap to champion climate responsibility, harness emissions reduction, and embolden individuals to be architects of change within their own borders.

Harmonizing Responsibility: The Policy Framework

Our first note resounds in the form of a Personal Carbon Pricing Mechanism. The NCDEP establishes a groundbreaking approach, individualizing carbon pricing with a tag of $30 per ton. This ingenious move captures the essence of carbon accountability, enveloping both historical emissions and future projections within its cadence. It's not just a price tag; it's an invitation to internalize environmental costs on a personal level.

As our symphony builds, we introduce the Enforcement and Carbon Contribution element. A domestic regulatory body, backed by the authority of governing institutions, takes center stage. Its role is clear: oversee enforcement, ensure fair contribution. For those whose carbon footprints exceed their budgets, financial contributions are on the horizon, serving as a stark reminder that climate action is a shared endeavor, transcending borders and echoing within our very homes.

Direct Empowerment: The Dividend Distribution Movement

But now, the crescendo: Direct Dividend Distribution. Funds collected from those who exceed their carbon budgets are not lost in the bureaucratic labyrinth. No, they are channeled directly to the citizens operating below their carbon thresholds. This is empowerment in its purest form, a dividend of responsibility rewarded with financial incentives. No intermediaries, no red tape – just direct dividends, fueling a movement of change from within.

Equity in Harmony: The Dividend Allocation Approach

As the symphony evolves, the Equitable Dividend Allocation movement takes the stage. It's a carefully orchestrated dance of equity and incentive. Dividends are calculated based on each individual's carbon budget performance. Those who dance below their budgets receive dividends, their percentage under the budget dictating the melody of rewards. It's an anthem of progress, a testament to the collective impact of personal actions.

Navigating Non-Compliance: The Encore of Incentives and Penalties

But what of non-compliance, you ask? Fear not, for our symphony has an encore. Carbon Contribution and Penalties emerge as a rhythm of accountability. Excess carbon emissions come at a cost, a financial penalty that feeds the dividend pool. Compliance is now interwoven with personal benefit, a harmony of collective responsibility and personal gain.

Reflecting on Strengths and Weaknesses: A Melodic Assessment

As with any masterpiece, our symphony has strengths and weaknesses. It empowers citizens, enhancing transparency, and fosters a sense of personal responsibility. Yet, it wrestles with equity challenges, behavioral changes, and administrative complexities. It's a composition that recognizes both the potential and the challenges, inviting us to navigate its harmonies with wisdom and determination.

The Final Cadence: A Call to Action

In closing, the National Carbon Dividend Equity Program isn't just a proposal; it's an anthem for change. It's an invitation to each one of us, a symphony that echoes the power of individuals to transform our relationship with the environment. Together, as we unleash the Green Pioneers within, let us embrace this visionary program and conduct a harmonious movement of climate responsibility. Thank you.

(Curtains close, the applause resonates)

And now, amidst this gathering, the echoes of skepticism reverberate among us, as if in harmonious resonance with the lyrical notes of a cherished French Chansonier, the great Joe Dassin, who once mused, "Ça va pas changer le monde" – "It won't change the world." To these doubters, I extend a response borrowed from the melodies of yet another beloved artist, the immortal John Lennon: "You may say I'm a dreamer, but I'm not the only one. I hope someday you'll join us, and the world will be as one."

Yes, it's true, these two policy proposals, like any human creation, may not be flawless; they may bear their share of imperfections. Yet, I implore you to unite with us, to transform these visions into reality. For in a world where those who orchestrate the planet's devastation are held accountable, only through this reckoning can we hope to usher in transformative behavioral change. It won't emanate from decrees and prohibitions, nor from state-imposed levies that seem to exist solely to fund the pet projects of politicians.

Now, the time has arrived, an opportune moment to concretize the profound vision of the legendary Bob Marley: "…You can fool some people sometimes, but you can't fool all the people all the time. So now we see the light (watch you gon' do?). We gonna stand up for our rights (yeah, yeah)…" The harmonious chorus of change beckons – will you heed its call?

III. Artistic Echoes Of Climate Change: Urban Uprising

CHAPTER *Eleven*

My Eyes Of Hope: Witnessing The Promised Land, Forging A Future Steeped In Possibilities

"Well, I don't know what will happen now. We've got some difficult days ahead. But it doesn't matter with me now. Because I've been to the mountaintop. And I don't mind. Like anybody, I would like to live a long life. Longevity has its place. But I'm not concerned about that now. I just want to do God's will. And He's allowed me to go up to the mountain. And I've looked over. And I've seen the promised land. I may not get there with you. But I want you to know tonight, that we, as a people will get to the promised land. And I'm happy, tonight. I'm not worried about anything. I'm not fearing any man. Mine eyes have seen the glory of the coming of the Lord…"

– Martin Luther King, Jr. I See the Promised Land
April 3, 1968 Memphis, Tennessee

I. Africa On My Mind: The Whys And Wherefores

In the great theater of our planet's climate saga, an often overlooked but pivotal character steps into the spotlight – Africa. Amidst the narratives of towering emissions and global responsibility, the continent's role is more than a mere subplot; it's a narrative strand that could determine the grand finale of this climatic tale.

Shift your focus from the usual protagonists, and you'll find that the world's most influential emitters hold a significant sway over the climate narrative. Yet, their endeavors, however well-intentioned, may fall short without a crucial piece in the puzzle – Africa's energy story.

For too long, Africa has been painted as a canvas of climate victimhood, absorbing the wrath of changing weather patterns while contributing the least to historical emissions. However, hidden in plain sight is a narrative that pulses with potential, one that could shape the trajectory of global emissions reduction – or undermine it.

The story begins with Africa's burgeoning population and its dance towards urbanization. It stands at the intersection of potential and peril. Energy deficiency looms large, yearning to be bridged by the flicker of electricity, the rumble of transportation, and the hum of industries.

This juncture reveals both promise and peril. It's a crossroads where Africa can opt for fossil fuel dependency, emulating the path of predecessors, or can pave an unprecedented trail where renewable energy reigns supreme.

Herein lies the heart of the matter: Africa's economy is intertwined with fossil resources. Oil, natural gas, and coal have long fueled national budgets, fed families, and offered livelihoods to countless individuals. It's a reality that cannot be brushed aside.

South Africa stands as a notable example, its economic engine propelled by coal. But beyond its borders, Africa's collective carbon footprint remains relatively modest. This is not merely a footnote; it's a turning point that reverberates through global climate negotiations.

In this context, Africa's evolution is a symphony of growth, aspiration, and the need for energy. It's an orchestra that depends heavily on the tune of fossil resources. This is not a mere energy transition; it's a transformation of economies, a shift that must be carefully orchestrated to avoid discord.

As world leaders set targets and emissions goals, it's imperative to understand that Africa's energy narrative is not solely a regional chapter. It's a thread woven into the fabric of our global climate saga. Its oilfields, gas reserves, and coal mines, while contributors to economies, also stand as markers of vulnerability in a changing world.

It's easy to prescribe renewable remedies from distant shores, but the truth lies in the fine print – replacing fossil resources requires more than just vision; it demands alternatives that can sustain economies and secure livelihoods.

Yet, Africa's leaders recognize the intersection of necessity and opportunity. Solar panels glint under the African sun, and wind turbines pierce the skies. The costs of renewable energy, once prohibitive, now bow to the might of economies of scale. These alternatives have the potential to fuel growth without stoking the flames of carbon emissions.

As discussions on emissions unfold, Africa's energy narrative must be embraced with pragmatism, understanding that a clean energy transition carries the weight of economies and the livelihoods of millions. It's about finding substitutes that can light up homes, power industries, and secure futures.

The tale of Africa's energy transition holds within it the potential to mold a narrative of triumph. It's about more than numbers on balance sheets; it's about steering economies and societies toward greener pastures. The future of global emissions hinges on how Africa balances its energy needs with its commitment to a sustainable world.

In this global climate chronicle, Africa emerges as a protagonist – not merely a player with a script handed down, but a playwright who shapes the narrative, who decides whether the ending will be one of harmony or discord. The outcome rests not just on reducing emissions, but on a symphony of sustainable choices that resonate through the ages, setting the stage for a world that thrives.

Henceforth, this segment shall be devoted exclusively to the cases I vision pertaining to the African continent. Despite the challenges that may lie ahead, uncertainty lingering like an enigmatic mist, I am acutely aware that my own role in Africa's journey is a question yet unanswered. Yet, within the tapestry of my aspirations, a vivid image unfurls – a sweeping vista unfolds, revealing ambitious and transformative undertakings within the cherished embrace of my envisioned African sanctuary. This vision, akin to the resonant verses of Ray Charles' hymn to his native Georgia, echoes within me for Africa, Africa on my mind:

"

...

Amidst outstretched arms, I am embraced.
Gentle gazes bestow their tender grace.
In tranquil dreams, the path unveils,
Leading me back to Africa's tales.

...

Oh Africa, my Africa, a tranquility eludes (a tranquility eludes).
Yet within the strains of a cherished melody.
Africa's essence, steadfastly pursued.

...

My thoughts and visions bind.
Keeps Africa on my mind.

...

"

II. Calling All Visionaries And Industrial Titans: Join The Transformative Journey To Build A Sustainable Future In Africa!

Greetings innovators, forward-thinkers, industrial leaders, and champions of change, Imagine the synergy of visionary investment and industrial might converging to shape the trajectory of Sub-Saharan Africa. As we stand at the cusp of transformation, we extend an invitation to those with funds, wealthy families, limited partners, and corporations – as well as to the industrial giants who can help build the very infrastructure that drives this change. Together, let's embark on a journey that transcends profit margins, fuels growth, and ushers in an era of sustainable progress.

Our Vision:

Imagine a Sub-Saharan Africa thriving on renewable energy, sustainable industries, and groundbreaking innovation. We're not just investing; we're co-founding an era defined by impact, innovation, and economic evolution.

The Promise of Transformation:

Our transformative initiatives, from integrated critical minerals to green data centers, hold the power to revolutionize industries while addressing pressing global challenges. We're seeking partners who don't just see an investment but envision a legacy.

Why Join Us?

First-Mover Advantage: In emerging markets, we're positioned for first-mover advantage, unlocking untapped potential and driving unparalleled growth.

Holistic Impact: We're offering more than financial returns; we're fostering positive social and environmental change, aligning with conscious consumers and communities.

Enduring Legacy: Your participation today will leave a lasting imprint on Africa's future, establishing a legacy that resonates through generations.

A Call to Industrial Titans:

Industrial leaders, your expertise and resources are transformative forces waiting to be unleashed. Partnering with us is an opportunity not just to expand your top and bottom line but to catalyze sustainable progress.

Act Now, Shape Tomorrow:

As we call out to funds, wealthy families, limited partners, and corporations, we also extend an invitation to industrial giants. Your involvement can redefine the trajectory of industries and economies and leave a legacy that reverberates across time.

Contact Us:

Reach out today to explore how your resources, innovation, and industrial capabilities can align with our transformative vision. Join us as co-founders of this historic journey, building infrastructure, driving growth, and shaping a sustainable future.

Together, We Thrive:

This is an opportunity not just to invest, but to lead with purpose, shape industries, and create a legacy that stands as a testament to visionary leadership.

Join us in this extraordinary journey, where visionary investment meets industrial prowess, and together, we carve a legacy that shines forever.

With anticipation and determination,

Jude S. Ngu' Ewodo

III. Sub-Saharan Africa's Metamorphic Expedition Unveiled

In the vast expanse of Sub-Saharan Africa, a transformation is brewing. It's a convergence of far-sighted investment and sustainable progress, a haven for potential that seems boundless. Here, ideas don't merely take shape; they ascend, crafting narratives of extraordinary impact. As we immerse ourselves in projects that possess the power to reshape economies, elevate lives, and nurture our planet, we embark on a voyage into a realm of profound change. Welcome to a landscape where the wisdom of our research seamlessly intertwines with the revolutionary spirit of climate solutions. This is an opportunity for those discerning investors, lenders, venture capitalists, infrastructure managers, and others who yearn to not only invest their funds but also to rescue the planet while reaping returns beyond the confines of conventional markets.

Integrated Critical Minerals Value Chain: The Bedrock of Resilience

Visualize a tapestry where minerals are sourced with a conscience, fueled by the vigor of renewable energy. Beneath the numerical fabric adorned with IRRs spanning 35% to 40% and ROIs that ascend to 8x to 10x, we uncover a narrative of sustainability meticulously woven into each thread. The equilibrium attained within 4 to 6 years isn't just a financial milestone; it's a proclamation of the creation of myriad direct and indirect jobs, the honing of skills, the stimulation of economic growth, and the elevation of countless communities.

The forecast heralds an estimated 15,000 direct jobs, subsequently empowering an approximate 75,000 livelihoods. And the monetary tapestry of yearly revenue extends from billions to tens of billions $. The funds required for this venture will emerge from infrastructure investment and venture capital.

Green Steel Manufacturing: Forging a Carbon-Neutral Legacy

Step into the realm of green steel, a space where innovation dances with ecological consciousness. Within the radiant sphere of IRRs ranging from 30% to 45% and ROIs ascending to 8x to 12x, a story unfolds – a narrative of burgeoning employments, refined skills, and the reduction of carbon emissions. The equilibrium achieved within 4 to 7 years isn't just a mark of financial stability; it's the birth of an enduring legacy. Lives transform through direct employment, empowering communities to stride towards brighter vistas.

This narrative envisions the creation of 10,000 direct jobs, nurturing an additional 50,000 livelihoods in its embrace. The yearly revenue panorama spans from hundreds of millions to billions $. The funding will find its roots in infrastructure investment and venture capital.

Green Hydrogen Production and Export: Igniting Emission-Free Futures

Embark on a voyage through the realm of green hydrogen, where returns flourish at 6x to 8x, and net profit margins shimmer at 30% to 35%. Beyond the numerical veneer lies a story of green jobs, cleaner energy, and a diminished carbon footprint. As we tread the path towards equilibrium within 5 to 8 years, we're not merely erecting an industry – we're crafting a greener world. The ripple effect extends through societies, offering opportunities and disseminating the gift of cleaner energy access.

This case forecast the generation of 10,000 direct jobs, indirectly supporting over 50,000 livelihoods. The yearly top line landscape extends from billions to tens of billions $. The catalysts for this transformation will stem from energy project financing and strategic partnerships.

Renewable Energy Regional Hubs: Beacons of Progress

Picture landscapes adorned with renewable energy hubs, their brilliance casting IRRs of 25% to 35%, and ROIs illuminating a realm of 5x to 6x. Beneath these numbers flourishes a narrative of assured energy access, burgeoning employments, and the promise of reduced dependence on fossil fuels. Equilibrium achieved within 4 to 6 years isn't just a financial milestone; it resonates with the energy landscape of an entire region. The impact ripples through lives empowered with clean energy, fanning the flames of a thriving sustainable job market.

The projection stands at an estimated 12,000 direct jobs, nurturing a potential 60,000 livelihoods indirectly. The yearly revenue spectrum blankets from hundreds of millions to billions $. The funding avenues will stem from renewable energy funds, infrastructure development, and public-private partnerships.

Circular Economy Initiatives: Weaving Sustainability into Progress' Fabric

Step into the embrace of the circular economy, where IRRs intertwine with ROIs ranging from 25% to 30%. Yet, beneath these fiscal façades, a tale unfolds – a tale of waste reduction, rejuvenated recycling, and an ecologically affirmative footprint. The equilibrium achieved within 3 to 5 years isn't just about charting a profitable course; it's about steering towards a cleaner, healthier future. Thousands will reap the rewards of waste reduction, recycling prospects, and a more vibrant environment.

The canvas of opportunity envisions the creation of approximately 7,000 direct jobs, fostering the livelihoods of over 35,000 souls indirectly. The yearly revenue panorama sweeps from hundreds of millions to billions $. The funding tendrils will extend from venture capital and impact investments to climate-conscious initiatives.

Climate-Smart Agriculture Innovation: Cultivating Resilient Tomorrows

Within this chapter, IRRs sway between 30% to 40%, while ROIs beckon with returns of 6x to 8x. However, beneath these figures resides a narrative of empowered farmers, fortified food security, and sustainable cultivation. Equilibrium reached within 4 to 6 years doesn't merely denote financial stability; it heralds the dawn of an agricultural renaissance. The social impact reverberates through empowered farmers, fortified food security, and the cultivation of sustainability.

The projection unveils an expectation of 25,000 direct jobs, indirectly nurturing an approximate 125,000 livelihoods. The yearly revenue terrain spans from hundreds of millions to billions $. The funding orchestration will draw from venture capital, agricultural and food security funds, and climate-conscious initiatives.

Affordable Clean Water Solutions: Quenching Thirst, Elevating Lives

The figures suggest returns that span from 4x to 5x, with IRRs hovering around 20% to 25%. Yet, beneath these numerical representations resides a narrative of amplified public health, access to clean water, and the decline of waterborne ailments. Achieving equilibrium within 5 to 7 years isn't just about financial gains; it's about realizing a fundamental human entitlement. Thousands will gain access to clean water, a transformation that elevates public health and overall well-being.

The employment forecast for this project envisages the creation of 6,000 direct jobs, in turn nurturing the livelihoods of approximately 30,000 individuals. The yearly revenue vista spans from hundreds of millions to billions $. The funding strands will intertwine from sources like public-private partnerships, impact investments, and infrastructure investment.

Eco-Tourism and Conservation: Confluence of Humanity and Nature

Amidst a backdrop of ROIs that glisten at 5x to 7x, while IRRs emanate from 22% to 30%, unfolds a narrative of conserving biodiversity, fostering local employments, and nurturing cultural exchange. Equilibrium within 4 to 6 years isn't merely about financial poise; it epitomizes the harmony between human progress and the sanctity of nature. Communities flourish through opportunities for employment, cultural exchange, and thriving eco-systems.

The projection envisions the provision of approximately 10,000 direct jobs, indirectly nurturing over 50,000 livelihoods. The yearly revenue landscape stretches from hundreds of millions to billions $. The investment currents will flow from avenues like tourism infrastructure, climate-conscious investments, carbon credit programs, and public-private partnerships.

Green Transportation Infrastructure: Paving the Route to a Greener Commute

Within these pages, ROIs whisper between 25% to 35%, and IRRs hum from 5x to 6x. Yet beneath these numerical façades lies a narrative of curtailing air pollution, reinvigorating urban mobility, and sowing the seeds of green employments. Achieving equilibrium within 3 to 5 years isn't just a financial milestone; it's a pivotal juncture towards sustainable mobility. The impact resonates with reduced pollution, heightened urban mobility, and fresh avenues of employment.

In terms of the forecasted employment, this project is poised to generate 15,000 direct jobs, indirectly supporting an additional 75,000 positions. The yearly revenue horizon varies from hundreds of millions to billions $. The conduits for investment here encompass infrastructure development, electric vehicle charging networks, and climate-conscious initiatives.

Green Data Centers: Where Innovation Embraces Sustainability

Roam through the realm of data centers, where IRRs range from 25% to 30%, while ROIs span 4x to 5x. Beyond these numerical revelations, a narrative unfurls – one of sustainable digital expansion interwoven with technological strides. Equilibrium attained within 3 to 5 years isn't just about financial robustness; it symbolizes the emergence of a greener, tech-centric era. The impact reverberates through leaps in technology, sustainable growth, and the creation of fresh avenues for employment.

In terms of employment forecast, this venture envisions the creation of approximately 5,000 direct jobs, indirectly nurturing the livelihoods of around 25,000 individuals. The yearly revenue tableau ranges from hundreds of millions to billions $. The channels for investment will stem from data center development, technological investment, and infrastructure development.

As we navigate through these transformative narratives, remember that the numbers are not mere statistics; they are gateways to tales of profound impact, innovation, and metamorphosis. Each project carries the potential to reshape the contours of Sub-Saharan Africa and redefine the global climate narrative.

Join us as we inscribe history together, forging a legacy of growth, innovation, and enduring impact.

IV. Artistic Echoes Of Climate Change: The Promised Lands

CHAPTER *Twelve*

Stewardship in Action: Ask Not What Your Land Can Do For You, But What You Can Do For Your Land

"In the long history of the world, only a few generations have been granted the role of defending freedom in its hour of maximum danger. I do not shrink from this responsibility – I welcome it. I do not believe that any of us would exchange places with any other people or any other generation. The energy, the faith, the devotion which we bring to this endeavour will light our country and all who serve it – and the glow from that fire can truly light the world.

And so, my fellow Americans: ask not what your country can do for you – ask what you can do for your country.

My fellow citizens of the world: ask not what America will do for you, but what together we can do for the freedom of man.

Finally, whether you are citizens of America or citizens of the world, ask of us the same high standards of strength and sacrifice which we ask of you. With a good conscience our only sure reward, with history the final judge of our deeds, let us go forth to lead the land we love, asking His blessing and His help, but knowing that here on earth God's work must truly be our own."

– John F. Kennedy's Inaugural Address Ask Not What Your Country Can Do For You January 20, 1961 United States Capitol in Washington, D.C.

I. The Twist: We Got The Power

Picture this, my fellow readers: an age where personal actions take center stage, where science and sustainable investments face relentless attacks a la Rumpelstiltskin from a few "conservatives," and are shamelessly defamed as "woke" endeavors. It's a tale of tumultuous times, where the battle for a greener world is fought not only in large investment firms, laboratories and boardrooms but also on the very battleground of public perception.

In this theater of climate crusaders, we witness a paradoxical struggle. On one hand, the march of science and innovation propels us towards a more sustainable future. Bright minds work tirelessly to unravel the mysteries of climate change and devise ingenious solutions to combat it. Sustainable investments, like mighty warriors, emerge to champion the cause, rallying the support of conscientious investors eager to make a difference.

But alas, in the shadows lurk those who seek to stifle progress. A cabal of "conservatives" emerges, painting the canvas of sustainable investments with broad strokes of skepticism and disdain. They view science as a threat to their status quo, an enemy to be vanquished. They fear the rise of a new era, where green ideas challenge the old order, and economic success is intertwined with environmental stewardship.

As these climate champions face the barrage of attacks, they find themselves at a crossroads. Do they retreat, abandoning the fight for a brighter future, or do they stand tall, resolute in their pursuit of change? The stakes have never been higher, and the outcome will determine the fate of our planet.

But lo and behold, a new player strides onto the stage, a maverick in the theater of climate crusaders. A bold activist movement rises like a phoenix, armed to the teeth not with swords, but with an arsenal of crowd funding, investment clubs, and do-it-yourself shared knowledge. Their weapons of choice? Intelligence and data, wielded like a masterful blade to prove the worth of sustainable practices and investments.

This intrepid breed of activists doesn't cower in the shadows of doubt or succumb to the whispers of naysayers. No, they charge forth into the battlefield of "just do it" and "can do" attitudes, determined to make their mark. These warriors know that their actions bear more weight than mere symbolic gestures. Their mission is clear – to rattle the very foundations of the status quo and rewrite the script of the future.

In the influencers market, they stand firm, unwavering in their resolve. Armed with data-driven arguments and disruptive ideas, they challenge the dark forces of climate change denial head-on. These modern-day visionaries recognize that the path to victory lies in shaking the pillars of the establishment and defying the odds stacked against them.

Their battle cry echoes through the halls of power, and their impact reverberates far beyond the confines of traditional activism. This is a movement of action, of tangible change, and of seismic shifts in the landscape of climate consciousness.

So, heed the call, my fellow readers, and witness the rise of these climate champions. For they are not bound by the limitations of the past; instead, they are pioneers of a new era, where intelligence and innovation are their most potent weapons. The stage is set, and the script is theirs to write. Let us join hands with these audacious warriors and step into a future defined by sustainable practices, transformative investments, and a resolve that knows no bounds. The journey is arduous, but the rewards are boundless – a greener, more resilient world that stands as a testament to their unwavering spirit.

II. Understand Your Carbon Footprint

Ladies and gentlemen, we embark on a journey of self-discovery, one that not only enlightens us but also empowers us to take meaningful action in the fight against one of the most pressing challenges of our time: climate change.

Understand Your Carbon Footprint:

Let's begin with the cornerstone of our mission – understanding your carbon footprint Picture, if you will, a grand symphony, where each note played represents a single action you take in your daily life. From the moment you wake up to the instant your head hits the pillow, you're composing a carbon composition that influences the climate's melody.

Why It Matters:

Now, why should you care about this composition? Because, my friends, understanding your carbon footprint is akin to understanding the very score of our future. It's the blueprint for informed decision-making, and it lays the foundation for change. By quantifying your impact, you gain the power to pinpoint precisely where your actions resonate most with the environment.

How to Calculate Your Carbon Footprint:

To measure your role in this grand orchestra, there are tools at your disposal. Imagine them as tuning forks, each finely calibrated to assess different aspects of your life. These digital wizards take into account the energy you consume, the way you move about, the food you put on your plate, and how you handle waste.

Key Areas to Consider:

Home Energy Use: Think of your dwelling as a stage where energy consumption takes center stage. Swap out those old incandescent bulbs for LEDs, ensure your home's insulation is topnotch, and seal those drafts that let precious energy escape. In this act, the encore is a smaller carbon footprint and lower utility bills.

Transportation: Our next movement takes us on a journey through your transportation choices. Do you saunter, cycle, carpool, or ride public transport? Perhaps you've embraced the electric vehicle revolution. Each step you take, every pedal you push, contributes to our collective harmony with the planet.

Food Choices: Ah, the dining room – the stage for our culinary choices. Here, plant-based diets are the virtuoso, playing a tune of lower carbon emissions. Opt for locally sourced and seasonal foods to reduce the carbon footprint of your meal.

Waste Generation: Our fourth act, a bit less glamorous but equally vital. Take a bow for your efforts to reduce, reuse, and recycle. These actions ensure that less waste makes its way to the landfill, thus reducing emissions.

Consumer Habits: In the fifth act, we contemplate our consumption patterns. Do you choose products designed for longevity? Are you inclined to repair rather than replace? Supporting companies with strong sustainability practices makes you a key player in the climate concerto.

Keep Track of Your Progress:

And as the performance unfolds, don't forget to record your notes. Keeping a journal of your efforts and their impact is like preserving a record of our symphony's evolution. Periodically revisit your carbon footprint, and, like a composer fine-tuning a masterpiece, adjust your actions to create a more harmonious future.

In Conclusion:

In closing, understanding your carbon footprint is not just a task; it's a journey of self-discovery and empowerment. It's about recognizing that every note in your life's composition can be fine-tuned for a more sustainable, beautiful future. It's about contributing to a healthier planet, where our actions, like the best symphonies, resonate for generations to come.

III. Reduce Energy Consumption

Ladies and gentlemen, this time, let's embark on a journey through the world of energy consumption. Imagine energy as the lifeblood of our modern world, coursing through the veins of our homes, fueling our vehicles, and powering our countless devices. Yet, it's a double-edged sword, for it's also a primary source of greenhouse gas emissions, shaping our climate's destiny.

Why It Matters:

Why should we care about reducing our energy consumption? This is where the story gains depth. The energy we consume often comes from fossil fuels, like coal, oil, and natural gas. When burned, these fuels release carbon dioxide (CO_2) into the atmosphere, contributing to global warming by trapping heat.

Reducing Our Energy Consumption: A How-To Guide:

Now, let's delve into the practicalities of reducing our energy consumption. I'm about to equip you with the tools to become energy-conscious virtuosos, conducting the symphony of your life in harmony with the environment.

Switch to Energy-Efficient Appliances: Modern, energy-efficient appliances consume less electricity and often perform better. Look for the ENERGY EFFICIENCY labels within your country.

Improve Insulation and Sealing: Invest in proper insulation for walls, ceilings, and floors to retain heat in the winter and cool air in the summer.

Use a Programmable Thermostat: A programmable thermostat allows you to efficiently regulate your home's temperature.

Unplug Devices and Use Smart Power Strips: Devices that draw power when off waste energy. Unplug or use smart power strips to cut power to inactive devices.

Natural Lighting and Ventilation: Make the most of natural light during the day and open windows for natural ventilation on mild days.

Regular Maintenance: Regular maintenance ensures your heating and cooling systems operate efficiently.

Consider Renewable Energy: Investing in solar panels or wind turbines reduces your carbon footprint and promises long-term energy savings.

The Key Takeaway:

As we conclude, remember that reducing energy consumption is not just a noble endeavor; it's a tangible and transformative journey. It's about harmonizing our lives with the environment, where every action lessens our ecological footprint and enriches our financial well-being. By adopting these practices, you contribute to the fight against climate change and set an example for a more sustainable future.

IV. Adopting Renewable Energy Sources As Consumers

As consumers, each of us holds a unique superpower - the power to shape the energy landscape. By embracing renewable energy sources in our daily lives, we become catalysts for a cleaner, more sustainable future. Allow me to shed light on how individuals and households can make this transformative shift across various aspects of their energy needs.

Why It Matters:

Our transition to a sustainable energy future hinges significantly on consumer-adoption of renewable energy. This choice reduces harmful greenhouse gas emissions, bolsters the renewable energy sector, and paves the way for positive environmental change.

Solar Power for Home Electricity:

Imagine the sun's rays on your rooftop not only providing warmth but also generating electricity. That's the promise of solar power. Installing solar panels on your roof allows you to harness the sun's energy to reduce your reliance on fossil fuels. Furthermore, many regions offer financial incentives and tax credits, making solar installations a financially viable option for homeowners.

Wind Energy for Residential Use:

In regions blessed with a consistent breeze, small wind turbines can be a game-changer for residential energy needs. These turbines gracefully harvest clean, wind-generated power, reducing your carbon footprint. Local regulations and incentives can provide guidance and support for wind energy installations.

Geothermal Heating and Cooling:

Let's delve into the Earth's natural stability for heating and cooling your home. Geothermal heat pumps efficiently tap into this energy, reducing your overall energy consumption while keeping your indoor environment comfortable. The best part? They can seamlessly integrate into your existing HVAC system.

Biomass and Biogas for Cooking and Heating:

Cooking and heating your home can also go green. Consider alternatives like biomass stoves that utilize wood pellets or biogas generated from organic waste. Modern biomass heating systems are not only efficient but also emit fewer pollutants compared to traditional wood-burning stoves.

Solar Water Heaters:

In sunnier regions, solar water heaters are a shining example of sustainable choices. They use sunlight to warm water for various household uses, providing an eco-friendly alternative to conventional water heaters.

Electric Vehicles (EVs):

Imagine your daily commute with zero emissions and a reduced reliance on fossil fuels. Electric vehicles (EVs) powered by renewable electricity are a significant stride toward sustainability. They help cut down greenhouse gas emissions from transportation, paving the way for cleaner, greener roads.

Supporting Renewable Energy Providers:

Choose energy providers that champion renewable sources. Many regions offer green energy options, allowing consumers to support clean energy production through their electricity bills.

Home Energy Storage:

Harnessing excess energy generated from renewables is a smart move. Solutions like lithium-ion batteries or emerging technologies like hydrogen-based storage allow you to store surplus energy for use during periods of low generation or grid outages.

Government Incentives and Rebates:

Governments often offer a helping hand in your journey toward renewables. Explore incentives, tax credits, and rebates for adopting renewable energy technologies. These incentives make the transition more affordable and accessible for consumers.

Community Federation and Investment:

Communities, municipalities, and villages can unite their purchasing and investment power to build and operate mini-grids for energy independence. This collective effort not only reduces environmental impact but also fosters community resilience.

ESG and Climate-Friendly Financial Products:

Investing in ESG (Environmental, Social, and Governance) and climate-friendly financial products is another step toward sustainability. However, be vigilant for greenwashing – ensure your investments genuinely align with green and sustainable objectives.

Key Takeaway:

In conclusion, consumers are the architects of a sustainable energy future. By embracing renewable sources for electricity, heating, cooling, and transportation, individuals and households wield the power to reduce their carbon footprint and support a greener world. The transition to renewables is not just a responsible choice; it's a transformational journey that leads to a brighter, healthier, and more sustainable future for all of humanity.

V. Transportation Choices

Ladies and gentlemen, now, we embark on a journey through the vital realm of transportation choices, an area of our daily lives that holds the power to shape our existence and impact the very planet we call home.

Transportation Choices:

Picture, if you will, your daily commute as a symphony of decisions. Each choice, like selecting an instrument for an orchestra, creates a unique tone, a unique impact. How you move about not only shapes your personal experience but also resonates deeply with the environment we all share.

Why It Matters:

Transportation is the heartbeat of modern life, connecting us to work, leisure, and loved ones. Yet, it's also a major player in the grand narrative of climate change. Vehicles, powered by fossil fuels such as gasoline and diesel, are the culprits releasing CO_2 and other pollutants into the atmosphere, intensifying the specter of climate change. The choices we make in this arena hold the power to rewrite this story.

Evaluating Your Transportation Options:

Let's take a stroll through the orchestra of transportation choices.

Walking and Biking: These modes of transportation are the gentle notes of a flute. They produce no emissions, serenade you with health benefits, and are especially harmonious for short trips in urban landscapes.

Public Transportation: Buses, trams, and trains form the rhythm section of our transportation ensemble. They transport large numbers efficiently, reducing traffic congestion and emissions. Opting for public transportation is a commendable choice, reducing the cacophony of individual cars.

Carpooling and Ridesharing: Carpooling and ridesharing perform a well-coordinated duet. They reduce the number of vehicles on the road, saving fuel and curbing emissions. Sharing rides with colleagues or utilizing ridesharing apps embodies convenience and Eco friendliness.

Electric Vehicles (EVs): EVs may represent the future of transportation, where the melody of sustainability meets innovation. They produce zero tailpipe emissions and are increasingly accessible as charging infrastructure expands. Consider transitioning to an EV to harmonize your daily commute with environmental consciousness.

Telecommuting and Remote Work: Think of this as the "virtual concert." Telecommuting and remote work have come to the forefront, reducing the need for daily commutes and significantly lowering your carbon footprint.

Active Transportation: Walking and biking, the notes of an environmentally friendly flute, contribute not only to a healthier lifestyle but also to a cleaner planet. Incorporating more physical activity into your daily routine reduces emissions and enhances your well-being.

Key Takeaway:

In conclusion, transportation choices are not mere pragmatic decisions; they are a symphony of options, a harmony with our environment. The choices you make while navigating your world resonate deeply in the grand composition of climate change. By opting for sustainable modes of transportation, whether it's the gentle notes of walking or biking, the harmonious rhythms of public transit, or the futuristic melody of electric vehicles, you not only reduce your carbon footprint but also play a vital role in creating a world with cleaner air, less congested streets, and a future in harmonious coexistence with our precious planet.

VI. Sustainable Food Choices

Ladies and gentlemen, this time, we venture into the profound realm of "Sustainable Food Choices." Picture your daily meals not as mere sustenance but as a symphony of flavors, a harmonious experience that extends its influence far beyond your taste buds. Each morsel you select, each ingredient you savor, is akin to a musical note in this grand symphony of sustenance, a symphony that carries its melody into the very heart of our environment.

Why It Matters:

Food, as you may know, is more than just nourishment; it's a cornerstone of our culture, identity, and daily life. But herein lies a paradox: the way we produce and consume food wields significant environmental consequences. Agriculture, livestock, and food processing contribute to the emission of greenhouse gases (GHGs), lead to habitat destruction, and pollute our precious waterways. Sustainable food choices are not merely desirable; they're essential because they hold the power to alleviate these environmental impacts and contribute to a healthier planet.

Evaluating Your Food Choices:

Now, let's embark on a culinary journey, exploring the world of sustainable food choices.

Plant-Based Diets: Think of plant-based diets as the soothing melody of a flute. They inherently possess a lower carbon footprint compared to meat-heavy diets. By scaling down your meat consumption and embracing a rich tapestry of fruits, vegetables, legumes, and grains, you not only reduce emissions but also nurture better health.

Local and Seasonal Foods: Picture local and seasonal foods as the dependable rhythm section of our culinary orchestra. These foods are fresher, demand less energy for transportation, and bolster local economies. When you favor foods that are in sync with the seasons and cultivated nearby, you actively reduce the carbon emissions tied to long-haul transportation.

Reduce Food Waste: Cutting down on food waste is akin to fine-tuning the harmony in your food symphony. Careful meal planning, proper food storage, and the art of creatively repurposing leftovers not only save you money but also diminish the volume of discarded food sent to landfills, where it generates the problematic methane emissions.

Support Sustainable Farming Practices: Supporting sustainable farming practices is akin to endorsing a virtuoso soloist in our gastronomic ensemble. Seek out foods labeled as organic, fair trade, or certified by sustainability organizations. These choices not only advocate for environmentally friendly farming methods but also uphold ethical treatment of the hands that till the land.

Mindful Seafood Choices: Seafood choices are like the delicate notes of a violin in your culinary orchestra. Overfishing and destructive fishing practices imperil our marine eco-systems. Opt for sustainably sourced seafood to contribute to the preservation of our oceans' delicate balance.

Meat Consumption: Now, for the note that requires delicate handling – meat consumption. If parting ways with meat entirely is a challenge, fret not; moderation is key. Consume meat consciously by scaling back the quantity and scrutinizing the quality and production methods. Opt for meat from sources that prioritize animal welfare and practice sustainable farming. By doing so, you can savor meat responsibly while diminishing its environmental footprint.

Key Takeaway:

In conclusion, sustainable food choices transcend mere dietary decisions; they resonate deeply with the environment. Every meal you relish, every ingredient you select, holds the potential to shape a healthier planet. By embracing plant-based diets, favoring local and seasonal foods, minimizing food waste, championing sustainable farming, making mindful seafood choices, and approaching meat consumption with consciousness, you become a conductor in the symphony of sustainability. Your choices not only nourish your body but also nurture our planet, paving the way for a harmonious world where food serves as both sustenance and a testament to our commitment to a greener, more sustainable future.

VII. Waste Generation

Ladies and gentlemen, as we venture further into our exploration of environmental stewardship, let us now turn our attention to the intricate topic of "Waste Generation."

Waste Generation:

Imagine the life cycle of your possessions, from their entrance into your world to their eventual exit. Every item, every package, every disposable product follows a trajectory, a journey that often concludes in waste. Waste generation orchestrates this journey, playing a vital role in the environmental symphony.

Why It Matters:

Waste generation isn't solely about the objects we discard; it's about the profound ecological consequences of our consumption and disposal habits. The waste we create ripples through our environment, from overflowing landfills to the contamination of natural habitats. Understanding the impact of waste is crucial for reducing our ecological footprint.

Evaluating Your Waste Generation:

Now, let's explore the art of responsible waste generation, with a special emphasis on avoiding plastics and single-use packaging.

Reduce: The act of waste reduction is like composing a minimalist masterpiece. It begins with mindful consumption and a commitment to steer clear of single-use plastics and excessive packaging. Challenge yourself: do you truly need that item encased in layers of plastic? Seek alternatives that generate less waste and are packaged in materials that are easily recyclable or biodegradable. By curbing excessive purchases and avoiding plastic-laden products, you actively reduce the volume of waste you produce.

Reuse: Reusing items you already possess is akin to crafting a timeless composition with familiar notes. Instead of hastily discarding something after a single use, discover ways to extend its life. Repair damaged items, repurpose containers, and explore the world of secondhand or vintage goods. The longer an item remains in service, the less waste it contributes to the environmental score.

Recycle: Recycling is the harmonious process through which discarded materials find new life in a grand symphony of renewal. Harmonize your recycling efforts by sorting recyclables with precision, adhering to local recycling guidelines, and actively supporting businesses that incorporate recycled materials into their products. Recycling not only conserves resources but also mitigates the use of plastics through the production of goods with recycled content.

Compost: The act of composting is akin to nurturing a lush garden of harmonious notes, all while avoiding plastic contamination in your compost pile. Organic waste, encompassing food scraps and yard trimmings, can be nurtured to create nutrient-rich soil. By composting these materials, you divert organic waste from landfills, where it would otherwise produce methane, a potent greenhouse gas.

Mindful Disposal: Responsible disposal stands as the grand crescendo in the symphony of waste generation. When handling hazardous waste, electronics, and other specialized items, it's imperative to follow precise guidelines. These materials demand unique handling to safeguard the environment and protect human health, all while avoiding plastic contamination in the disposal process.

Key Takeaway:

In conclusion, waste generation is an integral facet of our contemporary lives, a chapter in the grand composition of our existence. Yet, it's a chapter we possess the power to reshape into a more harmonious and sustainable melody. By tempering our consumption, embracing reusability, recycling with precision, nurturing organic waste through composting, and observing mindful disposal practices – all while avoiding plastics and single-use packaging – we can effectively reduce the environmental footprint of our waste generation.

Consider waste not as discord but as a composition waiting to be perfected. With each conscious action you take, you become a contributor to the creation of a world that resonates with sustainability. Your choices in waste generation hold the promise of protecting the environment, conserving invaluable resources, and crafting a cleaner, healthier planet for generations yet to come.

VIII. Water Conservation

Water, the essence of life itself, flows through the tapestry of our planet like a precious thread. As consumers, we wield the power to shape the destiny of this finite resource through actions that are simple yet profoundly effective.

Why It Matters:

The significance of water conservation cannot be overstated. It is the guardian of freshwater reserves for today and the safeguard for generations yet to come. Within the grasp of every individual are practical steps that can make an indelible mark on water conservation.

Fix Leaks and Embrace Water-Saving Devices:

Imagine the unseen drip of a leaky faucet – a seemingly insignificant occurrence that, over time, squanders vast quantities of water. Promptly identify and repair leaks in your home, whether in faucets or pipes. Embrace the marvel of low-flow fixtures like efficient showerheads and faucets. These unassuming devices reduce water consumption without compromising performance, making every drop count.

Conscious Daily Water Usage:

In the grand symphony of water conservation, each drop plays a role. Conserve by turning off the tap while brushing your teeth or engaging in the soothing rhythm of lathering your hands with soap. Shorten your showers, and employ a timer to keep a vigilant watch over the time spent under cascading water. Capture the cool embrace of cold shower water in a bucket, a virtuoso move that offers an encore as it waters your plants or refreshingly flushes toilets.

The Art of Landscape Management:

Envision your garden as a canvas of possibility. Choose drought-resistant plants that thrive on less water. Embrace the subtlety of efficient irrigation systems like the gentle notes of drip irrigation, which nourishes plants directly at their roots. Harness the grace of rainwater, collected in barrels, to nourish the lushness of your outdoor realm. Employ the technique of mulching to retain soil moisture, a harmonious act that diminishes the need for frequent watering.

The Symphony of Reduce, Reuse, and Recycle:

Let your creative spirit flow as you seek inventive ways to reuse water in your daily life. The roof above your head captures rainwater – a reservoir of possibility, ready to grace your toilets or breathe life into your garden. Repurpose water used for washing vegetables or simmering pasta to tend to the verdant chorus of your indoor plants. Through these creative endeavors, you weave a tapestry of water conservation, minimizing waste and reducing consumption.

Thoughtful Consumption Habits:

When traversing the marketplace, make selections that harmonize with water conservation. Deliberately choose water-efficient appliances and products, distinguished by certifications such as the WaterSense label. This emblem serves as a testament to their commitment to both water efficiency and performance standards.

Advocacy for Responsible Water Usage:

As you journey along the path of water conservation, invite your community to join the symphony. Engage with initiatives that champion responsible water usage, both locally and beyond. Encourage the enactment of policies that incentivize water conservation and discourage wasteful practices. Through your involvement with organizations dedicated to water conservation, your individual notes crescendo into a chorus of change.

Key Takeaway:

In conclusion, water conservation is not a solitary endeavor but a harmonious commitment that binds us all. Through these practical steps and daily mindfulness, consumers hold the power to protect this irreplaceable resource. Water conservation is more than a responsibility; it is a collective dedication to preserving Earth's liquid lifeline for a sustainable and harmonious future.

IX. Consumer Habits

Permit me to express a contemplation on the profound influence each of us wields in sculpting a world beneficial to the climate through the choices we make as conscientious consumers.

Imagine your daily routines as a symphony, a complex composition where each choice you make contributes to the melody of your life. These are your consumer habits, the rhythm and tempo of your existence, influencing not only your own well-being but also the world that surrounds you. Deeply embedded in our culture, these habits wield tremendous influence in the grand environmental symphony.

Why It Matters:

Consumer habits aren't mere routines; they are the unseen gears driving our economy and guiding the production and disposal of goods. Yet, when left unchecked, many of these habits become unwitting accomplices in resource depletion, waste generation, and environmental degradation. Understanding the profound impact of consumer habits is paramount to fostering sustainability.

Evaluating Your Consumer Habits:

Now, let's embark on a journey through the art of responsible consumer habits.

Mindful Consumption:

Mindful consumption is akin to conducting a symphony with intention. It starts with asking essential questions before making a purchase. Do you truly need this item, or is it merely a fleeting desire? How was it produced, and what environmental footprint does it carry? By pausing to consider the life cycle of a product, you can make more informed choices and reduce the unnecessary consumption that burdens our planet.

Longevity Over Disposability:

Prioritizing product longevity over disposability is akin to composing a timeless sonata. Instead of opting for cheap, short-lived goods, consider investing in high-quality items built to endure. Reflect on the lifespan of a product before acquiring it, and be open to the idea of repair rather than replacement whenever possible. Such choices not only reduce waste but also conserve precious resources.

Supporting Sustainable Brands:

Supporting sustainable brands is akin to endorsing virtuoso performers. Seek out companies and products committed to environmental responsibility and ethical production practices. Champion businesses dedicated to reducing their carbon footprint, utilizing eco-friendly materials, and ensuring fair treatment of workers.

Reduce, Reuse, Recycle:

The mantra of "reduce, reuse, recycle" runs through the core of every responsible consumer's symphony. It begins by reducing your consumption, embracing minimalism when circumstances permit. Reuse items whenever possible, and recycle meticulously to divert waste from overflowing landfills. Always consider the full lifecycle of products when making choices about recycling or disposal.

Product Lifecycle Awareness:

Understanding the complete lifecycle of products is akin to reading the musical score of sustainability. From the extraction of raw materials to production, transportation, usage, and ultimate disposal, each phase carries environmental implications. Be aware of these impacts and seek alternatives that leave a lighter footprint.

Consumer Activism:

Engaging in consumer activism is akin to leading a protest for change. Support initiatives and organizations that advocate for responsible production and consumption. Vote with your wallet by choosing products and services aligned with your values, sending a powerful message to businesses that sustainability matters.

Key Takeaway:

In conclusion, consumer habits are the notes that compose the grand symphony of our lives, a symphony that profoundly influences the health of our planet. By cultivating mindful consumption, prioritizing product longevity, supporting sustainable brands, adhering to the principles of reduce, reuse, recycle, and fostering product lifecycle awareness, you become a conductor in the orchestra of sustainability.

Consumer habits possess the power to either contribute to the degradation of our environment or serve as instruments of positive change. Your choices as a responsible consumer resonate not only in your own life but also in the world we collectively share. By harmonizing your consumer habits with sustainability, you compose a brighter future – one where responsible choices lead to a healthier planet for generations to come.

X. Education And Awareness

Imagine, if you will, a world where knowledge serves as the sword, and awareness becomes the armor in our battle against the formidable adversary known as climate change. As citizens, we bear the remarkable mantle of educating ourselves and illuminating the path for our communities. Today, we embark on a journey to uncover how we can wield these weapons to effect meaningful change in the fight against climate change.

Why It Matters:

The climate crisis is a global call to arms, demanding a united front. Raising awareness and educating ourselves and others about its causes, consequences, and solutions is pivotal to driving meaningful action.

Stay Informed and Educate Yourself:

Our journey begins with the quest for knowledge, the relentless pursuit of wisdom from credible sources. Dive into the works of respected scientists and organizations that illuminate the intricate tapestry of climate change. Stay abreast of the latest discoveries and revelations. Understanding the science behind climate change equips us to have informed conversations and make enlightened choices.

Engage in Climate Literacy:

Climate literacy is the bedrock of effective climate action, akin to mastering the notes in a symphony. Familiarize yourself with fundamental concepts such as the ebb and flow of greenhouse gas emissions, the delicate dance of the carbon cycle, and the profound impacts of climate change. Equip yourself with the knowledge needed to engage in constructive discussions and advocate for informed policies.

Share Knowledge and Raise Awareness:

Within the realm of climate change, your voice possesses the alchemical power to transmute awareness into action. Share the treasury of your knowledge about climate change with friends, family, and colleagues. Forge conversations that cast a spotlight on the urgency of our predicament and underscore the importance of collective endeavor. In the digital age, seize the opportunities of social media platforms to disseminate wisdom, champion sustainable practices, and amplify the voices of climate scientists and activists.

Support Climate Education Initiatives:

Advocate ardently for the inclusion of comprehensive climate education in the hallowed halls of our schools and universities. Plead the case for curricula that weave climate science, sustainability, and environmental ethics into the fabric of education. Collaborate with educators and policymakers, wielding your influence to ensure that climate change is instilled as an indomitable subject, arming the next generation with the knowledge and tools to confront this challenge.

Participate in Community Initiatives:

Engage in the heartbeats of local climate action groups, partake in the nurturing of community gardens, or lend your strength to the causes championed by environmental organizations. Join hands with initiatives that center on sustainability, clean energy, and the guardianship of nature. Contribute your sweat and passion to neighborhood cleanups, tree planting ceremonies, and awareness campaigns, nurturing a profound sense of stewardship toward our environment.

Reduce Your Own Carbon Footprint:

Become a beacon of inspiration through your own actions. Lead by the sheer force of example. Imbue your daily life with sustainable practices that rekindle the spirit of our planet. Lower your energy consumption, explore the virtues of public transportation or carpooling, and embrace eco-friendly habits as your daily creed. In your actions, you demonstrate the attainability of sustainable living.

Advocate for Policy Change:

Raise your voice and engage in dialogue with the custodians of policy at local, state, and national levels. Advocate fervently for policies that set us on the path to addressing climate change with courage and conviction. Extend your support to legislation that promotes renewable energy, embraces carbon pricing, and fortifies the shield of environmental protection. Stand shoulder to shoulder with your fellow citizens in peaceful demonstrations and rallies, exclaiming that the time for climate action is now.

Support Climate-Focused Organizations:

Consider bestowing your support upon non-profit organizations and initiatives wholly dedicated to the cause of climate change mitigation and adaptation. Whether through the benevolence of your donations or the dedication of your time as a volunteer, your contributions resound as impactful notes in the grand symphony of change on the ground.

Key Takeaway:

In conclusion, education and awareness are catalysts for change in the fight against climate change. By arming ourselves with knowledge, disseminating that knowledge, and actively participating in efforts to combat climate change, we become informed and engaged citizens who can drive meaningful action. Climate change is not an isolated challenge but a shared responsibility. Through education and awareness, we chart a course toward a more sustainable and resilient future, where every note we strike reverberates in harmony with the Earth's own song of survival.

XI. Green Spaces and Biodiversity Conservation

Imagine a world where every patch of green, from sprawling national parks to the tiniest backyard gardens, is a sanctuary for biodiversity. As citizens, we hold the power to transform our homes, communities, and nations into havens for nature. Today, we embark on a journey to explore how we can build, strengthen, and expand green spaces, fostering biodiversity conservation in every corner of our lives.

Why It Matters:

Biodiversity is the heartbeat of our planet, a symphony of life that sustains eco-systems, cleans the air, and provides us with sustenance. Building, strengthening, and expanding green spaces are critical steps toward preserving biodiversity and securing a healthier future for all.

Create Green Havens at Home:

Our journey begins within the confines of our own homes, where every yard, balcony, or windowsill holds the potential to become a thriving green haven. Plant native species in your garden to attract local wildlife. Install bird feeders and birdhouses to welcome feathered friends. Set up a bee-friendly garden, planting wildflowers and flowering herbs that nourish pollinators. Create a composting system to recycle organic waste, enriching your soil naturally.

Support Community Green Initiatives:

Extend your reach to the community level, where collective efforts can amplify the impact. Join or initiate local green space projects that aim to rejuvenate parks, vacant lots, or urban areas with native flora. Participate in tree-planting drives that enhance green canopy cover in your neighborhood. Advocate for urban planning that prioritizes green corridors, wildlife corridors, and the protection of natural habitats within your city.

Engage in Citizen Science:

Become a citizen scientist, contributing to biodiversity research and conservation efforts. Participate in wildlife surveys, bird counts, or butterfly monitoring programs in your area. Your observations can provide valuable data for researchers and help track changes in local eco-systems. Through citizen science, you become a guardian of biodiversity.

Promote Responsible Land Management:

Advocate for responsible land management practices in your country and region. Support policies that safeguard natural habitats, wetlands, and critical eco-systems. Encourage sustainable farming practices that reduce the use of harmful pesticides and promote wildlife-friendly agriculture. Engage with policymakers to ensure that green spaces are integrated into urban and rural planning.

Embrace Sustainable Landscaping:

Choose sustainable landscaping practices for your green spaces. Implement rain gardens that capture and filter stormwater, preventing pollution of local water bodies. Install permeable pavements that allow rainwater to recharge groundwater. Reduce or eliminate the use of chemical fertilizers and pesticides, which can harm biodiversity. Prioritize a balanced eco-system where predators and prey coexist naturally.

Educate and Inspire Others:

Share your passion for biodiversity conservation with friends, family, and your community. Organize educational events, nature walks, or workshops to raise awareness about local flora and fauna. Encourage others to join in your efforts and discover the wonder of biodiversity. When you inspire others, the collective impact grows exponentially.

Support Biodiversity-Focused Organizations:

Consider contributing your time, resources, or donations to organizations dedicated to biodiversity conservation. These groups often work tirelessly to protect endangered species, restore eco-systems, and advocate for policies that prioritize nature. Your support can help fund crucial initiatives that safeguard biodiversity on a larger scale.

Key Takeaway:

In conclusion, green spaces and biodiversity conservation are not just endeavors for experts and ecologists. As citizens, we hold the power to transform our homes, communities, and nations into sanctuaries for nature. By creating green havens at home, supporting community initiatives, engaging in citizen science, promoting responsible land management, embracing sustainable landscaping, educating and inspiring others, and supporting biodiversity-focused organizations, we become stewards of biodiversity in every sense of the word.

Biodiversity conservation is not a luxury; it's a necessity for the well-being of our planet and future generations. When we nurture nature, we enrich our lives and secure a harmonious future where all species, including our own, thrive together.

XII. Artistic Echoes Of Climate Change: Healing Harmonies

CHAPTER

Thirteen

The Last Word

"The punishment which the wise suffer who refuse to take part in the government, is to live under the government of worse men."
- Plato, "The Republic"

In the languid twilight of late summer 2023, as I sat with my neighbor in his impeccably manicured French-style garden, a sense of disillusionment washed over me. The world seemed to teeter on the precipice of catastrophe, and I had just finished chronicling the amoral, nasty, hypocritical, and greed-driven leaders of the climate change denial movement.

Our conversation was momentarily interrupted by the buzz of our smartphones, heralding the news of a young scion of Indian immigrants gunning for the highest office in the land. It was a political tale that would have been unremarkable if not for the sinister twist that marred his campaign. This young man, armed with an Ivy League pedigree and a trail of business successes, was shamelessly courting the MAGA electorate. And his chosen weapon? Denial. Denial of climate change, the undeniable reality looming before us like a foreboding storm.

The summer of 2023 bore witness to Earth's hottest temperatures since we began keeping records in 1880, a stark warning etched in the scorched earth and sweltering air. NASA's Goddard Institute of Space Studies confirmed what should have been undeniable: June, July, and August combined were hotter than any other summer in recorded history. August alone surpassed the average by a blistering 2.2 degrees Fahrenheit. The evidence was incontrovertible.

Weather and climate disasters raged across the United States, shattering previous records. The National Oceanic and Atmospheric Administration somberly reported 23 separate billion-dollar disasters, a damning testament to the climate's wrath. Over 250 lives were claimed by these cataclysms, and the economic toll soared above $57.6 billion. Torrential rains, flash floods, and unforgiving wildfires wrought havoc upon Europe, sparing few nations from nature's fury. It was a combination of climate change and the unruly El Niño that fueled these calamities, a relentless duo conspiring to unleash chaos.

Central Europe became a battleground for floods, with Germany, Austria, Hungary, Slovenia, the Czech Republic, and Georgia bearing the brunt of relentless downpours. Greece witnessed a year's worth of rain in just 18 hours. Meanwhile, Spain, Italy, Greece, Cyprus, Algeria, and China mourned the heat deaths of their citizens. Southern regions of the United States languished under heat alerts, straining power grids and decimating vital crops.

Yet, amidst this maelstrom of irrefutable evidence, there existed a cabal of amoral opportunists, hungry for money, attention, and political power. They twisted reality to suit their ambitions, echoing the denialist rhetoric that had become all too familiar. They dared to claim that climate change was a hoax, their words a dagger in the heart of reason.

Allow me to be unequivocal: If you choose to follow these charlatans, who place their self-interest above all else, including the future of their own progeny, you are sealing your fate and that of our beleaguered planet. Like the hapless children of the German legend "Rattenfänger von Hameln," you will be led to the abyss, with no savior to rescue you. Your father may succumb to the unrelenting heat, your mother to some zoonotic malady born of climate upheaval, and your children… who can predict their destiny?

In this dire scenario, you will have no one to blame but yourself. Cease following these amoral figures, dissect their lust for wealth, political ascendancy, and unquenchable greed. Their concern for your well-being doesn't amount to a solitary penny. Instead, turn your gaze to science, to facts, and accept the somber truth that we must change our course. The alternative is a collective descent into death and destruction.

I employ hyperbole not to shock but to emphasize the urgency of our situation. Climate deaths are mounting, an ominous precursor of what lies ahead. These are the final words of this book, a clarion call to action, a plea for reason in the face of madness. The choice, ultimately, is yours.

Bibliography

- ADAC. (2023, June 28). Dieselskandal bei Audi, Seat, Skoda, Porsche, VW. Retrieved from https://www.adac.de/rund-ums-fahrzeug/auto-kaufenverkaufen/abgasskandal-rechte/vw/.
- abcNEWS (Camp Fire).(2018). This was a firestorm: Deadly California wildfire leaves entire Paradise town council homeless. Retrieved from https://abcnews.go.com/beta-story-container/US/deadly-camp-fire-leaves-entireparadise-town-council/story?id=59159481.
- Alami, A. (2019, July 20). In Spain, Workers in Strawberry Fields Speak Out on Abuse. Retrieved From https://www.nytimes.com/2019/07/20/world/europe/spainstrawberry-fields-abuse.html.
- Alley, R.., Clark, P., Huybrechts, P., & Joughin, I. (2005). Ice-Sheet and Sea_level Changes. Science. 310(5747).
- AMAP (2017). Snow, Water, Ice and Permafrost in the Arctic (SWIPA) 2017. Arctic Monitoring and Assessment Programme (AMAP), Oslo, Norway.
- American Association for the Advancement of Science (AAAS). (2014). THE REALITY, RISKS, AND RESPONSE TO CLIMATE CHANGE. Retrieved from https://whatweknow.aaas.org/wp-content/uploads/2014/07/whatweknow_website.pdf.
- American Scientific Societies (ACS). (2019). Scientific Consensus: Earth's Climate Is Warming. Retrieved from https://climate.nasa.gov/scientific-consensus/.
- Anderson, K., Peters, G. (2016). The trouble with negative emissions - Reliance on negative-emission concepts locks in humankind's carbon addiction. SCIENCE, Vol 354(6309), pp. 182-183.
- Arvesen, A., Hertwich, E. G., & Suh, S. (2011). On the Importance of Metals in the Carbon Footprint of Vehicle Manufacturing. Environmental Research Letters, 6(3), 034007.
- Aurand, T. W., Finley, W., Krishnan, V., Sullivan, U. Y., Abresch, J., Bowen, J., Rackauskas, M., Thomas, R., & Willkomm, J. (2018). The

VW Diesel Scandal: A Case of Corporate Commissioned Greenwashing. Journal of Organizational Psychology, 18(1).

- Balch, O. (2012, May 4). Nike reveals a new, innovative game plan for sustainability (The Guardian). Retrieved from https://www.theguardian.com/sustainablebusiness/nike-sustainability-report-social-environmental-impact.
- Balch, O. (2013, May 3). H&M: can fast fashion and sustainability ever really mix? (The Guardian). Retrieved from https://www.theguardian.com/sustainable-business/hand-m-fashion-sustainability-mix.
- Balch, O. (2020, December 8). The curse of 'white oil': electric vehicles' dirty secret. The Guardian. Retrieved from https://www.theguardian.com/news/2020/dec/08/thecurse-of-white-oil-electric-vehicles-dirty-secret-lithium.
- BBC. (2020). Brazil's Amazon: Deforestation high in January despite rainy season. Retrieved from https://www.bbc.com/news/world-latin-america-51425408.
- Beckert, S. (2014). Empire of Cotton: A Global History. Vintage; Reprint edition (December 2, 2014).
- Bintanja, R., van der Linden, E. C. (2013). The changing seasonal climate in the Arctic. Sci. Rep. 3, 1–8.
- Birnbaum, C., Remme, K.(2010). Beyond Petroleum. Deutschlandfunk. Retrieved from https://www.deutschlandfunk.de/bp-beyond-petroleum-100.html
- Blazin , N., M. B. Gavrilov, M.B. (WTO). (2014). The tsunami of 26th December 2004. Retrieved from https://www.witpress.com/Secure/elibrary/papers/EID14/EID14015FU1.pdf.
- Bleakley, D. (2023). Toyota could face $50 million "greenwashing" fine after referral to consumer watchdog. Retrieved from https://thedriven.io/2023/03/06/toyota-could-face-50-million-greenwashingfine-after-referral-to-consumer-watchdog/.
- Bombardi, L.M., Changoe, A. (2022). TOXIC TRADING – The EU pesticide lobby's offensive in Brazil. Retrieved from https://friendsoftheearth.eu/wpcontent/ uploads/2022/04/Toxic-Trading-EN.pdf.
- Bontron, C. (2012, August 7). Rare-earth mining in China comes at a heavy cost for local villages. The Guardian. Retrieved from. https://

www.theguardian.com/environment/2012/aug/07/china-rare-earthvillage-pollution.

- Briscoe, N. (2021, 8 July). The Irish Times. Car industry accused of 'greenwashing' as we wait for electric revolution. Retrieved from https://www.irishtimes.com/life-and-style/motors/car-industry-accused-of-greenwashing-as-wewait-for-electric-revolution-1.4614887.
- Briscoe, N. (2023, June 28). Electric car batteries: Just how bad are they for the environment. Irish Times. Retrieved from https://www.irishtimes.com/motors/2023/06/28/electric-car-batteries-just-how-badare-they-for-the-environment/.
- Broughton, E. (2005). The Bhopal disaster and its aftermath: a review. Environmental Health. Retrieved from https://www.researchgate.net/publication/7859596_The_Bhopal_disaster_and_its_aftermath_A_review.
- Bucciere, C. (2014, November 4). Syngenta, CropLife America Lobby for Policy Change in EU Neonicotinoid Use. Retrieved from https://www.agribusinessglobal.com/agrochemicals/syngenta-croplife-america-lobbyfor-policy-change-in-eu-neonicotinoid-use/.
- Bundesministerium für Wirtschaft und Klimaschutz (BMWK). (2023). Rohstoffe und Ressourcen – Fracking. Retrieved from https://www.bmwk.de/Redaktion/DE/Artikel/Industrie/fracking.html.
- Bundesministerium Wirtschaft und Klima (BMWK). (2023, May 5). Retrieved from https://www.bmwk.de/Redaktion/DE/Pressemitteilungen/2023/05/20230505-habecklegt-arbeitspapier-zum-industriestrompreis-vor.html.
- Butler, S. (2022). Dirty greenwashing: watchdog targets fashion brands over misleading claims. The Guardian. Retrieved from https://www.theguardian.com/business/2022/jan/14/dirty-greenwashing-watchdogtargets-fashion-brands-over-misleading-claims.
- C40 Cities (C40). (2018). How Climate Change Could Impact the World's Greatest Cities. Retrieved from https://www.c40.org/wp-content/uploads/2023/04/1789_Future_We_Dont_Want_Report_1.4_hires_120618.original-compressed.pdf.
- Canadell, J.G. (2022). Global Carbon and other Biogeochemical Cycles and Feed-backs. In Climate Change 2021: The Physical Science Basis. Contribution of Working Group I to the Sixth Assessment Re-

port of the Intergovernmental Panel on Climate. Cambridge University Press.

- Caradonio, J. (2022, July 19). The Hotel Industry's Big Carbon Lie. Bloomberg. Retrieved from https://www.bloomberg.com/news/articles/2022-07-19/embodiedcarbon-is-dirty-secret-of-hotel-industry-greenwashing-eco-ratings.
- Carter, D. (2022, October 26). EU car lobby blocks tougher emissions standards. The Brussels Times. Retrieved from https://www.brusselstimes.com/312566/eu-car-lobbyblocks-tougher-emissions-standards.
- Chancel, L., Bothe, P., Voituriez, T. (2023) Climate Inequality Report 2023, World Inequality Lab Study, 2023(1).
- Chapman, S.C., Watkins, N.W., & Stainforth, D.A. (AGU). (2019). Warming Trends in Summer Heatwaves. Geophysical Research Letters. 46(3), 1634-1640.
- Chason, R., Sharrock, C. (2023, April 27). On frontier of new 'gold rush,' quest for coveted EV metals yields misery. The Washington Post. Retrieved from https://www.washingtonpost.com/world/interactive/2023/ev-battery-bauxite-guinea/.
- Christensen, C.M. (2016). The Innovator's Dilemma: When New Technologies Cause Great Firms to Fail (Management of Innovation and Change). HARVARD BUSINESS REVIEW PRESS.
- Ciais, P. et al.. (2013). Carbon and Other Biogeochemical Cycles. In: Climate Change 2013: The Physical Science Basis. Contribution of Working Group I to the Fifth Assessment Report of the Intergovernmental Panel on Climate. Cambridge University Press, Cambridge, United Kingdom and New York, NY, USA, pp. 465–570, doi:10.1017/cbo9781107415324.015.
- Clawson, L. (2020, January 21). Marriott's 'green choice' isn't so green, and it's hurting workers. Daily Kos. Retrieved from https://www.dailykos.com/tories/2020/1/20/1911782/-Marriott-s-green-choice-isn-tso-green-and-it-s-hurting-workers.
- ClientEarth. (2023, June 7). Landmark greenwashing lawsuit against KLM airline granted court permission. Retrieved from https://www.clientearth.org/latest/pressoffice/press/landmark-greenwashing-lawsuit-against-klm-airline-granted-courtpermission/

- Climate Change Laws of the world. (2021). The Climate Change Act Fiji 2021.Retrieved from: https://climatelaws. org/document/climate-change-act-2021_8bf7.
- CNBC (Statistics South Africa). (2018). Cape Town is running out of water, and no one knows what economic impact that will have. Retrieved from https://www.cnbc.com/2018/03/06/south-africa-cape-town-drought-economicimpact.html.
- Corby, S. (2022, March 22). Are hybrid cars really better for the environment? The Alps - Challenges and Potentials of a Brand Management. Retrieved from https://www.carsguide.com.au/ev/advice/are-hybrid-cars-really-better-for-theenvironment-86067.
- Creutzig, F., Roy, J., Lamb, W.F. et al. (2018). Towards demand-side solutions for mitigating climate change. Nature Clim Change 8, 260–263.
- Daszak, P., Olival, K. J., Li, H., & Han, H. (2020). A strategy to prevent future epidemics similar to the 2019-nCoV outbreak. Biosafety and Health, 2(1), 6-8.
- DeConto, R., Pollard, D. (2016). Contribution of Antarctica to past and future sea level rise. Nature 531, 591–597.
- Deforestation in the Amazon rainforest continues to plunge (Mongabay). Retrieved from https://news.mongabay.com/2023/09/de-forestation-in-the-amazon-rainforest-continues-to-plunge/.
- Delpey, R. (1985). Affaires centrafricaines. Paris : J. Grancher, c1985.
- de Trenqualye, M. (2023). It's inequality that kills': Naomi Klein on the future of climate justice. https://www.theguardian.com/books/2023/feb/13/its-inequality-that-kills-naomiklein-on-the-future-of-climate-justice.
- Die ZEIT (Zeit). (2015, November 27). Manipulierte Stickoxid-Messungen, getrickste CO_2-Werte: Der VW-Skandal zieht weltweit Kreise. Wir beantworten die wichtigsten Fragen zum Fall Volkswagen. Retrieved from https://www.zeit.de/wirtschaft/2015- 09/vw-abgase-manipulation-faq.
- Dr. Leu, A., et al. (2023). Degenerative Agriculture: Bayer/Monsanto's and Syngenta's Toxic Greenwashing Deception. Regeneration International. Retrieved from https://regenerationinternational.org/2023/06/29/degenerative-agriculture-bayer-monsantos-and-syngentas-toxic-greenwashing-deception/.

- Duff, I. (2011, February 23). Ecologist Informed by Nature. Retrieved from https://theecologist.org/2011/feb/23/greenpeace-asia-pulp-and-papers-pr-greenwashtheir-sustainability-claims-joke.
- Duff, I. (2011, February 23). Greenpeace: Asia Pulp and Paper's PR is greenwash, their sustainability claims a joke. ECOLOGIST INFORMED BY NATURE. Retrieved from https://theecologist.org/2011/feb/23/greenpeace-asia-pulp-and-paperspr-greenwash-their-sustainability-claims-joke.
- Duncombe, J. (2023). (EOS) The Unequal Benefits of California's Electric Vehicle Transition. Retrieved from https://eos.org/articles/the-unequal-benefits-of-californias-electric-vehicle-transition.
- EGR.(2023). (EGR20). EGR Environmental Policy. Retrieved from: https://www.egrgroup.com/egr_environment.html.
- Eltis et al. (2013). Transatlantic Trade. Cambridge University Press.
- Environmental Investigation Agency. (2021) (EEA). Retrieved from: https://eia-international.org/report/ikeas-dirtysecret/.
- Environmental Pollution. (2020). Retrieved from ScienceDirect.
- Environmental Science and Pollution Research. (2016). Retrieved from Science-Direct.
- European Environment Agency (EEA). (2021). Observed snow depth trends in the European Alps: 1971 to 2019. 5(3), 1343–1382.
- Fairhead, J., Leach, M., and Scoones, I. (2012). Green Grabbing: A New Appropriation of Nature (Critical Agrarian Studies) (English Edition). Routledge.
- Farquhar, G.D., Prentice, I.C. (2018). The Carbon Cycle and Atmospheric Carbon Dioxide. Retrieved from https://www.ipcc.ch/site/assets/uploads/2018/02/TAR-03.pdf.
- Feldenkirchen, M. (2022, April 15). SPD-Ministerpräsidentin Schwesig und der Kreml – Mecklenburg-Gazprommern. DER SPIEGEL. Retrieved from https://www.spiegel.de/politik/deutschland/stiftung-klima-und-umweltschutz-mvmecklenburg-gazprommern-kolumne-a-1a8a57ee-581d-4eda-be89-6a147744412d.
- Fetting, C. (2023). The European Green New Deal. Retrieved from https://www.esdn.eu/fileadmin/ESDN_Reports/ESDN_Report_2_2020.pdf.

- Fischer, K. (2023, June, 23). Investorenstudie entlarvt die grünen Versprechen der Ölindustrie. Wirtschaftswoche. Retrieved from https://www.wiwo.de/unternehmen/industrie/erderwaermung-investorenstudieentlarvt-die-gruenen-versprechen-der-oelindustrie/29230922.html.
- Fogel, R.W., EngermaN, S.L. (1985). Time on the Cross: The Economics of American Negro Slavery. University Press of America.
- Food and Agriculture Organization of the United Nations (FAO). (2023). Mapping ways to reduce methane emissions from livestock and rice. Retrieved from https://www.fao.org/newsroom/detail/mapping-ways-to-reduce-methane-emissionsfrom-livestock-and-rice/en.
- Fortune Business Insights. (2023). (Fortune). Cement Market Size. Retrieved from https://www.fortunebusinessinsights.com/industry-reports/cement-market-101825.
- Fraser, B. (2014). Deforestation: Carving up the Amazon. Nature 509, 418–419 (2014).
- Gardner, T. (2015, October 24). U.S. oil export battle heats up as drillers group to fight ban. REUTERS. Retrieved from https://www.reuters.com/article/us-oil-exportslobbying-idUSKCN0ID23020141024.
- Gibbs, S. G., Gibbs, E. P., Anderson, T. C., & Young, C. C. (2015). Wild animals and the risk of Ebola virus outbreaks in humans. Journal of Emerging Infectious Diseases, 21(5), 845-849.
- Girardi, A. (2018). The Ski Slopes in the Dolomites. Retrieved from https://www.forbes.com/sites/annalisagirardi/2018/12/27/the-ski-slopes-in-the-dolomitesbetween-tradition-and-innovation/.
- GLOBAL COVENANT of MAYORS for CLIMATE & ENERGY (2023). (Mayors). Retrieved from: https://www.globalcovenantofmayors.org.
- Goldenberg, S. (2015, February 19). World's biggest PR firm calls it quits with American oil lobby – reports. The Guardian. Retrieved from https://www.theguardian.com/business/2015/feb/19/edelman-public-relations-endsrelationship-american-petroleum-institute.
- GovInfo (2011, January 11). Deep Water: The Gulf Oil Disaster and the Future of Offshore Drilling. Report to the President (BP Oil Spill Commission Report).

- Green Infrastructure and Biodiversity Plan. (2020). (BGIBP). Retrieved from https://climateadapt. eea.europa.eu/en/metadata/case-studies/barcelona-trees-tempering-the-mediterranean-cityclimate/11302639.pdf.
- Hahn, J. (2021, March 24). Many Banks Committing to Climate Goals Are Engaging in Greenwashing. The Magazine Of The Sierra Club. Retrieved from https://www.sierraclub.org/sierra/many-banks-committing-climate-goals-areengaging-greenwashing-banking-on-climate-chaos.
- Hansen, J., Fung, I., Lacis, A., Rind, D., Lebedeff, S., Ruedy, R., Russell, G., and Stone, P. (1988). Global Climate Changes as Forecast by Goddard Institute for Space Studies Three-Dimensional Model. Geophysical Research Letters, 15(9), 867-870.
- Hansen, J., Sato, M., Ruedy, R., Lo, K., Lea, D. W., and Medina-Elizade, M. (2010). Global Temperature Change. Proceedings of the National Academy of Sciences, 107(21), 9552-9555.
- Harvey, F. (2012, April 2). Multinationals vow to boycott APP after outcry over illegal logging. (The Guardian). Retrieved from https://www.theguardian.com/environment/2012/apr/02/boycott-app-illegal-logging.
- Harvey, F. (2020, June 23). Timber from unsustainable logging allegedly being sold in EU as ethical. The Guardian. Retrieved from https://www.theguardian.com/environment/2020/jun/23/timber-unsustainablelogging-allegedly-sold-eu-ethical.
- Hausfather, Z. (2023, May 1). Will global temperatures exceed 1.5C in 2024?. The Climate Brink. Retrieved from https://www.theclimatebrink.com/p/will-globaltemperatures-exceed-15c.
- Heilprin, A. (2011). Mont Pelee and the Tragedy of Martinique: A Study of the Great Catastrophes of 1902, with Observations and Experiences in the Field (Cambridge Library Collection – Earth Science) Taschenbuch.
- Henley, J. (2023). The Guardian. Many Europeans want climate action – but less so if it changes their lifestyle, shows poll. Retrieved from https://www.theguardian.com/environment/2023/may/02/many-europeans-want-climateaction-but-less-so-if-it-changes-their-lifestyle-shows-poll.

- Hochschild, A. (2020). King Leopold's Ghost: A Story of Greed, Terror, and Heroism in Colonial Africa. Marina Books.
- Hosan, B. (2021). Magnitogorsk: Die Stadt aus Stahl und Wahnsinn. Retrieved from: https://www.capital.de/wirtschaft-politik/history-crime/magnitogorsk-die-stadt-aus-stahl-und-wahnsinn-30414276.html.
- Hotten, R. (2015, December 10). Volkswagen: The Scandal Explained. BBC. Retrieved from https://www.bbc.com/news/business-34324772.
- Howell, B. (2023, January 26). What Happens To Dead Electric Car Batteries? Retrieved from https://www.theecoexperts.co.uk/electric-vehicles/what-happens-todeadbatteries#:~:text=Once%20your%20electric%20vehicle%20battery,life%2C%20or%20recycle%20the%20parts.
- Huxley, A. (1932). Brave New World. Harper Perennial (2006, October 18).
- Ighobor, K., Bafana, B. (2014). Financing Africa's massive projects. Retrieved from https://www.un.org/africarenewal/magazine/december-2014/financing-africa's-massiveprojects
- Influence Map. (2023). Analyzing the Automotive Sector on Climate Change -January 2023 Update. Retrieved from https://automotive.influencemap.org.
- Inikori, J. E. (2004). Africans and the Industrial Revolution in England. Cambridge University Press.
- Instituto Nacional de Pesquisas Espaciais (INPE). (2023, September 8).
- Intergovernmental Panel on Climate Change (IPCC). (2014). Climate Change 2014: Impacts, Adaptation, and Vulnerability. Cambridge University Press.
- Intergovernmental Panel on Climate Change (IPCC). (2014). Climate Change 2014: Synthesis Report. Contribution of Working Groups I, II and III to the Fifth Assessment Report of the Intergovernmental Panel on Climate Change. IPCC, Geneva, Switzerland, 151 pp.
- Intergovernmental Panel on Climate Change (IPCC). (2021). Climate Change 2021: The Physical Science Basis. Contribution of Working

Group to the Sixt Assessment Report of the Intergovernmental Panel on Climate Change. Cambridge University Press.

- Intergovernmental Panel on Climate Change (IPCC). (2022). IPCC Sixth Assessment Report (AR6): Climate Change 2021 - The Physical Science Basis, Impacts, Adaptation, and Vulnerability, Mitigation of Climate Change. IPCC, Geneva, Switzerland.
- Intergovernmental Panel on Climate Change (SR15). (2019). Global Warming of 1.5 °C. Retrieved from https://www.ipcc.ch/sr15/.
- Intergovernmental Panel on Climate ChangeAR4). (2007). Fourth Assessment Report – Climate Change 2007. Retrieved from https://www.ipcc.ch/assessment-report/ar4/.
- Intergovernmental Report on Climate Change (IPCC). (2013). Climate Change 2013: The Physical Science Basis. Cambridge University Press.
- International Atomic Energy Agency (IAEA). (2020). Advanced Reactors: Technology Options for Generation IV and Small Reactors. Retrieved from https://aris.iaea.org/Publications/IAEA_SMR_Booklet_2014.pdf.
- International Energy Agency' (IAE). (2021). Global Energy Review. Retrieved from https://www.iea.org/reports/global-energy-review-2021.
- International Energy Agency (IEA). (2021). Net Zero by 2050: A Roadmap for the Global Energy Sector. Retrieved from https://www.iea.org/reports/net-zero-by-2050.
- International Energy Agency (IEA). (2023). What is the role of cement in clean energy transitions? Retrieved from https://www.iea.org/energysystem/industry/cement.
- International Renewable Energy Agency. (IRENA). (2020). Innovative Sweden Power. Retrieved from: https://www.irena.org/-/media/Files/IRENA/Agency/Publication/2020/Jan/IRENA_Innovative_power_Sweden_2020_summary.pdf?la=en&hash=9FC47DCAD97F5001B07663FD7D246872DBC0F868.
- Johnson, G. (2005, 22 Aug.). Los Angeles Times. 'Greenwashing' Leaves a Stain of Distortion. Retrieved from https://www.latimes.com/archives/la-xpm-2004-aug-22-op-johnson22-story.html.

- Jones, K. E., Patel, N. G., Levy, M. A., Storeygard, A., Balk, D., Gittleman, J. L., &

  Daszak, P. (2008). Global trends in emerging infectious diseases. Nature, 451(7181), 990-993.

- Jones, T., Awokoya, A. How the Italian mafia makes millions by exploiting migrants. (The Guardian). Retrived from https://www.theguardian.com/world/2019/jun/20/tomatoes-italy-mafia-migrant-labour-modern-slavery.

- Joshi, K. (2021, February 16). Plastics: A carbon copy of the climate crisis. ClientEarth. Retrieved from https://www.clientearth.org/latest/latestupdates/stories/plastics-a-carbon-copy-of-the-climate-crisis/.

- Kálmán, A., Pena, P. (2022, June 24). In Brussels, the multi-billion-euro pesticide business is everywhere. Investigate Europe. https://www.investigateeurope. eu/posts/pesticide-business-lobby-in-brussels.

- Kamins, A. (2023, February 2023). The Impact of Climate Change on U.S. Subnational Economies. Moody Analytics. Retrieved https://www.moodysanalytics.com/articles/pa/2023/the-impact-of-climate-change-onus-subnational-economies.

- Kell, G. (2022, December 5). From Emissions Cheater To Climate Leader: VW's ourney From Dieselgate To Embracing E-Mobility. Forbes. Retrieved from https://www.forbes.com/sites/georgkell/2022/12/05/from-emissions-cheater-toclimate-leader-vws-journey-from-dieselgate-to-embracing-emobility/?sh=-555442d68a55.

- Kinkartz, S. (2023, March 7). Nord Stream: Der lange Arm Putins in die deutsche Politik? DW. Retrieved from https://www.dw.com/de/nord-stream-der-lange-armputins-in-die-deutsche-politik/a-64896116.

- Klawans, J. (2023, June 20). How the cruise industry is pivoting to sustainability. THE WEEK. Retrieved from https://theweek.com/environmental-news/1024392/how-the-cruise-industry-is-pivoting-to-sustainability.

- Kolstad, E. W., Johansson, K. A., & Ebi, K. L. (2020). Healthy people in a healthy world: An assessment of the co-benefits of climate change mitigation for human health. Environment international, 134, 105279.

- Koven, R. (1981, May 8). Angry Ex-Emperor Bokassa Seeks to Thwart Giscard's Reelection. The Washington Post. Retrieved from https://www.washingtonpost.com/archive/politics/1981/05/08/angry-ex-emperorbokassa-seeks-to-thwart-giscards-reelectio/8b-4700ca-d5e5-4c51-a669-9a280c42e9af/.
- Kulp, S. A., & Strauss, B. H. (2019). New Elevation Data Triple Estimates of Global Vulnerability to Sea-Level Rise and Coastal Flooding. Nature Communications, 10(1), 4844.
- Kwasniewski, N. (2013, October 31). Greenpeace wirft Adidas und Nike Schönfärberei vor. Der Spiegel. Retrieved from https://www.spiegel.de/wirtschaft/service/greenpeace-bezeichnet-adidas-und-nike-inranking-als-greenwasher-a-930973.html.
- KYOTO PROTOCOL TO THE UNITED NATIONS FRAMEWORK CONVENTION ON CLIMATE CHANGE (Kyoto). (1198). United Nations. Retrieved from https://unfccc.int/resource/docs/convkp/kpeng.pdf#page=12.
- Landrigan, P.J., et al. (2023, March 21). The Minderoo-Monaco Commission on Plastics and Human Health. Ann Glob Health. 89(1):23. doi: 10.5334/aogh.4056. PMID: 36969097; PMCID: PMC10038118.
- Lappé, A. (2014, October 28). Could Bhopal happen here? Aljazeera America. http://america.aljazeera.com/opinions/2014/10/bhopal-indiachemicalplantsafetyepaenvironment. html.
- Laville, S. (2019, October 10). Exclusive: carmakers among key opponents of climate action. The Guardian. Retrieved from https://www.theguardian.com/environment/2019/oct/10/exclusive-carmakersopponents-climate-action-us-europe-emissions.
- Le Figaro. (2021). Vacances d'été : succès massif en vue pour la montagne. Retrieved from: https://www.lefigaro.fr/voyages/montagne/vacances-d-ete-succes-massif-envue-pour-la-montagne-20210618.
- Le Quéré, C., et al. (2018). Global Carbon Budget 2018. Earth System Science Data, 10(4), 2141-2194.
- Leavitt, S. (2021, June 18). ESG Spotlight: Banking on Net Zero – Fact or Fiction? Amenity. Retrieved from https://www.amenityanalytics.com/blog-articles/esgspotlight-banking-on-net-zero-fact-or-fiction.

- Lennard, N. (2011, August 5) Is this the most anti-environment House in history? Salon. Retrieved from https://www.salon.com/2011/08/05/anti_environment_gop_house/
- Lenton, T. M., Held, H., Kriegler, E., Hall, J. W. (2008). Tipping elements in the Earth's climate system. Proc. Natl. Acad. Sci. U.S.A. 105, 1786–1793.
- Lenton, T.M., Rockström, J. (2019) Climate tipping points - too risky to bet against. Nature, 575, 592–595.
- Lis Cunha (2023, April 19). Stellungnahme zur öffentlichen Anhörung im Wirtschaftsausschuss zum EU-Mercosur-Abkommen. Greenpeace – Ausschussdrucksache 20(9)237- Deutscher Bundestag. Retrieved from https://www.bundestag.de/resource/blob/943246/1629ab7636a295de84e505a41ec83b 58/Stellungnahme_Greenpeace-data.pdf.
- Malhi, Y., Aragão, L. E. O. C., Galbraith, D., Huntingford, C., Fisher, R., Zelazowski, P., ... & Meir, P. (2009). Exploring the likelihood and mechanism of a climatechange-induced dieback of the Amazon rainforest. Proceedings of the National Academy of Sciences, 106(49), 20610-20615.
- Marcus, L. (2021, March 30). Your eco-friendly hotel might not be so green after all. CNN. Retrieved from https://edition.cnn.com/travel/article/eco-friendly-hotelsgreenwashing-cmd/index.html.
- Markey, E. (2011, July 27). REPS. WAXMAN AND MARKEY RELEASE REPORT DETAILING MOST ANTI-ENVIRONMENT HOUSE IN HISTORY. Retrieved from https://www.markey.senate.gov/news/press-releases/reps-waxmanand-markey-release-report-detailing-most-anti-environment-house-in-history.
- Masengarb, C. (2023). Philipps Beteiligung an Africa GreenTec – Klappt der Plan, macht er Habecks Staatssekretär zum 15-fachen Millionär. FOCUS. Retrieved from https://www.focus.de/finanzen/habecks-staatssekretaer-africa-greentec-kann-ihn-zummulti-millionaer-machen_id_194666201.html.
- Maslin, M. (2014). Climate Change: A Very Short Introduction. Oxford University Press.
- Maslin, M. (2023). Professor Mark Maslin, How to save Our Planet. Retrieved from https://twitter.com/profmarkmaslin?lang=de.

- Mayer, J. (2019). Kochland Examines the Koch Brothers' Early, Crucial Role in Climate-Change Denial. The NEW YORKER. Retrieved from https://www.newyorker.com/news/daily-comment/kochland-examines-how-the-koch-brothers-made-theirfortune-and-the-influence-it-bought.
- McGreal, C. (2021, October 26). Revealed: 60% of Americans say oil firms are to blame for the climate crisis. The Guardian. Retrieved from https://www.theguardian.com/environment/2021/oct/26/climate-change-poll-oil-gascompanies-environment.
- McGuire, J. (2021, October 5).. McDonald's Rebuked for Greenwashing Climate Pledge. (EcoWatch). Retrieved from https://www.ecowatch.com/mcdonaldsgreenwashing-2655223811.html.
- McMichael, A. J., Lindgren, E., & Nykvist, B. (2018). Climate change: Heatwaves hit the elderly hardest. Nature, 567(7748), 178-180.
- McVeigh, K. (2022, 28 Jan.). The Guardian. West accused of 'climate hypocrisy' as emissions dwarf those of poor countries. Retrieved from https://www.theguardian.com/global-development/2022/jan/28/west-accused-ofclimate-hypocrisy-as-emissions-dwarf-those-of-poor-countries.
- Meyer, R. (2019). Trump Isn't a Climate Denier. He's Worse. The Atlantic. Retrieved from https://www.theatlantic.com/science/archive/2019/11/ideologybehind-donald-trumps-paris-withdrawal/601462/.
- Mishra, S. (2023, January 24). Fossil fuel lobby waged $4m disinformation campaign during climate summit, report finds. INDEPENDENT. Retrieved from https://www.independent.co.uk/climate-change/news/climate-disinformation-fossilfuel-cop27-b2265216.html.
- Mogensen, J.F. (2018, 25 Jan.). Grist.org . Sierra Club's new video slams Ford for greenwashing its public image. Retrieved from https://grist.org/technology/sierra-clubs-new-video-slams-ford-for-greenwashing-its-public-image/.
- NASA Vital Signs (2022). (NASA). Facts – Methane. Retrieved from https://climate.nasa.gov/vital-signs/methane/.
- National Aeronautics and Space Administration (NASA). (2022). Global Climate Change: Vital Signs of the Planet. Retrieved from https://climate.nasa.gov/

- National Oceanic and Atmospheric Administration (NOAA). (2022). Climate.gov. Retrieved from https://www.climate.gov/.
- National Snow and Ice Data Center (NSIDC). (2020). Sea Ice Index. National Snow and Ice Data Center. Retrieved from https://nsidc.org/arcticseaicenews/2023/09/arctic-sea-ice-minimum-at-sixth/.
- Nepstad, D., McGrath, D. G., Stickler, C. M., Alencar, A., Azevedo-Ramos, C., & Swette, B. (2014). Slowing Amazon deforestation through public policy and interventions in beef and soy supply chains. Science, 344(6188), 1118-1123.
- Neue Zürcher Zeitung (NZZ). (2023, February 24). BASF Investiert Milliarden in China. Retrieved from https://www.nzz.ch/meinung/basf-investiert-milliarden-inchina-unbelehrbar-nein-pragmatisch-europa-sollte-das-dennoch-ernst-nehmenld. 1727678.
- New International (Newint). (2011, July 1). BIOFUELS – THE GOOD, THE BAD AND THE UGLY. Retrieved from https://newint.org/features/2011/07/01/biofuelsfacts-future-environmental-greenwashing-2655223811.html.
- McMichael, A. J., Lindgren, E., & Nykvist, B. (2018). Climate change: Heatwaves hit the elderly hardest. Nature, 567(7748), 178-180.
- McVeigh, K. (2022, 28 Jan.). The Guardian. West accused of 'climate hypocrisy' as emissions dwarf those of poor countries. Retrieved from https://www.theguardian.com/global-development/2022/jan/28/west-accused-ofclimate-hypocrisy-as-emissions-dwarf-those-of-poor-countries.
- Meyer, R. (2019). Trump Isn't a Climate Denier. He's Worse. The Atlantic. Retrieved from https://www.theatlantic.com/science/archive/2019/11/ideologybehind-donald-trumps-paris-withdrawal/601462/.
- Mishra, S. (2023, January 24). Fossil fuel lobby waged $4m disinformation campaign during climate summit, report finds. INDEPENDENT. Retrieved from https://www.independent.co.uk/climate-change/news/climate-disinformation-fossilfuel-cop27-b2265216.html.
- NASA Vital Signs (2022). (NASA). Facts – Methane. Retrieved from https://climate.nasa.gov/vital-signs/methane/.
- National Aeronautics and Space Administration (NASA). (2022). Global Climate Change: Vital Signs of the Planet. Retrieved from https://climate.nasa.gov/

- National Oceanic and Atmospheric Administration (NOAA). (2022). Climate.gov. Retrieved from https://www.climate.gov/.
- National Snow and Ice Data Center (NSIDC). (2020). Sea Ice Index. National Snow and Ice Data Center. Retrieved from https://nsidc.org/arcticseaicenews/2023/09/arctic-sea-ice-minimum-at-sixth/.
- Nepstad, D., McGrath, D. G., Stickler, C. M., Alencar, A., Azevedo-Ramos, C., & Swette, B. (2014). Slowing Amazon deforestation through public policy and interventions in beef and soy supply chains. Science, 344(6188), 1118-1123.
- Neue Zürcher Zeitung (NZZ). (2023, February 24). BASF Investiert Milliarden in China. Retrieved from https://www.nzz.ch/meinung/basf-investiert-milliarden-inchina-unbelehrbar-nein-pragmatisch-europa-sollte-das-dennoch-ernst-nehmenld. 1727678.
- New International (Newint). (2011, July 1). BIOFUELS – THE GOOD, THE BAD AND THE UGLY. Retrieved from https://newint.org/features/2011/07/01/biofuelsfacts-future-environmental-green.
- Nova, A. (2023). (CNBC). Climate change could impose 'substantial financial costs' on U.S. household finances, Treasury warns. Retrieved from https://www.cnbc.com/2023/10/02/climate-change-could-devastate-household-finances-ustreasury-warns.html
- Nuccitelli, D. (2016, July 25). The Guardian.These are the best arguments from the 3% of climate scientist 'skeptics.' Really. Retrieved from https://www.theguardian.com/environment/climate-consensus-97-percent/2016/jul/25/these-are-the-best-arguments-from-the-3-of-climate-scientist-skeptics-really.
- NYTimes. (2021). Behind Manchin's Opposition, a Long History of Fighting Climate Measures. Retrieved from https://www.nytimes.com/2021/12/20/us/politics/manchin-climate-change-coal.html.
- NYTimes. (2019). Who Keeps Europe's Farm Billions Flowing? Often, Those Who Benefit. Retrieved from https://www.nytimes.com/2019/12/11/world/europe/eu-farm-subsidy-lobbying.html.
- Oekoreich. (2022, September 9). So dreist ist das aktuelle Greenwashing von IKEA. Retrieved from https://www.oekoreich.com/medium/so-dreist-ist-das-aktuellegreenwashing-von-ikea.

- Open SECRETS (CRP). (2015 February 18). Final Tally: 2014's Midterm Was Most Expensive, With Fewer Donors. Retrieved from https://www.opensecrets.org/news/2015/02/final-tally-2014s-mid-term-was-mostexpensive-with-fewer-donors/.
- Paddison, L. (2023, January 24). Bill Gates backs start-up tackling cow burps and farts. Retrieved from https://www.cbs58.com/news/bill-gates-backs-start-up-tacklingcow-burps-and-farts.
- Pashley, A. (2016, April 7). Big oil spent $115m 'obstructing' climate laws in 2015, NGO says. CLIMATE HOME NEWS. Retrieved from https://www.climatechangenews.com/2016/04/07/big-oil-spent-115m-obstructingclimate-laws-in-2015-ngo-says/.
- Pearce, F. (2009, September 3). Monsanto? Sustainable? Water bully, I'd say… The Guardian. Retrieved from https://www.theguardian.com/environment/cifgreen/2009/sep/03/monsanto-water-green-wash.
- Pechlaner, H., Raich, F., Zehrer, A. (2007). The Alps - Challenges and Potentials of a Brand Management. Tourism Analysis 12(5). Retrieved from https://www.researchgate.net/publication/50518220_The_Alps_-_Challenges_and_Potentials_of_a_Brand_Management.
- Perkins, T. (2021, December 21). (The Guardian) Hold the beef: McDonald's avoids the bold step it must take to cut emissions. Retrieved from https://www.theguar-dian.com/environment/2021/dec/10/mcdonalds-emissions-beefburgers.
- Perkins-Kirkpatrick, S.E., Lewis, S.C. (2020). Increasing trends in regional heatwaves. Nature Communications, 11(3357).
- Peters, G.P. et al., 2020: Carbon dioxide emissions continue to grow amidst slowly emerging climate policies. Nat. Clim. Change, 10(1), 3–6, doi:10. 1038/s41558-019-0659-6.
- Phillips, D. (2020, September 23. Brazil meat giant JBS pledges to axe suppliers linked to deforestation. The Guardian. Retrieved from https://www.theguardian.com/environment/2020/jun/29/brazilian-meat-giant-jbslinked-to-deforestation-in-the-amazon.
- Plumer, B. (2017). What 'Clean Coal' Is – and Isn't. The New York Times. Retrieved from https://www.nytimes.com/2017/08/23/climate/what-clean-coal-is-and-isnt.html.
- Porter, M. (1990). The Competitive Advantage of Nations. New York: Free Press.

- Prentice, I. C., Sitch S., Meyer R., Hooss G., et al..200. Global warming feedbacks on terrestrial carbon uptake under the Intergovernmental Panel on Climate Change (IPCC) emission scenarios, GLOBAL BIOGEOCHEMICAL CYCLES, Vol: 15, Pages: 891-907, ISSN: 0886-6236.
- Prof. McCurdy, H.E.E. (2003). Faster, Better, Cheaper: Low-Cost Innovation in the U.S. Space Program. John Hopkins University Press.
- Reiners, P., Sobrino, J., Kuenzer, C. (2023). Satellite-Derived Land Surface Temperature Dynamics in the Context of Global Change – A Review, Remote Sensing, 15, (7).
- Renault Group (2019, September 13). The challenges of recycling electric car batteries. Retrieved from The Guardian. (2023). Retrieved from https://www.renaultgroup.com/en/news-on-air/news/the-challenges-of-recyclingelectric-car-batteries/.
- Rodney, W. (2018). How Europe Underdeveloped Africa. Verso.
- Rogers, D. (2011). GOP outlines new EPA cuts. Politico. Retrieved from https://www.politico.com/story/2011/07/gop-outlines-new-epa-cuts-058409.
- Roy, B., Penha-Lopes, G.P., Uddin, M.S., Kabir, M.H., Lourenço, T.C., Torrejano. A. (2022). Sea level rise induced impacts on coastal areas of Bangladesh and local-led community-based adaptation. The International Journal of Disaster Risk Reduction, 73(102905).
- Sasse, JP., Trutnevyte, E. A low-carbon electricity sector in Europe risks sustaining regional inequalities in benefits and vulnerabilities. Nat Commun 14, 2205 (2023). https://doi.org/10.1038/s41467-023-37946-3.
- Scaling Solar in Africa. (2023). (Scaling). Retrieved from https://www.ifc.org/content/dam/ifc/doc/2023-delta/infra-factsheet-scaling-solar-2023.pdf
- Schäfer, S. (2023, July 3). Focus: Spanier lösen Hauptproblem der Klimawendeund schonen sogar die Umwelt. Focus. Retrieved from https://www.focus.de/perspektiven/constructive-world-award/nominiertebeitraege/nominiert-fuer-constructive-world-award-spanier-loesen-hauptproblem-derenergiewende-mit-wasser_id_194592378.html
- Schellingerhout, M. (2023, February 13). Slippery slope? Alpine tourism in the faceof climate crisis – photo essay. The Guardian. Re-

trieved from https://www.theguardian.com/artanddesign/2023/feb/13/alpine-tourism-climate-crisisphotoessay#:~: text=The%20Alps%2C%20with%20their,%2C%20snowboarding%2C%20hiking%20and%20cycling.

- Schimel, D.,. Enting, L.G, et al. (1995). CO_2 and the Carbon Cycle. Climate Change 1994: Radiative Forcing of Climate Change and An Evaluation of the IPCC IS92 Emission Scenarios, 1995. hal-03384881
- Schmidt, R., Abdelilah, A. (2022). Ikea made in Belarus. Taz. Retrieved from https://taz.de/Material-aus-Strafgefangenen-Lagern/5892344/.
- Schumpeter, J.A. (1947). Capitalism, Socialism and Democracy. Kessinger Publishing (2007).
- Screen, J. A., Deser, C., & Sun, L. (2018). Projected changes in regional climate extremes arising from Arctic Sea ice loss. Environmental Research Letters, 13(5), 054019.
- Shabecoff, P. (1988, June 24). Global Warming Has Begun, Expert Tells Senate. The New York Times. Retrieved from https://www.nytimes.com/1988/06/24/us/globalwarming-has-begun-expert-tells-senate.html.
- Silkenat, D. (2022). Scars on the Land: An Environmental History of Slavery in the American South. Oxford University Press.
- Sillmann, J., et al. (Environ). (2014). Observed and simulated temperature extremes during the recent warming hiatus. Environmental Research Letters, 9, 064023.
- Skinner, C. (2021, May 12). Greenwashing the system: which banks are worst? Retrieved from https://thefinanser.com/2021/05/greenwashing-the-system-whichbanks-are-worst.
- Smith, C. (2021). The Earth's energy budget, climate feedbacks, and climate sensitivity supplementary material. Contribution of Working Group I to the Sixth Assessment Report of the Intergovernmental Panel on Climate Change, Cambridge University Press (2021).
- Sovacool, B.K., Turnheim, B., Hook, A., Brock, A., Martiskain, M. (2021). Dispossessed by decarbonisation: Reducing vulnerability, injustice, and inequality in the lived experience of low-carbon pathways. World Development, 137(2021).

- Sovacool, B.K., Tan-Mullins, M., Ockwell, D., & Newell, P. (2017). Political economy, poverty, and polycentrism in the Global Environment Facility's Least Developed Countries Fund (LDCF) for climate change adaptation (Version 1). University of Sussex. https://hdl.handle.net/10779/uos.23440658.v1.
- Sovacool, B.K.., and Linnér , B.O. (2016). The Perils of Climate Diplomacy: The Political Economy of the UNFCCC. The Political Economy of Climate Change Adaptation, pp 110-135.
- Statista. (2023 Weltweite und europäische Kunststoffproduktion in den Jahren von 1950 bis 2022. Retrieved from https://de.statista.com/statistik/daten/studie/167099/umfrage/weltproduktion-von-kunststoff-seit-1950/.
- Statistic South Africa. (2008). South African Statistics, 2008. Retrieved from https://www.statssa.gov.za/publications/SAStatistics/SAStatistics2008.pdf.
- Sterling, T., Plucinska, J. (2023, September 13). Fly responsibly? Airlines face a storm over climate claims. Reuters. Retrieved from https://www.reuters.com/business/aerospace-defense/fly-responsibly-airlines-facestorm-over-climate-claims-2023-09-13/
- Survival International (Survival). (2023, May). Survival International. Retrieved from https://www.survivalinternational.org/about/deforestation.
- Symons, A. (2023, April 6). KLM has pulled its 'Fly Responsibly' ads – but environmental lawyers fear future greenwashing campaigns. Euronews.green. Retrieved from https://www.euronews.com/green/2023/04/06/klm-airline-accused-ofgreenwashing-adverts-which-encourage-responsible-flying-in-the-neth.
- Talbot, H. (2021). How this Swiss city is using green roofs to combat climate change. Retrieved from https://www.euronews.com/green/2021/07/09/how-this-swiss-city-is-using-green-roofs-to-combatclimate-change.
- Tamme, K. (2023, March 8). Klimaschutzstiftung MV: Skandale und Salamitaktik.NDR. Retrieved from https://www.ndr.de/nachrichten/mecklenburgvorpommern/Klimaschutzstiftung-MV-Skandale-und-Salamitaktik,klimastiftunghintergrund100.html.
- Tark, J., Oh, W.Y. (2021). Social and Sustainability Marketing. Productivity Press.

- TAZ. (2021). German Engineered Klimatschutz. taz. Retrieved from https://taz.de/FDP-Chef-Lindner-ueber-Klimapolitik/5797246/
- The World Business Council for Sustainable Development. (2023). Cement Sustainability Initiative. Retrieved from: https://www.wbcsd.org/Sector-Projects/Cement-Sustainability-Initiative/Cement-Sustainability-Initiative-CSI.
- The Conflict and Environmental Observatory (CEOBS). (2016, November 30). The environmental consequences of Iraq's oil fires are going unrecorded. Retrieved from https://ceobs.org/the-environmental-consequences-of-iraqs-oil-fires-are-goingunrecorded/.
- The Gartner Hype Curve. (2017, August 15). Top Trends in the Gartner Hype Cycle for Emerging Technologies, 2017. Retrieved from https://www.gartner.com/smarterwithgartner/top-trends-in-the-gartner-hype-cycle-foremerging-technologies-2017.
- The Global Slavery Index. (2023). Retrieved from https://www.walkfree.org/globalslavery-index/.
- The Government of Costa Rica. (2023).The National Decarbonization Plan – Costa Rica. Retrieved from https://unfccc.int/sites/default/files/resource/NationalDecarbonizationPlan.pdf.
- The Guardian. (2023). The Challenges of Recycling Electric Batteries. Retrieved from https://www.renaultgroup.com/en/news-on-air/news/the-challenges-of-recyclingelectric-car-batteries/
- The Guardian (2023). Toyota accused of greenwashing in Greenpeace complaint filed to ACCC. Retrieved from https://www.theguardian.com/business/2023/mar/03/toyota-accused-of-greenwashing-in-greenpeace-complaintfiled-to-accc.
- The Japan times (2023, March 24). Japan's endless summer pushes some toward cooler places. Retrieved from https://www.japantimes.co.jp/environment/2023/09/24/climate-change/katsuurasummer-heat-escape/.
- The National Aeronautics and Space Administration (NASA). (2023). Vital Signs: Global Temperature. Retrieved from https://climate.nasa.gov/vital-signs/globaltemperature/#:~: text=Earth%27s%20global%20average%20surface%20tempe-rature,average%20from%20 1951%20to%201980.
- The New York Times (NYT). (2013). Ford's Green Ads Challenged as Misleading in Britain. Retrieved from https://www.nytimes.

com/2013/06/27/business/global/fordsgreen-ads-challenged-as-misleading-in-britain.html.

- The New York Times (NYTimes). (2022, October 19). E.V.s Start With a Bigger Carbon Footprint. Retrieved from https://www.nytimes.com/2022/10/19/business/electric-vehicles-carbon-footprintbatteries.html.
- Thornhill, G.D., et al. (2021). The Earth's energy budget, climate feedbacks, and climate sensitivity supplementary material. Atmos. Chem. Phys., 21(2).
- Touboul, S.; Glachant, M.; Dechezleprêtre, A.; Fankhauser, S.; Stoever, J. Invention and global diffusion of technologies for climate change adaptation: A patent analysis. Rev Env. Econ Policy 2023, 17, 316–335.
- Tracy, B., Novak, A. (2023). Cement industry accounts for about 8% of CO_2 emissions. One startup seeks to change that. CBS NEWS. Retrieved from https://www.cbsnews.com/news/cement-industry-co2-emissions-climate-changebrimstone/
- Traufetter, G., & Weiland, S. (2023, May 19). So geriet Habeck in das Heizungschaos. Der Spiegel. Retrieved from https://www.spiegel.de/politik/deutschland/gebaeudeenergiegesetz-wie-roberthabeck-in-das-heizungschaos-geriet-a-266353ad-c7b3-4444-8c91-0873 8404548b.
- Traufetter, G. (2023, April 14). Dieselgate: VW kämpft gegen Golf-Stilllegungen. Der Spiegel. Retrieved from https://www.spiegel.de/wirtschaft/unternehmen/volkswagen-diesel-skandal-vwkaempft-gegen-golf-stilllegungen-a-13c127ed-1128-403f-9dde-4a373f-f8e5f7.
- U.S. Geological Survey (USGS). (2023). Cement Statistics and Information. Retrieved from https://www.usgs.gov/centers/national-minerals-informationcenter/cement-statistics-and-information.
- UNICEF. (2015). The impact of climate change on children. Retrieved from https://www.unicef.org/media/60111/file.
- Union of Concerned Scientists (UCS). (2018). Underwater: Rising seas, chronic floods, and the implications for US coastal real estate. Retrieved from https://www.ucsusa.org/resources/underwater-impacts-climate-change-us-coastalreal-estate-0.

- United Nations. (2022). (UN). Climate Action. Retrieved from https://www.un.org/en/climatechange/net-zerocoalition
- United Nations Climate Change Conference (ccacoalition). (2021). New global methane pledge aims to tackle climate change. Retrieved from https://www.unep.org/news-and-stories/story/new-global-methane-pledge-aimstackle-climate-change.
- United Nations Environment Programme (UNEP). (2021, May 6). Global Assessment: Urgent steps must be taken to reduce methane emissions this decade. Retrieved from https://www.unep.org/news-and-stories/press-release/globalassessment-urgent-steps-must-be-taken-reduce-methane.
- United Nations Framework Convention on Climate Change (UNFCCC). (2023). What is the clean development mechanism? Retrieved from https://cdm.unfccc.int/about/index.html.
- University of Chicago (AQLI). (2023). How much longer you live if you breathed clean air? Retrieved from https://aqli.epic.uchicago.edu.
- Urban Climate Change Research Network. (2023). (UCCRN). Retrieved from https://uccrn.ei.columbia.edu
- Usborne, S.. (2022, January 12). More people is the last thing this planet needs': the men getting vasectomies to save the world. The Guardian. https://www.theguardian.com/lifeandstyle/2022/jan/12/more-people-is-the-last-thingthis-planet-needs-the-men-getting-vasectomies-to-save-the-world.
- Verisk Maplecroft. (2021). The Climate Change Vulnerability Index. Retrieved from https://www.maplecroft.com/risk-indices/climate-change-vulnerability-index/.
- Vieira, I. C., Gardner, T. A., Ferreira, J., Lees, A. C., Barlow, J., & Viana, V. M. (2018). The Amazon's carbon balance: the contribution of pristine and disturbed forests. Ambio, 47(4), 380-394.
- Waters, J. (2015). Barack Obama singles out Koch brothers over fossil fuel lobbying .The Guardian. Retrieved from https://www.theguardian.com/us-news/2015/aug/25/barack-obama-warns-fossil-fuelled-opponents-he-has-cleanenergy-to-spare.
- Watson, R.T., Rhode (1990). Greenhouse gases and aerosols. In: Climate Change: the IPCC Scientific Assessment, J.T. Houghton, G.J.

Jenkins and J.J. Ephraums (eds.), Cambridge University Press, Cambridge, UK, ppl-40.

- Watts, N., Amann, M., Ayeb-Karlsson, S., Belesova, K., Bouley, T., Boykoff, M., …& Costello, A. (2019). The Lancet Countdown on health and climate change: from 25 years of inaction to a global transformation for public health. The Lancet, 394(10211), 1836-1878.
- Weir, D. (1998). The Bhopal Syndrome: Pesticides, Environment and Health. Earthscan Ltd.
- Welt. (2022, August 5). Lindner bat Porsche-Chef um "argumentative Unterstützung" bei E-Fuels. Welt. Retrieved from https://www.welt.de/politik/deutschland/article240328945/Christian-Lindner-bat-Porsche-Chef-um-argumentative-Unterstuetzung-bei-E-Fuels.html.
- Williams, L. (2023). Where's the Justice in Net Zero? Retrieved from https://www.bloomberg.com/opinion/articles/2023-05-18/climate-change-policy-where-s-thejustice-in-net-zero
- Williamson, P. (2016, February 10). Emissions reduction: Scrutinize CO_2 removal methods. Nature 530, 153–155 (2016).
- Willmroth, J. (2021, Januar 13). Umweltschützer werfen Blackrock Greenwashing vor. Sueddeutsche. Retrieved from https://www.sueddeutsche.de/wirtschaft/blackrock-nachhaltigkeit-brief-kritik-1.5172295.
- Wimmers, A., Madlener, R. (2020). The European Market for Guarantees of Origin for Green Electricity: A Scenario-Based Evaluation of Trading under Uncertainty. FCN Working Papers 17/2020, E.ON Energy Research Center, Future Energy Consumer Needs and Behavior (FCN).
- Woody, T. (2023). Bloomberg. EVs are cleaning up California's air, but mostly for the affluent. Retrieved from https://www.denverpost.com/2023/05/03/electric-vehicles-california-air-pollution-wealthy/.
- World Resources Institute (WRSI). (2022, March 2). World Resources Institute.

  (2022). World Greenhouse Gas Emissions: 2019. Retrieved from https://www.wri.org/insights/interactive-chart-shows-changes-worlds-top-10-emitters.

- World Travel and Tourism Council (WTTC). (2005). Impact of the Tsunami on Travel and Tourism: An Economic Analysis. Retrieved from https://www.traveldailynews.com/associations/wttc-tsunami-impact-on-traveltourismis-significant-but-limited/.
- World Wild Fund For Nature (WWF). (2023). Amazon – Facts. Retrieved from https://www.worldwildlife.org/places/amazon.
- Wu, F., Zhao, S., Yu, B., Chen, Y. M., Wang, W., Song, Z. G., … & Zhang, Y. Z. (2020). A new coronavirus associated with human respiratory disease in China. Nature, 579(7798), 265-269.
- Yale School of Environment (2008, June 2). The Myth of Clean Coal. The Yale School of the Environment. Retrieved from https://e360.yale.edu/features/the_myth_of_clean_coal.
- Zhou, Y., et al. (2018). Projected changes in temperature and precipitation extremes over Africa from CMIP5 models. Environmental Research Letters, 13(6), 0650

About This Book

In this groundbreaking exploration of our planet's most urgent crisis, we journey through the complex tapestry of climate change. From the meticulous research that unveils its chilling consequences to the inspiring stories of individuals and communities driving change, this book offers a panoramic view of our climate's past, present, and uncertain future.

With clarity and urgency, it delves into the science, politics, and human dimensions of the climate crisis. It illuminates the innovative solutions and sustainable strategies that promise hope and renewal for our fragile Earth.

This is the definitive guide for those seeking to understand, act, and advocate for a more sustainable, resilient, and equitable world. As the call to address climate change intensifies, this book is your indispensable companion on the path to a brighter, cooler, and more compassionate future.

About The Author

Jude S. Ngu'Ewodo is a highly accomplished Climate, Natural Resources, and Technology Investor and Entrepreneur from Germany, with a distinguished global portfolio of successful ventures. His exceptional proficiency spans a wide spectrum, encompassing the establishment, growth, and strategic success of pioneering enterprises and business opportunities on a global scale. Beyond his investment and financial acumen, Jude is an unwavering advocate for sustainability. As a farmer himself, he leads the charge in introducing water-efficient, pesticide-free, and circular farming practices in his country of birth, Cameroon. His multifaceted journey has endowed him with a unique perspective, positioning him as an inspired visionary at the intersection of commerce and environmental stewardship.